GIS

for

Coastal Zone Management

GIS
for
Coastal Zone Management

Edited by
Darius Bartlett and Jennifer Smith

CRC PRESS

Boca Raton London New York Washington, D.C.

Library of Congress Cataloging-in-Publication Data

CoastGIS '01 Conference (2001 : Halifax, N.S.)
GIS for coastal zone management / edited by Darius J. Bartlett and Jennifer L. Smith.
p. cm.
Includes updated and edited presentations made to the CoastGIS '01 Conference in Halifax, Canada 18-20 June 2001.
Includes bibliographical references and index.
ISBN 0-415-31972-2
1. Coast changes—Congresses. 2. Environmental mapping—Congresses. 3. Coastal zone management—Congresses. 4. Geographic information systems—Congresses. I. Bartlett, Darius J., 1955- II. Smith, Jennifer L. III. Title.

GB450.2.C625 2001
333.91'7'0285-dc22 2004050302

Direct all inquiries to CRC Press LLC, 2000 N.W. Corporate Blvd., Boca Raton, Florida 33431.

Visit the CRC Press Web site at www.crcpress.com

International Standard Book Number 0-415-31972-2
Library of Congress Card Number 2004050302
Printed in the United States of America 1 2 3 4 5 6 7 8 9 0
Printed on acid-free paper

Contents

Foreword

The material presented in this volume comprises updated and edited presentations first made to the CoastGIS'01 Conference conducted in Halifax, Canada between the 18th and 20th June 2001 together with chapters commissioned by the Editors.

The CoastGIS series of conferences have been the outcome of a fruitful collaboration between the International Geographical Union's Commission on Coastal Systems and the International Cartographic Association's Commission on Marine Cartography. Generally entitled "International Symposium on GIS and Computer Mapping for Coastal Zone Management" we have seen five successful CoastGIS conferences held over this decade-long collaboration.

These conferences were held in Ireland (Cork, 1995), Scotland (Aberdeen, 1997), France (Brest, 1999) Canada (Halifax, 2001) and Italy (Genoa, 2003). A closely allied CoastalGIS conference was conducted in Wollongong, Australia in July 2003. Future conferences are planned for Scotland, Australia and Barbados.

At the first meeting in Cork, we had the honour of being addressed in a keynote presentation by Lord Chorley, who referred in his address to the House of Commons Environment Select Committee's 1992 Report on coastal zone protection and planning. Reflecting on the findings of that Report, Lord Chorley was then struck by three main points.

"First, it is only in recent years that the coastal zone has been recognised as one important topic in its own right. Second, the huge range of relevant aspects or considerations. (Thirdly): the huge number of agencies involved, often with overlapping and perhaps incompatible responsibilities, jurisdictions and objectives."

These themes have recurred throughout the conferences that followed.

Halifax 2001

CoastGIS'01 was convened in Halifax, Canada, at Saint Mary's University between 18th – 20th June 2001. The conference attracted over 150 delegates from the Americas, Europe, Africa, Asia and Australia who presented 36 oral presentations and live demonstrations in a single stream format and 50 posters. The theme selected by the 2001 Science Committee was "Managing the Interfaces," a theme with a multitude of possible interpretations. Overall, the shift in emphasis towards *integration* of systems for coastal management and the growing interest in coastal spatial data infrastructures were especially in evidence at this meeting. So too was the international dimension of coastal GIS.

Notably, for the first time, financial assistance from the Canadian International Development Agency and the Geomatics Association of Nova Scotia permitted CoastGIS 2001 to fund ten delegates from the developing world to participate in the conference. Two chapters in the book result from this initiative (those by Nwilo and Euán-Avila *et al.*).

CoastGIS 2001 also instituted demonstrations of live GIS systems. Three chapters in this volume arose from this innovation (Laflamme *et al.*, Mosbech *et al.* and Bourcier). There were field trips to Nova Scotia Community College's Centre for Geographical Sciences in the Annapolis Valley and to the Bay of Fundy, where we considered the coastal issues facing a region that experiences the highest tides in the world.

This setting drove home the dynamic nature of the coastal zone interface of land, sea and air. Within this framework, several of the conference presentations that evolved into chapters in this volume deal with the dynamics of the coastal zone, while others address approaches to bridging the land-sea divide.

Many presentations at the Halifax gathering focused on the need for an effective interface amongst the range of participants and stakeholders involved in coastal management. The chapters in this volume that describe applications and case studies and those that include traditional ecological knowledge demonstrate the impact of effective communication between these parties.

The use of increasingly advanced technologies in the coastal zone (notably remote sensing, web mapping and mobile application technologies, visualization techniques, and LiDAR) to support research and management was a highlight of the conference and is detailed in several chapters.

The reader will no doubt be well aware of the amazing developments of GIS capability over the last decade in particular. Nonetheless, the development of standards, formats and data models together with the sheer genius of GIS technological developments and ultimate cost effectiveness are perhaps still hampered by the paucity of available data sets. At CoastGIS 2001, the development of spatial data infrastructures was highlighted. Susan Lambert, then the Executive Director of the Kentucky Office of Geographic Information and now with the United States Geological Survey, presented a keynote lecture on the development of the GeoData Alliance, a nonprofit organization open to all individuals and institutions committed to using geographic information to improve the health of our communities, our economies and the Earth. A presentation was also made on Canada's Marine Geospatial Data Infrastructure. The Editors of this volume invited Roger Longhorn to summarize the progress in the development of coastal spatial data infrastructures for this volume.

Genoa 2003 and beyond

One advantage the authors have in writing this Foreword is being able to do it immediately after the successful conclusion of this successful conference, which ran from October 16 to 18, 2003. Exploration of the SDI theme continued at the 2003 CoastGIS conference held in Genoa, Italy. At this gathering emphasis was placed upon many of the non-science and non-technology issues that continue to adversely impact the success and long-term sustainability of many coastal zone projects and wider coastal zone monitoring initiatives at national and regional levels due to barriers to the access to data and information.

In summarizing the outcomes of the Genoa conference on behalf of the Scientific Committee, Roger Longhorn noted that virtually all presentations had covered coastal zone research, monitoring or management work in a *single* nation, often in a single sub-national region. Very few therefore faced the added difficulties that can arise when trying to locate, access, understand and agree on the usage and dissemination terms for data from owners outside not only one's own discipline, but outside one's national legislative infrastructure for information use. The ocean, as Longhorn pointed out, has a "nasty" way of connecting one piece of coastline to another, and neither the ocean nor the physical coastline show any respect for national boundaries and differing jurisdictions.

Some of the key points highlighted by different members of the Scientific Committee at a meeting held on the last day of the conference included the following:

1. We need political champions to help guard our interests in seeing that coastal information needs are not forgotten as larger national and regional (trans-national) spatial data infrastructure (SDI) frameworks are created.

2. GIS in the coastal zone is certainly about supporting "science work", but there are also non-scientific and non-technical issues to be considered, hence the need for a policy level of collaboration.

3. Data usability is a key concern and continues to require both research and information management focus for continued development of ways to harmonise data for wider use.

4. Information infrastructure developments are needed that permit easier discovery of existing data and use of data once located, in a variety of forms, from multiple data owners or custodians.

5. We need to find ways to engage stakeholders (data creators, custodians and users) even more widely in the data management and access issues.

With the launch of the EU Water Framework Directive in 2000, to be fully implemented by December 2003 in all EU Member States, we have seen the first institutionalized, regional (trans-national) legal requirement that GIS be used in monitoring the implementation of a major EU policy, and one of extremely high importance on a global level - i.e., maintaining good quality water resources in river basins, groundwater, coastal zones and the off-shore transitional waters leading to the coastal zone. In all likelihood this is only the first such legal requirement that we will see coming from major international institutions for use of spatial information and GIS tools for planning and monitoring purposes.

Therefore, coastal GIS practitioners need to address their next efforts towards effective usability of coastal knowledge (not just coastal data) as a major contribution to regional planning and monitoring, even at transnational level. In regard to this perceived need, two issues arise:

1. The landscape/seascape paradigm offers a comprehensive perspective of both the physical and human/cultural aspects and their interaction, defining the present state and heritage. At the European level, the European Landscape Convention (2000) may be assumed as a reference for Coastal GIS attuned for Administration in the governance, planning and design phases. Definition of relationships with non-European landscape policies should be sought.

2. The operation of data, jointly with the implementation of data infrastructures, may be regarded as a chief subject for GIS optimisation. It is hoped to create a link with the running global and pan-European initiatives and/or policies by offering a contribution for Data Infrastructure Profiles suitable for coastal GIS and or promoting these achievements towards non-GIS and non-ICAM specialists, addressing the concerned stakeholders in public administration and industry.

The legal requirement to use GIS for monitoring the Water Framework Directive, and the implied directive to use GIS to monitor the EU ICZM Recommendation both focus on primarily physical data, i.e., coastal or benthic flora and fauna, geomorphology, etc. Yet for wider planning and monitoring purposes, many administrative and non-physical data sources will be needed. These must somehow be accommodated by the evolving coastal SDI.

As we write, it is fairly clear that CoastGIS as a gathering will be around for the foreseeable future. Perhaps the main intangible, but nonetheless very real, benefit from the series of gatherings has been the camaraderie and consequent networking of many of the main contributors. However, as researchers and practitioners we all need more tangible records of these significant events. We trust that the presentation of part of the ongoing record in this volume will contribute to the development and improvement of coastal zone management around the globe.

Ron Furness – *Chair, International Cartographic Association Commission on Marine Cartography*
Andy Sherin – *Chair, CoastGIS 2001 Science Committee, Co-chair of the CoastGIS 2001 Organizing Committee and Coastal Information Specialist, Geological Survey of Canada*

Sydney and Ottawa, 13th November 2003

Preface

Darius Bartlett and Jennifer Smith

This book has arisen out of a decade-long collaborative initiative between the Commission on Marine Cartography of the International Cartographic Association and the Commission on Coastal Systems of the International Geographical Union, and manifested in the series of conferences known as the CoastGIS Symposia. The first CoastGIS meeting was held in Cork, Ireland, in February 1995. Since then, successive events have taken place in Aberdeen, Scotland (1997), Brest, France (1999) and Halifax, Nova Scotia (2001). The majority of chapters presented in the pages that follow had their origins in papers presented at the Halifax meeting, supplemented by a selection of additional contributions commissioned by the editors specifically for this volume.

Previous volumes have focused on GIS research in the marine and coastal realms (Wright and Bartlett, 2000) and on the application of GIS to oceanography and fisheries (Valavanis, 2002). The current volume is, to the best of our knowledge, the first to focus specifically on the role of GIS in integrated coastal zone management. We hope it will provide guidance, inspiration, encouragement and, where merited, a degree of caution, for all those tasked with the stewardship of the world's coasts, as well as for those whose interests are more academic and research-oriented.

The wide diversity of perspectives that can and must be brought to bear on the challenge of coastal zone management is reflected in the range and organisation of chapters in this book. Thus the opening chapters focus on technical issues, ranging from the incorporation of GIS within wider information infrastructures to techniques of visualisation, the importance of error and uncertainty in coastal databases, and the interfacing of GIS with simulation and process models. This is followed by a number of chapters that step back from technology, and which seek to put coastal zone GIS into a more human context, particularly through examination of cultural issues and exploration of techniques for incorporating traditional ecological knowledge within GIS-enabled coastal management regimes; and, finally, attention focuses on the use of GIS to historic shoreline change analysis, the application of geomatics to estuary management, and to better understanding and management of environmentally sensitive shorelines.

We are particularly delighted that contributions to this volume have come from each of the inhabited continents of the world, namely from Africa, Asia, Australia, Europe and North and South America. The diversity of perspectives on coastal management arising from the cultural and professional backgrounds of the authors, and also from the range of geographic locations used in the case studies and applications reported on, underscores the truly international dimension of coastal management today.

As always, compilation of an edited collection of papers depends on the support, encouragement and assistance of a vast number of people who have worked "behind the scenes." It is, of course, a pleasure to thank the authors who have contributed chapters to the book, and who have borne with cheerful patience the many demands – some reasonable, some perhaps less so – of the editors. We also acknowledge with gratitude the support of the International Geographical Union and the International Cartographic Association.

On an individual level, to merely "thank" Ron Furness and Andy Sherin seems woefully inadequate: it is no exaggeration to say that, without the sustained friendship and cheerful encouragement of both these gentlemen, this volume simply would never have seen the light of day. No less valued was the encouragement of our friends and colleagues on the International CoastGIS Scientific Committee, past and present.

Closer to home, Darius Bartlett wishes to thank friends and colleagues within the Geography Department and the Coastal and Marine Resources Centre at University College Cork; his postgraduate students for their lively discussions and thought-provoking questions; and, above all Mary-Anne, Becky and dog Jessa for putting up with my irregular hours, my absences from home and my all-too-frequent neglect of domestic duties and responsibilities. For her part, Jennifer Smith would like to thank Andy Sherin and the Canadian CoastGIS committee who facilitated her involvement in this project.

Finally, both authors acknowledge with gratitude the assistance, support and guidance of Tony Moore at Taylor & Francis in London and Randi Cohen and Jay Margolis at CRC Press in Florida, who helped steer production of this volume from conception through all stages of publication to its final appearance on booksellers' shelves.

REFERENCES

Valavanis, V. D., 2002, *Geographic Information Systems in Oceanography and Fisheries*, (London: Taylor and Francis).

Wright, D.J. and Bartlett, D.J., 2000, *Marine and Coastal Geographical Information Systems*. (London: Taylor and Francis).

Contributors

THE EDITORS

Darius Bartlett
Department of Geography, University College Cork, Cork, Ireland. Phone: +353 21 4902835;
Fax: +353 21 4902190; e-mail: djb@ucc.ie

Darius Bartlett first encountered GIS as a postgraduate student at Edinburgh University in about 1982, and has been researching and writing on conceptual, institutional and related issues arising out of coastal zone applications of GIS since the mid-1980s. More recently, he has started investigating the incorporation of marine and coastal areas into SDI initiatives, issues surrounding the diffusion to and use of GIS in the Developing World, and use of GIS by community groups, voluntary organisations and Non-Governmental Organisations (NGOs). An avid traveller, he has so far visited over 65 countries around the world, and looks forward to visiting the remainder in due course. He is a Member of the International Geographical Union's Commission on Coastal Systems; and was one of the founder organisers of the CoastGIS series of biannual conferences.

Jennifer L. Smith
World Wildlife Fund Canada, Atlantic Office, Suite 1202, Halifax, Nova Scotia, Canada, B3J 1P3. Phone: (902) 482-1105; Fax: (902) 482-1107; e-mail: jsmith@wwfcanada.org

Jennifer Smith manages the application of GIS in conservation planning for World Wildlife Fund Canada's Marine Program. She holds an Honours degree in Geography from McGill University, Montreal. Her interests in work and studies have focused on environmental monitoring, change in ecological systems, seagrass ecosystems, developing areas and GIS-based decision support for protected areas network design.

THE AUTHORS

Roger A. Longhorn
ral@alum.mit.edu

Roger Longhorn is an independent ICT policy consultant who holds B.Sc. and M.Sc. degrees in Ocean Engineering and Shipping Management from M.I.T, Cambridge, MA, USA. After more than a decade of implementing marine information systems for global maritime clients, in 1989 Roger became an external ICT expert to the European Commission, where since 1992 he focused on GIS technology and markets in the evolving Information Society. His special area of interest is GIS applied to the coastal zone. Currently a Ph.D. candidate in Information Policy at City University, London, his research focuses on regional spatial data infrastructures.

Simon Gomm
Ordnance Survey, Romsey Road, Southampton. SO16 4GU. UK. Phone: (+44)023 80305149

Simon has worked for Ordnance Survey in a variety of roles including geodetic surveying and data quality assurance and is now a Senior Research Leader where he is responsible for coordinating research on topics related to the capture, maintenance and use of spatial data.

Rongxing (Ron) Li
Department of Civil & Environmental Engineering and Geodetic Science, The Ohio State University, 470 Hitchcock Hall, 2070 Neil Avenue, Columbus, OH 43210-1275, Tel. (614) 292-6946, Fax. (614) 292-2957; e-mail: li.282@osu.edu, http://shoreline.eng.ohio-state.edu

Dr. Ron Li is a professor at the Department of Civil & Environmental Engineering and Geodetic Science of The Ohio State University. His research interests include digital mapping, coastal and marine GIS, spatial data structure, Mars Rover localization and landing site mapping.

Kaichang Di
Department of Civil & Environmental Engineering and Geodetic Science, The Ohio State University, 470 Hitchcock Hall, 2070 Neil Avenue, Columbus, OH 43210-1275, Tel. (614) 292-4303, Fax. (614) 292-2957; e-mail: di.2@osu.edu

Dr. Kaichang Di is a research associate at the Department of Civil & Environmental Engineering and Geodetic Science of The Ohio State University. His current research interests are coastal mapping using high-resolution satellite imagery, Mars Rover localization and landing site mapping.

Ruijin Ma
Department of Civil & Environmental Engineering and Geodetic Science, The Ohio State University, 470 Hitchcock Hall, 2070 Neil Avenue, Columbus, OH 43210-1275, Tel. (614) 292-4950, Fax. (614) 292-2957; e-mail: ma.106@osu.edu

Mr. Ruijin Ma is a Ph.D. candidate at the Department of Civil & Environmental Engineering and Geodetic Science of The Ohio State University. His current research interests are coastal mapping and GIS, 3D model reconstruction from LiDAR and photographs, and remote sensing applications.

Paul Pan
62 Llanishen Street, Cardiff, CF 14 3QD, United Kingdom; e-mail: pan@cardiff.ac.uk

Paul Pan was the Principal Investigator for a number of innovative coastal monitoring projects in South Wales. He lectured and researched on GIS with the University of Wales, Swansea and Cardiff University for over 10 years. Paul now works as an independent consultant

Eleanor Bruce
School of Geosciences, University of Sydney, Sydney, NSW, 2006, Australia. Phone: +61 2 9351 6443; Fax: +61 2 9351 3644; e-mail: ebruce@geosci.usyd.edu.au

Eleanor Bruce is a senior lecturer in the School of Geosciences at the University of Sydney. She teaches GIS, coastal management and advanced spatial data analysis. Her research interests include the use of GIS for marine park zoning, nearshore habitat mapping and coastal system modelling. Currently, she is Assistant Director of the Spatial Science Innovation Unit at the University of Sydney.

Sam Macharia Ng'ang'a
Department of Geodesy and Geomatics Engineering, University of New Brunswick, P.O. Box 4400, Fredericton, New Brunswick, Canada, E3B 5A3. Phone: (506) 447-3259 or (506) 455-7073; e-mail: sam.nganga@unb.ca

Sam Macharia Ng'ang'a obtained a Bachelor's degree in surveying from the University of Nairobi, Kenya and a Master's degree (Land Information Systems) from the Department of Geodesy and Geomatics Engineering, University of New Brunswick (Canada). He is a part time lecturer at UNB and is currently completing his Doctorate degree on marine protected area information systems. He holds memberships in (among other institutions) the Canadian Institute of Geomatics (CIG) and the Canadian Hydrographic Association (CHA).

Chul-sue Hwang
Department of Geography, Kyung Hee University, Seoul 131-701, Republic of Korea. Phone: +82-2-961-9313; e-mail: hcs@khu.ac.kr

Chul-sue Hwang is assistant professor of geography at Kyung Hee University, Korea, and a member of the editorial board of the *Journal of the Geographical Information System Association of Korea*. His recent research focuses on uncertainty of remote sensing data, spatial data mining, and exploratory spatial data analysis.

Cha Yong Ku
Department of Geography, Sangmyung University, Jongro-gu, Seoul 110-743, Republic of Korea. Phone: +82-2-2287-5043; e-mail: koostar@smu.ac.kr

Cha Yong Ku is assistant professor of geography at Sangmyung University. His research interests include the integration of GIS and remote sensing, classification accuracy assessment and scale effects in remote sensing, and land use/land cover information extraction and modelling for coastal wetlands. He received his Ph.D. in Geography from Seoul National University.

Simon Jude
School of Environmental Sciences, University of East Anglia, Norwich, Norfolk, United Kingdom, NR4 7TJ. Phone: +00 44 1603 591360; e-mail: s.jude@uea.ac.uk.

Dr. Simon Jude is a research associate in the School of Environmental Sciences at the University of East Anglia UK. His research involves developing the use of GIS, virtual reality and visualisation techniques for coastal decision-making.

Andrew Jones
School of Environmental Sciences, University of East Anglia, Norwich, Norfolk, United Kingdom, NR4 7TJ. Phone: +00 44 1603 593127; e-mail: a.p.jones@uea.ac.uk

Dr. Andrew P. Jones is a lecturer in the School of Environmental Sciences at the University of East Anglia UK.

Julian Andrews
School of Environmental Sciences, University of East Anglia, Norwich, Norfolk, United Kingdom, NR3 7TJ. Phone: +00 44 1603 592536.

Dr. Julian E. Andrews is a sedimentologist at the School of Environmental Sciences (University of East Anglia, Norwich UK) with special interest in modern and Holocene coastal sediments.

Roberto Mayerle
Coastal Research Laboratory (Corelab), Christian Albrechts University, Otto Hahn Platz 3, 24118 Kiel, Germany; e-mail: rmayerle@corelab.uni-kiel.de

Professor Mayerle is a specialist in numerical modelling of waves, currents, sediment transport and morphological changes in rivers, coastal areas and near hydraulic structures, with extensive experience in physical modelling of hydroelectric power schemes. He graduated as a civil engineer at the Federal University of Paraná in Brazil in 1979 where he worked for an energy concern involved in the construction and operation of several major hydropower schemes such as Itaipu. In 1988 he got his Ph.D. degree at the University of Newcastle-upon-Tyne in England. From 1989 till 1992 he was in-charge of a team working on the development of numerical models at the National Centre for Computational Hydroscience and Engineering in Oxford, USA. From 1992 till 1996 he worked at the Institute of Fluid Mechanics of the University of Hanover in Germany, being in charge of several research projects dealing with the investigation of the impact of climate changes on the morphological development both on the German North and Baltic Seas. Since 1996 he heads the Coastal Research Laboratory (Corelab) of the University of Kiel. Corelab is a research and teaching unit established to foster research in coastal environments. The Laboratory is engaged in applied research using a combination of in situ measurements and investigations as well as databases and numerical models embedded into decision support systems to help in the management of coastal areas.

Fernando Toro
Wilrijkstraat 37, 2140 Antwerpen, Belgium; e-mail: toro.gunst@skynet.be

Fernando Toro was born in Medellin, Colombia. He received his degree in civil engineering in 1993. He worked in Colombia in the construction of the Metro in the city of Medellin and in an engineering consultant company for three years. He pursued a Master of Science in computational hydrosciences, in the National Center for Computational Hydroscience and Engineering, at The University of Mississippi, USA, from 1996 to 1998. In 1998, he moved to Germany and did his Doctorate studies in the Coastal Research Laboratory, at the Christian Albrechts University in Kiel, until Summer 2003. He is presently working in an engineering consultant company in Antwerp, Belgium. His interests are numerical models and GIS applied to engineering problems.

Jacques Populus
Service des Applications Opérationnelles DEL/AO, IFREMER, BP 70, 29280 Plouzané, France
Phone: 0298224310; Fax : 0298224555; e-mail: jpopulus@ifremer.fr

Jacques Populus is a civil engineer who originally specialised in applications of high resolution remote sensing to coastal studies. His current activities concern the handling of geo-information for coastal applications, with a view to making it available to practitioners, in both developed and developing countries. This concerns GI as output of data analysis, remote sensing and hydrodynamic modelling. Current applications deal with water and its use for aquaculture sustainability. More recently, he has focused on acquisition and processing of physical data (topography and bathymetry) in the coastal zone by way of innovative techniques such as the LiDAR.

Lionel Loubersac
Head of the Coastal Environment and Living Resources Laboratory of Ifremer in Sète (France). He is in charge of the organisation, coordination and planning of scientific and technical programs dealing with a) the monitoring of coastal environment and living resources quality along the Mediterranean shores, b) the development, interfacing and transfer of tools in the field of coastal oceanography; i.e. environmental monitoring networks and databases, hydrodynamic modelling, geographical data bases and their integration within standardized coastal GIS, new technologies of information and communication for transferring to the public scientific results on coastal environment quality. He is involved in training courses, scientific committees of symposiums as well as numerous projects at national, European and international levels in the field of the applications of Remote Sensing and Information Systems to coastal management. He has been selected as European evaluator and expert in the framework of the DG INFSO Programme Information Society Technology, Key Action "Systems and Services for the Citizen; Applications related to environmental protection."

Jean-François Le Roux
Jean-François is a computer engineer with a Master's Degree in computer science. He has been a software engineer at Simulog, a high-technology service company specializing in technical software engineering, between 1997 and 2000, detached at Thomson CSF Optronics, involved in several projects of technical and operational simulation (real-time 3D simulation). Now he is a system and software engineer at Ifremer for the operational applications service of the Environment and Coastal Planning Division (DEL/AO). He works on interfacing GIS, web and hydrodynamic models.

Franck Dumas
A coastal ocean circulation modeler, Franck graduated from Ecole Nationale Supérieure des Techniques Avancées. For the past seven years, he has been using the MARS-2D-3D modelling system intensively in various frameworks: in the context of pure physical oceanography and circulation along the French coast, but also in pluridisciplinary contexts for the understanding of complex coastal ecosystems such as that of the Baie du Mont Saint-Michel. He is now in charge of developing Ifremer's coastal ocean modelling system, and integrating all coastal ecosystem components developed over the past years within the Institute.

Valerie Cummins
Coastal and Marine Resources Centre, Environmental Research Institute, University College Cork, Naval Base, Haulbowline Island, Cork Harbour, Ireland. Phone: +353 (21) 4703100; Fax: +353 (21) 4703132; e-mail: v.cummins@ucc.ie

Valerie Cummins (B.Sc., M.Sc.) received her B.Sc. in marine geography from University of Wales, College of Cardiff, and her M.Sc. through the Department of Zoology and Aninmal Ecology in University College Cork. After several years working in the field of spatial data analysis/GIS in the UK at the British Oceanographic Data Centre, and subsequently at Landmark Information Group, Valerie returned to Ireland in 1999 to join the Coastal & Marine Resources Centre. She is currently manager of the CMRC with responsibility for the co-ordination of nineteen national and European funded research programs, with a current staff of 17.

Gerry Sutton
Coastal and Marine Resources Centre, Environmental Research Institute, University College Cork, Naval Base, Haulbowline Island, Cork Harbour, Ireland. Phone: +353 (21) 4703100; Fax: +353 (21) 4703132; e-mail: gerry.sutton@ucc.ie

Gerry Sutton graduated from University of Wales, Bangor, with Joint Honours degree in marine biology and zoology in 1984, following which he worked for two years as a fisheries officer with the Department of Fisheries in Sabah, East Malaysia. Here he focussed on research, development and subsequent local adoption of macro-algal cultivation techniques (*Eucheuma* spp.). Following his return to Ireland Gerry joined Hydrographic Surveys Ltd where, as senior hydrographic surveyor between 1991 and 1998, he was responsible for planning, conducting, processing and charting high precision coastal surveys associated with national and international civil engineering, dredging, and oceanographic projects. Since joining the CMRC in 1998 Gerry has been actively conducting research within the CMRC team, contributing to a number of EU and nationally funded projects, and consultative reports. Gerry recently received his M.Sc through the Department of Geography at University College Cork, in the field of marine resources and GIS. His primary research interests are currently in the fields of seabed mapping (specialising in multibeam sonar acquisition and processing); geophysics; oceanography; marine resources and technology; and marine geographic information systems. Gerry has been a professional member of the Irish Institution of Surveyors (IIS) since 1997.

Françoise Gourmelon
Laboratoire Géomer, LETG UMR6554 CNRS, European Institute for Marine Studies (IUEM), Technopôle Brest-Iroise, 29280 Plouzané, France. Phone: + 00 33 2 98 49 86 83; Fax: + 00 33 2 98 49 86 86; e-mail: francoise.gourmelon@univ-brest.fr

Françoise Gourmelon is a senior research worker at the CNRS (French National Center for Scientific Research). She is a member of the Géomer laboratory (LETG UMR 6554 CNRS), part of the European Institute of Marine Studies. She has been using GIS in ICZM studies since 1989, when she met François Cuq (Director of the CNRS Géosystèmes laboratory). She has been working with coastal and marine applications such as landuse and landcover changes in protected islets, habitat of marine fauna, in temperate and tropical coastal zones. Her current focuses are on the coupling of models with GIS for better knowledge of social/natural dynamics; on the use of remote sensing data for the long term monitoring of protected areas; and on the design of geomatic infrastructures dedicated to long term environmental observation.

Iwan Le Berre
Laboratoire Géomer, LETG UMR6554 CNRS, European Institute for Marine Studies (IUEM), Technopôle Brest-Iroise, 29280 Plouzané, France. Phone: 00 33 2 98 49 86 80 ; Fax: 00 33 2 98 49 86 86; e-mail: iwan.leberre@univ-brest.fr

Iwan Le Berre acquired his first experience with GIS when he joined the Géosystèmes laboratory as a Master of Science student in 1990 (and has stayed ever since!). His research interests deal with the implementation of GIS for synthetic mapping of coastal and marine environment. In 1992, he got a grant from the MaB committee of UNESCO for the achievement of a synthetic map of the Iroise Sea Biosphere Reserve. He defended his Ph.D., focused on the implementation of a coastal and marine GIS for the Iroise Sea, in 1999 and has been awarded by the Regional Council of Brittany in 2001. After several studies with different organisations in France (National Park Service, National coastal zone heritage agency, Ministry of Equipment) he got his position at the Western Brittany University in 2001 and he now teaches cartography, GIS and remote sensing for graduate, master and Ph. D. students.

R. Sudarshana
RI227, National Remote Sensing Agency, Dept. of Space, Govt. of India, Balanagar, Hyderabad 500 037, India; e-mail: sudarshana_r@nrsa.gov.in

R. Sudarshana conducted his doctoral research on seabed ecology in the Arabian sea and taught in Karnatak University, India for a while. He then joined the Department of Space, Government of India, wherein he developed an academic niche in marine science, developing and propagating remote sensing application skills. After several years of work in the field of marine applications of remote sensing, he now heads the programme planning affairs of the national remote sensing activities. He has been a consultant to

UNESCO in Africa, Arabia and Iran, besides being a visiting professor in institutions in Europe and Japan.

John Lindsay

John A. Lindsay, NOAA Pribilof Project Office, 7600- Sand Point Way NE, Building 3, Seattle, WA 98115. Phone: (206) 526-4560; e-mail: john.lindsay@noaa.gov

John A. Lindsay is the director of the Pribilof Islands Environmental Restoration Project. He began work with NOAA more than fifteen years ago following a long career as a marine invertebrate taxonomist and ecologist. At NOAA, Mr. Lindsay implemented the agency's natural resource trustee responsibility nationwide at hazardous waste sites, and chemical and oil spills. He also represented the U.S. Department of Commerce on the Joint U.S./Canada Atlantic Regional Environmental Emergencies Team.

Thomas J. Simon

NOAA Environmental Compliance, Health & Safety, and Security Office, 7600- Sand Point Way NE, Building 1, Seattle, WA 98115. Phone: (206) 526-6295; e-mail: Tom.Simon@NOAA.gov

Tom Simon is a geographer and GIS specialist who helped develop the NOAA Pribilof Project Office GIS project. His interests include the use of GIS to integrate varying technologies and develop interactive presentation tools.

Aquilina D. Lestenkof

Aleut Community of St. Paul Island, Tribal Government, P.O. Box 86, St. Paul Island, AK 99660. Phone: 907-546-2641; Fax: 907-546-2655; E-mail: aquilina@tdxak.com

Aquilina D. Lestenkof of St. Paul Island continually seeks ways to balance and blend the cultural knowledge of her people – Unangan (Aleut) – with present day life. Aquilina currently co-directs the Ecosystem Conservation Office of the Aleut Community of St. Paul Island's Tribal Government.

Phillip A. Zavadil

Aleut Community of St. Paul Island, Tribal Government, P.O. Box 86, St. Paul Island, AK 99660,
Phone: 907-546-2641; Fax: 907-546-2655; e-mail: pazavadil@tdxak.com

Phillip A. Zavadil of St. Paul Island co-directs the Ecosystem Conservation Office of the Aleut Community of St. Paul Island's Tribal Government. Mr. Zavadil enthusiastically seeks means to understand and improve the quality of life on St. Paul Island.

Peter Nwilo
Department of Surveying and Geoinformatics, University of Lagos, Lagos, Nigeria.

Dr. Nwilo is an associate professor of surveying and geoinformatics at the University of Lagos, Lagos, Nigeria, and is currently the sub-dean of the Postgraduate School of the University. Dr. Nwilo has a B.Sc. and M.Sc. in Surveying from the University of Lagos and a Ph.D. in environmental resources from the University of Salford, United Kingdom. His special research interests cover sea level variations and the impacts along the coast of Nigeria; GIS and its applications in coastal management; subsidence phenomenon along the coast of Nigeria; management of the navigable rivers of Nigeria; and management of Nigerian coastal areas. He has published extensively on these topics.

Tim Webster
Email: Timothy.Webster@nscc.ca

Tim Webster is a research scientist with the Applied Geomatics Research Group (AGRG) at the Centre of Geographic Sciences (COGS), a component of the Nova Scotia Community College. He currently is also pursuing his Ph.D. in earth sciences from Dalhousie University where he is applying DEM technologies including LIDAR to mapping geological landforms.

Charles Sangster
Email: sangstcw@gov.ns.ca

Charles Sangster is a graduate from the Applied Geomatics Research Program at COGS, and works in the field of conservation GIS. Currently, he is the GIS specialist for the Protected Areas Branch of the Nova Scotia Department of Environment and Labour.

Montfield Christian
Email: c_montfield@hotmail.com

Montfield Christian is a graduate from the Applied Geomatics Research Program at COGS, and works in Toronto, Ontario as an independent GIS analyst. His company is Green Eminent Consulting and they specialise in customised GIS projects.

Dennis Kingston
Email: Dennis.Kingston@nscc.ca

Dennis Kingston is the department head of the Information Technology Department, Annapolis Valley Campus, NSCC. Formerly, he taught GPS technology at COGS in the Survey Program.

Courtney Schupp
Department of Physical Sciences, Virginia Institute of Marine Science, Rt. 1208 Greate Road, Gloucester Point, VA 23062

Courtney Schupp focuses her research on shoreline behavior and its relationship to nearshore processes and underlying geology. After earning her B.S. in geology from Duke University, she worked as a GIS analyst at the U.S. Geological Survey and as a consultant to the Maryland Geological Survey. She is currently working towards an M.S. in marine science at the Virginia Institute of Marine Science at the College of William and Mary.

Rob Thieler
U.S. Geological Survey, Coastal and Marine Geology Program, 384 Woods Hole Road, Woods Hole, MA, 02543-1598, USA. Phone: (508) 457-2350; Fax: (508) 457-2310; e-mail: rthieler@usgs.gov; http://marine.usgs.gov

Rob Thieler is a research geologist with the U.S. Geological Survey in Woods Hole, MA. His research includes continental margin sedimentation, late Quaternary sea-level change, coastal hazards, and development of software to quantify rates of shoreline change.

James F. O'Connell
Woods Hole Oceanographic Institution Sea Grant Program, 193 Oyster Pond Road, MS#2, Woods Hole, MA, 02543-1525, USA. Phone: (508) 289-2993; e-mail: joconnell@whoi.edu

Jim is presently the coastal processes specialist with the Woods Hole Oceanographic Institution Sea Grant Program and Cape Cod Cooperative Extension (past 5 years). Prior to joining WHOI, Jim was the coastal geologist and coastal hazards specialist with the Massachusetts Office of Coastal Zone Management for 13 years, followed by a year with the Cape Cod Commission as a marine resources specialist. Jim's speciality is shoreline change analysis, analyzing human effects on coastal processes and coastal landforms, coastal floodplains, erosion control alternatives, and coastal hazard mitigation issues.

Jorge I. Euán-Avila,
Department of Marine Resources, CINVESTAV, Mérida, México; e-mail: euan@mda.cinvestav.mx.

Jorge I. Euán-Avila, Ph. D., is an engineer and researcher in the Department of Marine Resources, CINVESTAV, Mérida, México.

María de los Angeles Liceaga-Correa
Department of Marine Resources, CINVESTAV, Mérida, México; e-mail: liceaga@mda.cinvestav.mx.

María de los Angeles Liceaga-Correa,Ph.D, is a mathematician and researcher in the Department of Marine Resources, CINVESTAV, Mérida, México.

Héctor Rodríguez-Sánchez
Department of Marine Resources, CINVESTAV, Mérida, México; e-mail: hrs@mda.cinvestav.mx.

Héctor Rodríguez-Sánchez, M.Sc., is a biologist and research assistant in the Department of Marine Resources, CINVESTAV, Mérida, México.

David R. Green
Centre for Marine and Coastal Zone Management (CMCZM) / Aberdeen Institute of Coastal Sciences and Management (AICSM), Department of Geography and Environment, School of Geosciences, College of Physical Sciences, University of Aberdeen, Elphinstone Road, Aberdeen, AB24 3UF, Scotland, UK. / Tel. +44 (0)1224 272324 /Fax. +44 (0)1224 272331; e-mail: d.r.green@abdn.ac.uk Internet. www.abdn.ac.uk.cmczm or www.abdn.ac.uk/aicsm

David R. Green was educated in geography at the Universities of Edinburgh, Pennsylvania and Toronto. He is currently director of the Centre for Marine and Coastal Zone Management (CMCZM), and assistant director of the Aberdeen Institute for Coastal Sciences and Management (AICSM) at the University of Aberdeen in Scotland. He lectures in the environmental applications of the geospatial technologies (including remote sensing and GIS) with special research interests in online Internet-based GIS, mobile mapping, and user interfaces for use in integrated coastal zone management. David is currently president of the EUCC-The Coastal Union, and GIS Editor of the *Journal of Coastal Conservation* (JCC). He is also chief assessor and examiner for the ASET GIS Programme, chairman of the AGI Marine and Coastal Zone Management GIS Special Interest Group, and deputy chair of the ICA Commission on Marine Cartography (International Cartographic Association). David has presented papers and workshops at a number of national and international conferences, and has published a number of books on the use of GIS in landscape ecology, school education, marine and coastal applications, and use of remote sensing for coastal applications. He is currently involved in organising the Littoral 2004 and CoastGIS 2005 conferences.

Stephen D. King
Centre for Marine and Coastal Zone Management, Department of Geography and Environment, University of Aberdeen, Elphinstone Road, Aberdeen, AB24 3UF, UK.; Tel. +44 (0)1224 272324, Fax +44 (0)1224 272331; e-mail: s.d.king@abdn.ac.uk

Stephen is currently a research assistant in the Department of Geography and Environment at the University of Aberdeen. He has a B.A. in geography

(University of Liverpool) and an M.Sc. in environmental remote sensing (University of Aberdeen). His main interests are the application of remote sensing and GIS to coastal zone management and, in particular, the development of Internet-based coastal information systems and the use of mobile communications technology and GIS for field data collection.

Jean-Côme Bourcier
8 Rue Georges Laroque, Appartement 1021, 76300 Sotteville-Les-Rouen FRANCE. Phone: +33 02 63 87 30 or +33 06 88 43 54 95; e-mail: jean-come.bourcier@wanadoo.fr or geoinformatique@wanadoo.fr

J.-C. Bourcier obtained his Ph.D. at University of Le Havre, France, and works as a GIS specialist, environmental engineer and territorial planning engineer. His research concerns the use of GIS as a tool designed to create a better understanding of coastal processes, functioning and change and, more specifically, better management and development of coastal spaces such as estuaries. He is currently the president of GeoInformatique, an association which helps people to better use geographical information.

Anders Mosbech
National Environmental Research Institute, Department of Arctic Environment, Frederiksborgvej 399, P.O. Box 358, DK-4000 Roskilde, Denmark. Phone: +45 4630 1934; Fax: +45 4630 1914; e-mail: amo@dmu.dk

Anders Mosbech, Ph.D., is a senior scientist who performs ecological research in Greenland on marine birds, geese, muskoxen, marine mammals and vegetation. He is an advisor to the regulatory agencies in the Greenlandic and Danish governments concerning environmental impacts of oil exploration and development in Greenland.

David Boertmann
National Environmental Research Institute, Department of Arctic Environment, Frederiksborgvej 399, P.O. Box 358, DK-4000 Roskilde, Denmark. Phone: +45 4630 1934; Fax: +45 4630 1914; e-mail: dmb@dmu.dk

David Boertman is a senior research biologist with long experience in arctic ornithology and Arctic natural history, and has worked with environmental issues related to oil exploration in the arctic. He is currently associated with the National Environmental Research Institute, Denmark.

Louise Grøndahl
National Environmental Research Institute, Department of Arctic Environment, Frederiksborgvej 399, P.O. Box 358, DK-4000 Roskilde, Denmark. Phone: +45 4630 1934; Fax: +45 4630 1914; e-mail: lgr@dmu.dk

Louise Grøndahl holds an M.Sc. in physical geography. She has worked as research assistant at NERI with GIS and is now doing a Ph.D. on carbon flux in Northeast Greenland at the National Environmental Research Institute, Denmark.

Frants von Platen
Geological Survey of Denmark and Greenland (GEUS). E-mail: fph@geus.dk

Frants von Platen is a geographer and GIS-specialist who has been employed since 1983 at GEUS, where he works with geological databases and GIS development.

Niels Nielsen
Institute of Geography, University of Copenhagen, Denmark. Phone: +45 35322508; Fax: +45 35322501; e-mail: nn@geogr.ku.dk; www.geogr.ku.dk

Niels Nielsen is an associate professor in physical geography (coastal geomorphology) at the Institute of Geography of the University of Copenhagen.

Søren Stach Nielsen
Greenland Institute of Natural Resources, P.O. Box 570, DK-3900 Nuuk, Grønland. Phone: +299 328095 /ext. 243; Fax: +299 325957; email: SorenSN@natur.gl

Søren Stach Nielsen holds a Master of Technological and Sociological Planning degree from Roskilde University and is at present a project-researcher at the Greenland Institute of Natural Resources, investigating socioeconomic relations between hunting and hunting-gear and hunting and household economy, and incorporation of local knowledge into scientific research.

Morten Rasch
Danish Polar Center, Strandgade 100H, DK-1401 Copenhagen K, Denmark. Direct phone: +45 32880110 Mobile phone: +45 23490645; Secretary: +45 32880100; Fax: +45 32880101; e-mail: mr@dpc.dk; homepage, Zackenberg: www.zackenberg.dk, homepage, Danish Polar Center: www.dpc.dk

Morten Rasch, Ph.D, is a coastal geomorphologist educated at the Institute of Geography of the University of Copenhagen. He is a specialist in arctic coastal geomorphology and Holocene relative sea-level changes in Greenland. He has been working as station manager of Zackenberg Station, North East Greenland and is also head of logistics at Danish Polar Center.

Hans Kapel
SILA, The Greenland Research Center at The National Museum of Denmark, Frederiksholms Kanal 12, DK 1220 Copenhagen K. Phone: +45 33473256; e-mail. hans.kapel@museum.dk

Hans Kapel is a curator and archaeologist who has conducted archaeological research in Greenland since 1969. His field of responsibility includes The Central Register of Protected Monuments in Greenland.

André Laflamme
Environment Canada - Environmental Emergencies Section, Dartmouth, Nova Scotia, CANADA. Phone: 902-426-5324; Fax: 902-426-9709; e-mail: Andre.laflamme@ec.gc.ca

André Laflamme graduated from University of Sherbrooke, Quebec, in physical geography in 1994. He has worked since with Environment Canada Atlantic Region with the environmental emergencies section as an environmental emergencies officer. André is also the regional sensitivity mapping coordinator for Environment Canada. He was a co-author on scientific papers produced during several conferences such as the International Oil Spill Conference (IOSC), and the Arctic and Marine Oil Spill conference (AMOP). André was also part of a team tasked to develop a national oil spill contingency plan for the Republic of Chad (Africa) where he spent some time.

CHAPTER ONE

Coastal Spatial Data Infrastructure

Roger A. Longhorn

1.1 INTRODUCTION

The term Spatial Data Infrastructure (SDI) is now in common use in countries around the world, although definitions for the term differ quite considerably. The stated objectives of SDI initiatives vary as much as do the definitions, legal mandates, types of organisation responsible for specifying and implementing SDI and actual progress achieved in creating national and regional SDIs.

One complication in specifying any SDI is the nature of spatial information, i.e. information with an important location attribute, often said to represent 80% of all information held, especially at government level. The visionaries and designers of SDI must accommodate the widely varying information needs of highly diverse disciplines and sectors of society, business and government. Health epidemiologists are seldom interested in the same spatial data as geological surveyors, air traffic controllers or coastal zone managers. Yet an important overlap in jurisdiction and information needs may arise, e.g. when a potential health epidemic is generated by toxic chemical concentrations in marine fauna later consumed by area residents. Then knowledge of the coastal zone flora and fauna, hydrography, tidal states, nearby land use practices of industry and agriculture and transport routes, fishing practices and zones all become intertwined. The complex relationships between different types of spatial information are one reason that countries take different routes to specify their SDI, ranging from visions to strategies to goals to detailed content (data and standards) and implementation plans (rules and regulations).

We all recognize that the coastal zone is a difficult geographical area to manage due to temporal issues (tides and seasons) and the overlapping of physical geography and hydrography (offshore, near shore, shoreline, inshore), of jurisdictions, legal mandates and remits of government agencies and the often competing needs of stakeholders. Typically, many different local, national and regional government agencies are responsible for different aspects of the same physical areas and uses of the coastal zone, e.g. fisheries, environment, agriculture, transport (inland and marine), urban planning and cadastre, national mapping agency and the hydrographic service.

Conflict resolution in this complex piece of real estate called the "coast" is sometimes used as justification for special attention for the coast within NSDI initiatives. As Bartlett (2000) expresses it: "Given the diversity of interest groups, stakeholders, managerial authorities and administrative structures that converge at

0-41531-972-2/04/$0.00+$1.50

the shore, conflicts are almost inevitable between and among coastal users, managers, developers and the wider public, as well as between human society and the natural environment."

Because of such complex physical and institutional relationships, it is not possible to develop a coastal SDI (CSDI) in isolation from the broader National SDI (NSDI) for a nation or Regional SDI (RSDI). CSDI will necessarily be a subset of a more comprehensive NSDI because the coastal zone covers multiple physical and institutional spaces included in the generic NSDI. For that reason, it is important that people and agencies with specific knowledge and experience of the coastal zone and marine offshore areas and information requirements be an integral part of the NSDI and RSDI planning process.

1.2 CSDI WITHIN NATIONAL, REGIONAL AND GLOBAL SDI

Few nations have specified separate SDI components for individual sectors such as the coastal zone. In the USA, a Coastal NSDI vision exists based on four goals that relate to the USA NSDI (NOAA, 2001). Practical implementation work is undertaken by the National Oceanic and Atmospheric Administration (NOAA) and the Federal Geographic Data Committee (FGDC).

Canada has proposed a Marine Geospatial Data Infrastructure initiative (MGDI) within the national Canadian Geospatial Data Infrastructure (CGDI) (Chopin and Costain, 2001). MGDI is seen as an extension to the CGDI in response to "the need for a comprehensive, integrated and common infrastructure of marine data and information … accessible to all stakeholders."

1.2.1 Regional SDI Initiatives

At regional level, the INSPIRE project (Infrastructure for Spatial Information in Europe) recognizes hydrographic data as one of its "selected topographic themes" for data content for a Regional SDI for Europe (INSPIRE, 2002). More detailed data requirements are presented in the European Commission's Water Framework Directive (European Commission, 2000), which underpins the need for INPSIRE. These two European regional projects are discussed more fully in a later section as they have legal mandates from a recognized regional government body—the European Commission—and have developed detailed descriptions of spatial data requirements, including marine and coastal zone data.

Work on specification for an Asia-Pacific SDI (APSDI) began in 1994 with formation of the Permanent Committee on GIS Infrastructure for Asia and the Pacific (PCGIAP) under the auspices of the UN Regional Cartographic Conferences (UNRCC). The APSDI model stresses institutional frameworks, technical standards, fundamental datasets and an access network (PCGIAP, 1998).

A new initiative started in 2000 for a Regional SDI for the Americas (PC-IDEA), but it is too new to have achieved anything more than creating a basic project organisational framework at this stage.

1.2.2 Global SDI Initiatives

Global level SDI initiatives include the Global Map Project (ISCGM, 2001), which began in 1996, the Global Spatial Data Infrastructure (GSDI) Association, which began as a series of international conferences in 1996, and the Digital Earth initiative sponsored by the US government (Evans, 2001). These initiatives intend to build on existing national and/or regional SDI work and existing data resources (primarily remote sensing imagery). They typically mention shoreline data, boundary data (including marine, ocean, inland waters) and references to cadastre that may or may not include marine cadastral data. GSDI and the Global Map project are discussed more fully later.

Of more immediate interest—and importance—to coastal and marine managers and researchers is the work of the Intergovernmental Oceanographic Commission (IOC) International Oceanographic Data and Information Exchange (IODE) working group mentioned later and described in some detail by Longhorn (2002).

1.3 CSDI IN THE US

The US NSDI initiative began formally in April 1994 with a presidential Executive Order (Clinton, 1994) in response to a national performance review of federal government. One of the main elements of the US NSDI is the creation, harmonisation and promulgation of technical standards by which all federal spatial information can be recorded, metadata being of highest importance so that existing spatial data resources can be identified, accessed and used. This major task fell to the Federal Geographic Data Committee (FGDC), which was created in 1990. Responsibilities with regard to NSDI implementation for all federal agencies in the US was set out in the revised OMB Circular A-16 in 2002 (OMB, 2002).

In the US, the Coastal SDI vision is based on four goals relating to the NSDI (NOAA, 2001). The US CSDI initiative is led by the Coastal Services Centre of NOAA (National Oceanic and Atmospheric Administration). The main goals of the US CSDI are:

- that the coastal management community should understand and embrace the vision, concepts and benefits of the NSDI,
- spatial coastal and marine framework data should be readily available to the coastal management community,
- technologies to facilitate discovery, collection, description, access and preservation of spatial data should be widely available to the coastal management community, and
- NOAA should help develop and implement spatial data applications to meet the needs of the coastal and marine communities.

Note the emphasis on support for coastal resources management, a role assumed by NOAA under the 1972 Coastal Zone Management Act, as opposed to

NOAA's historical focus and initial primary mission of ensuring safety of navigation.

The main elements of CSDI that NOAA promote are bathymetry, shoreline identification and marine cadastre, although other types of data are under review for attention as a formal part of CSDI, e.g. coastal imagery, marine navigation, tidal benchmarks and benthic habitats (Lockwood and Fowler, 2001). Development of technical standards to support the CSDI is undertaken by the FGDC's Marine and Coastal Spatial Data Subcommittee, under chairmanship of NOAA/CSC (FGDC, 2002). The committee's primary mission is to develop and promote the Marine and Coastal NSDI so that "current and accurate geospatial coastal and ocean data will be readily available to contribute locally, nationally, and globally to economic growth, environmental quality and stability, and social progress" (NOAA, 2003). In 2000, this committee assumed the remit of the FGDC Bathymetric and Nautical Charting Data Subcommittee, formed in 1993.

Bathymetric data is treated as a sub-layer of the Elevation layer data in the NSDI Framework. Marine cadastre is being examined within both the FGDC Marine Boundary Working Group (MBWG) which includes members of the FGDC Cadastral Subcommittee. The MBWG was formed in 2001 to address issues relating to the legal and technical aspects of marine boundaries, with the goal to alleviate cross-agency problems concerning marine boundaries, plus provide outreach, standards development, partnerships and other data development critical to the NSDI. A major product of the MBWG work to date is the FGDC's Shoreline Metadata Profile (FGDC, 1998).

Two further standards nearing completion for the CSDI are the National Hydrography Data Content Standard for Inland and Coastal Waterways (FGDC, 2000a) and Accuracy Standards for Nautical Charting Hydrographic Surveys (FGDC, 2000b).

The US CSDI initiative should help identify the basic reference data required to achieve goals for national programs such as those stated in the Clean Water Action Plan - Coastal Research and Monitoring Strategy, which represents "the first effort to integrate coastal monitoring and research activities on a national scale to provide thorough, cross-cutting assessments of the health of the nation's coastal resources" (CRMSW, 2000). These goals include:

- Improving monitoring programs for integration at national level
- Integrating interagency research efforts
- Conducting national and regional coastal assessments, and
- Improving data management.

Interestingly, a report from the US Commission on Ocean Policy found a tight connection between inland systems for development and agriculture to areas traditionally designated as coastal. "The coastal zone is not a narrow band. It's the whole country" (US CoOP, 2002). Regarding data collection and sharing, the US Ocean Policy report further found that "there is no marine equivalent to the networks of meteorological observation stations distributed on land on all continents. Ocean observation efforts are limited temporally and spatially." This leads to the conclusion that "there is a need for a better and more comprehensive way to link the work of different disciplines in a manner that offers a more integrated understanding of the marine environment and the processes that control it… there is a need for standardized practices and procedures" (US CoOP, 2002). These findings reinforce the premise that CSDI cannot and should not be developed in isolation from the broader NSDI of a nation or region. Actions like the Clean Water Action Plan mentioned earlier provide important input into the political process for securing added support for creating a CSDI.

1.4 CSDI IN CANADA

In Canada, the NSDI is called the Canadian Geospatial Data Infrastructure (CGDI) with the snappier, market-oriented title GeoConnections. From its beginning, the CGDI recognized that "governments have a responsibility to make geospatial information available … for developing a knowledge economy in response to the needs of citizens, industry and communities in support of economic, social and environmental well-being" (Labonte, Corey and Evangelatos, 1998). Marine navigation and charting for pollution control, coastal zone management and environmental monitoring were considered important applications to be fostered under CGDI. The CGDI has five technical and policy thrusts: access, data framework, standards, partnerships and creating a supportive policy environment. The CGDI vision is "to enable timely access to geo-info data holdings and services in support of policy, decision-making and economic development through a co-operative interconnected infrastructure of government, private sector and academia participants. From the outset, CGDI planners realized that "institutional issues will likely eclipse technology as an impediment to CGDI development and implementation" (Labonte *et al.*, 1998).

Within the framework of the CGDI, the Marine Geospatial Data Infrastructure (MGDI) is being developed in Canada, "to enable simple, third party access to data and information that will facilitate more effective decision-making" (Gillespie *et al.*, 2000) for anyone involved in coastal zone management. MGDI is described as comprising data and information products, enabling technologies as well as network linkages, standards and institutional policies. The concept for an MGDI-like information network was first proposed in 1988 as the "Inland waters, Coastal and Ocean Information Network (ICOIN)" (after Butler *et al.*, 1998 in Gillespie *et al.*, 2000).

In support of the five main elements of GeoConnections (the CGDI), a Marine Advisory Committee was created in 1999 at the time of publication of a draft concept report for the MGDI (CCMC, 1999). The committee's remit is to ensure the full functionality of the CGDI in providing service to all marine stakeholders. To aid in this, a Marine Advisory Network has been set up to act as the physical focal point for stakeholder outreach and consultation (GeoConnections, 2003).

The MGDI recognizes the need for common standards permitting data to be used seamlessly across disciplines and systems. Because of the principally land-based focus of most NSDI developments, MGDI also recognizes that standards that apply perfectly well to land-based applications and data are sometimes incompatible with the marine world. It is also understood that simply making spatial data available across the Internet does not always provide solutions to specific problems. Data is not information and information is not knowledge. MGDI also recognizes that data pricing and related policy issues dealing with intellectual copyright will be crucial to the success of both CGDI and MDGI and that, while these potential barriers may not fall easily, they are not insurmountable. The needs of users and potential stakeholders were identified more completely in an extensive requirements analysis study in 2001 (DFO, 2001).

The MGDI architecture includes:

- A common spatial data model
- An integrated process and data modelling environment
- A common spatial language and data exchange format
- Methods for managing, querying and delivering data with integrity, and
- Open source productivity tools ensuring access for all.

Progress in implementing the MGDI as a coherent system has been slower than anticipated, due mainly to two things that affect nearly all such initiatives known to this author, i.e. lack of resources (especially any new or extra resources) and institutional barriers inherent in government whether local or national. Note that even the Presidential Executive Order creating the USA NSDI did not offer any new money, rather all work to achieve the NSDI had to be conducted from within existing budgets.

1.5 CSDI ELEMENTS IN REGIONAL SDI

Regional SDI (RSDI) initiatives have started in Europe (INSPIRE), Asia-Pacific (PCGIAP) and the Americas (PC-IDEA). These projects all intend to build on National SDIs located within their geographic regions. Unfortunately, since few NSDIs have reached an advanced degree of completion—and many have not yet even been specified or received legal mandates—it is still early to see how coastal and marine SDI requirements will be accommodated within the RSDI envelopes.

The Permanent Committee for SDI for the Americas (CP-IDEA/PC-IDEA) was only formed early in 2000 under the auspices of the UN Regional Cartographic Conference. This new initiative has yet to produce any substantial documents relating to how SDI will be coordinated across the nations of Central and South America. Initially it is focusing on institutional strengthening, metadata, access policies, outreach and similar issues to those being pursued by PCGIAP and within the global GSDI Association.

1.5.1 INSPIRE—Infrastructure for Spatial Information in Europe

INSPIRE recognizes hydrographic data as one of its selected topographic themes for data content for the proposed Regional SDI for Europe (INSPIRE, 2002). Hydrography is defined as "surface water features such as lakes and ponds, streams and rivers, canals, oceans and shorelines." During the many years that European SDI consultations progressed from the GI2000 initiative (1995–1999) through ETeMII (European Territorial Management Information Infrastructure 1999–2001) to INSPIRE (2001–present), reference to specific themes for spatial data content in the proposed SDI specification was initially resisted. After much debate, the hydrographic component (including coastal zones) was included, along with only two other "topographic themes" - transport and height. Hydrography is defined in INSPIRE as "surface water features such as lakes and ponds, streams and rivers, canals, oceans and shorelines." The "height" topographic theme includes "contour data showing heights by isolines, and including with the same data set spot heights, high and low water lines, breaklines and bathymetry" (INSPIRE, 2002). At the end of the consultation period, INSPIRE was extended to cover a wide range of coastal and marine data components, including "bathymetry, coastline, hydrography, surface water bodies, water catchments, oceans and seas, oceanographic spatial features, sea regions, aquaculture facilities, polluted areas and more (INSPIRE, 2003).

1.5.2 European Water Framework Directive

The EU Water Framework Directive (European Commission, 2000) represents the culmination of five years of consultation and negotiations for implementing a harmonized and integrated water policy for all European Union Member States (now 15, soon to be 25). The WFD places quite detailed—and some say onerous—monitoring and reporting requirements for the status of surface water in four regimes: rivers, lakes, coastal waters and transitional waters (estuaries and similar bodies of water which are partly saline but strongly influenced by freshwater flows).

The data relating to coastal and transitional waters required in order for EU Member States to report to the European Commission on water condition include: detailed location (boundary) information, various biological data for aquatic flora and benthic fauna, hydromorphological data including depth variations, tidal regimes, transparency, thermal conditions, oxygenation conditions, salinity, nutrient conditions and pollution.

In order to do this reporting, a very detailed GIS specification has been produced (European Communities, 2002). For coastal waters, the data to be captured include shape, name, various identifying codes, type of water body, status of the water body (artificial, heavily modified), salinity typology, depth typology, tidal typology, and more. The specification document is nearly 170 pages long—a far cry from the "hydrography" or "marine boundary" terms found in isolation, without further explanation, in other regional and all global SDI specifications. In fact, one worries if the European government authorities and their national oceanographic and coastal research institutions will be able to provide data at the level of detail required within the timeframe set by the Directive, which comes into force in December 2003.

The WFD Directive has been used by various Directorates General at the European Commission to justify the INSPIRE European Regional SDI initiative, on the basis that much harmonised, integrated and interoperable basic reference data will be required if WFD reporting requirements are to be met at least cost. It remains to be seen if the level of detail already specified in the INSPIRE working documents will in fact be sufficient to help fulfil that requirement.

1.5.3 Asia-Pacific Regional SDI (APSDI)

In the Asia-Pacific SDI (APSDI), what are termed "fundamental datasets" are only presented in the most general terms and with no detail. The discussion on what data should be included in the APSDI began in 1998 and has yet to reach a conclusion. Due to the large number of participants—55 nations across Asia and the Pacific are members—and very limited development resources, we should not expect the situation to change any time soon in regard to formal data specification. At the most recent meeting of the Working Group on Fundamental Data held in April 2002, the working group recommendations included (PCGIAP, 2002):

- The contents of the Asia-Pacific Regional Fundamental Dataset shall be defined, and further, the technical specifications of the Asia-Pacific Regional Fundamental Dataset shall be developed, with the Global Mapping Specifications and Administrative Boundary Pilot Project Specifications, as reference.
- For the first version of the Asia-Pacific Regional Fundamental Dataset, the vector data shall be no less accurate than 1:1 million in map scale, and the raster data shall be no less than 1KM in ground resolution.

At a 2001 meeting of the PCGIAP Working Group on Cadastre, an item was recommended for the work plan covering "identification of broad types of marine cadastres in use throughout the world as they become established" (PCGIAP, 2001). Unfortunately, the meeting also noted "concern about the apparent direction of the Cadastre WG in majoring in activities on marine cadastre, boundaries being a very sensitive issue in PCGIAP."

One surmises that little practical progress has been achieved in more detailed data specification in nearly six years of work and that the small scale of the data will be of minimal use to local coastal managers or researchers. Also, political issues may limit the degree of actual detail that will be agreed within the very broadly based group.

1.6 CSDI AT GLOBAL LEVEL

At the global level, the not-for-profit Global Spatial Data Infrastructure (GSDI) Association primarily offers a forum for discussion and exchange of experiences, especially in relation to creating SDI at national level. The GSDI's 'SDI Cookbook' (GSDI, 2001) does not specify what types of data nations should have in their SDI, but rather it concentrates on general, though still important, matters such as metadata creation, access policy and other administrative and institutional issues. In a general listing of the types of data that should be considered for including in a National SDI, the cookbook does mention cadastre and hydrography. No further details are given and there is no working group specifically targeting marine or coastal data requirements. A more thorough review of how coastal zone information is related to the GSDI can be found in Longhorn (2003b).

The Global Map specification (ISCGM, 2000) makes reference only to "coastline/shoreline" and "ocean/sea" boundary data, "ferry routes" under a "transportation" heading and edges of "islands" and "inland water" under "drainage (hydrography)." The remainder of the specification is devoted to identifying existing public domain data sources (all at scales of 1:1 million or smaller), data models for vector and raster data, precision and accuracy and metadata issues. The primary data sources listed are VMAP Level 0 from the US National Imagery and Mapping Agency (NIMA), the Global 30 Arc Second Elevation Data Set (GTOPO30) and Global Land Cover Characteristics Database, both available from the US Geological Survey and other sources. These resources have been available in the public domain for some years now, and include only small scale data, some of which are rather old compared to the time frames of interest for much coastal management and research work. For these reasons, there is little new in the Global Map project that would excite coastal managers and researchers.

What approximates a useful global marine SDI does exist for oceanographic data due to long established information management and data exchange programmes of the Intergovernmental Oceanographic Commission (IOC) (Longhorn, 2002). At its most recent meeting in June 2002, the IODE working group considered including coastal data in its remit, only to have the possibility removed from its future agenda by refusal or reluctance of the governments of some important developing country members to permit ready access to coastal zone data for national security and economic reasons.

1.7 PRINCIPAL COMPONENTS FOR CSDI

Having reviewed the few formal coastal SDI initiatives that exist, all at national level, it appears that a CSDI mainly comprises: data sources, standards, enabling technologies and institutional policies. Work in relation to the first two components needs to be carried out with a specific marine or coastal focus, which is sometimes missing from generic SDI initiatives. The latter two aspects apply to the wider requirements of any information infrastructure, not just that of coastal SDI or even for spatial data alone.

1.7.1 Basic Reference Data for the CSDI

The presence or absence of basic reference spatial data (also sometimes called 'framework' data) in National, Regional and Global SDI implementations that relate to the needs of a viable coastal SDI are shown in table 1.1. As one might suspect, the most common are bathymetry, shoreline and boundary data. For most Regional and Global SDI initiatives, there is not sufficient detail in specification of data elements to determine whether or not the needs of coastal and marine resource managers and researchers will be met (with the exception of INSPIRE/WFD). Since the basic data will be collected at national level, this might not appear to be a serious problem at the moment. Yet when data exchange is required for research purposes, resolving boundary disputes or to satisfy a nation's responsibilities regarding various international data exchange conventions, then the absence of regional and global agreement on SDI contents and access issues will become noticed.

Table 1.1 Coastal SDI components in national, regional and global SDIs

SDI Component	USA	Canada	INSPIRE-WFD	APSDI	GlobalMap
bathymetry	yes	yes	yes	maybe	maybe
shoreline	yes	yes	yes	yes	yes
marine cadastre	yes	yes	no	yes	maybe
coastal imagery	maybe	maybe	yes	no	no
marine navigation	maybe	yes	maybe	no	maybe
tidal benchmarks	maybe	maybe	yes	no	no
benthic habitats	maybe	maybe	yes-WFD	no	no

A 'yes' indicates that the component in the left-hand column is formally listed as an important data component in the definition of spatial data infrastructure at national, regional or global level. The number of 'no' and 'maybe' entries is disquieting. Fortunately, the 'nos' appear mainly in regional or global initiatives while 'maybes' indicate that detailed user requirements or specifications have been identified and published, along with existing data sources that might provide this data. However, no firm decisions have been made as to how or if this data will be included within the higher-level NSDI or not.

1.7.2 CSDI Is More Than Data

Other components of an SDI mentioned in most SDI descriptions cover: metadata creation and standards, technical guidance (including standards) on spatial precision, accuracy and data formats (both raster and vector), data access policies (some of which are more liberal than others) and intellectual property guidance.

Standards issues in the spatial data world are now much better addressed than a mere five years ago due to the extensive work of the International Standards Organization's Technical Committee 211 (TC/211) on Geographic Information/Geomatics which is creating "a structured set of standards for information concerning objects or phenomena that are directly or indirectly associated with a location relative to the Earth" comprising some 40 new GI-related standards (ISO, 2002). In parallel with ISO, the global reach and uptake of the GIS interoperability work of the Open GIS Consortium, Inc., and OGC Europe, Ltd., is providing a clear way forward in regard to integrating GI applications and data sources, especially using the Web as the service delivery machinery (OGC, 2002).

Other important standards developments relating to coastal and marine data include the S-57 (Special Publication No. 57) cartographic standard developed for and maintained by the International Hydrographic Organisation (IHO) and International Hydrographic Bureau (IHB) in Monaco (IHO, 1996). This standard is used for collection and exchange of hydrographic data among national Hydrographic offices globally. It is also very important for marine navigation as applied to the new Electronic Chart Display and Information Systems (ECDIS) now being introduced throughout the maritime industry. S-57 comprises a hydrographic data model, an object catalogue and an electronic nautical chart (ENC) product specification that are standard for ECDIS data.

Various shoreline and boundary data metadata standards have been developed at national level, as mentioned previously for the USA within the FGDC shoreline metadata working group. Looking to the future, two projects are underway that aim to specify a globally agreed standard for marineXML, a marine-specific implementation of the eXtensible Markup Language (XML) now used widely on the Internet for conveying semantic content of information as opposed to only the display specifications provided by HTML (Hypertext Markup Language). MarineXML is described as "an interoperability framework for global ocean observation systems" (ICES, 2003) which will encompass coastal zone elements as well. The development work on marineXML is undertaken by the European Union, via a part-funded project in the EU's Framework RTD programme (IOC, 2003a) in conjunction with the Intergovernmental Oceanographic Commission's Committee on International Oceanographic Data and Information Exchange (IODE) based at UNESCO headquarters in Paris (IOC, 2003b). National initiatives on creating marineXML specifications are also underway, for example in Australia and the USA (Sligoeris, 2002; Davis *et al.*, 2002).

More serious barriers remain in regard to harmonised data access policies and exploitation rights for spatial information, particularly that collected by public sector agencies. The European Union is trying to address this problem at a regional level via a new Directive setting out an agreed EU-wide framework for access to and exploitation of public sector information (European Commission, 2002). Similar initiatives are under discussion, in consultation or being implemented in both developed and developing nations across the globe. The impact that these initiatives will have on the coastal zone (and larger research) community cannot be underestimated. Even in countries with strong "freedom of information" cultures, such as the USA, some public sector marine information is not disclosed due to fear of liability actions against the data providers (Lockwood and Fowler, 2001). With intellectual property (IP) legislation in a state of flux across the globe while attempts are made to accommodate the prior IP regime and existing international IP conventions to the digital world, resolving many of the non-technical data-related problems is far easier than the institutional ones.

1.8 CONCLUSIONS

The CZM community should be aware that their input, based on knowledge and experience of the coastal zone, is imperative during specification of SDI initiatives, whether at national, regional or global level. What we can conclude from the above examination of various types of spatial data infrastructures that are under development—national, regional and global—is that, at regional and global level, the emphasis is virtually all on process, standards, metadata and institutional issues, and rarely on data specifications for the content of an SDI. Thus, for the most part, these initiatives will almost certainly never have a separate strong coastal or marine focus. The only exception at Regional SDI level is that of INSPIRE, justified partly by the legal data reporting requirements of the European Water Framework Directive, the latter of which focuses very heavily and in great detail on coastal and near-shore waters. As more such regional (non-European) environmentally focused initiatives evolve, we can expect a similar emphasis on the need to arise for wider access to and use of coastal and marine spatial data in regard to environmental monitoring.

At the global level, some of the most important work in regard to major elements of what comprise an SDI has been implemented within major international environmental monitoring and research programmes, such as those of IOC, the World Meteorological Organization (WMO), International Geosphere-Biosphere Program (IGBP) and others (Longhorn, 2003a).

At national level, where one tends to find more detail in specification of the SDI and a more advanced state of actual implementation of elements of SDI, there is evidence that only the more obvious data needs are being considered, e.g. shoreline, hydrography/bathymetry and (sometimes) cadastre/boundaries. Yet other important elements such as habitat and navigation do not fit well within the primarily land-focused sections of NSDI specifications. This may be due partly to the absence of coastal zone/marine professionals on the SDI development teams or for other reasons, e.g. a national decision that the SDI should focus on only the most basic 'reference data' that is considered commonly needed to underpin broadly based economic activities and governance.

Finally, one should not forget that spatial data is only one facet of an SDI implementation. Important institutional, jurisdictional, data policy and standards/interoperability issues also Figure high on the agenda. These appear to be the principal focal points for most regional and global SDI initiatives, rather than detailed basic reference data specifications.

1.9 REFERENCES

Bartlett, D.J., 2000, Working on the Frontiers of Science: Applying GIS to the Coastal Zone, In *Marine and Coastal Geographical Information Systems* edited by Wright, D. and Bartlett, D. (London: Taylor & Francis).

Butler, M.J.A., LeBlanc, C. and Stanley, J.M., 1998, *Inland Waters, Coastal and Ocean Information Network (ICOIN)*, Report submitted to the Canadian Hydrographic Service, Department of Fisheries and Oceans (DFO) by MRMS, Inc. pp. 85.

CCMC, 1999, *Draft Concept Outline Marine Geospatial Data Infrastructure (MGDI)*, Canadian Centre for Marine Communications, http://cgdi.gc.ca/english/geospatial/MGDI/pdf/mgdi.pdf (May 22, 2003).

Chopin, T. and Costain, K., 2001, *Beyond 2000: An Agenda for Integrated Coastal Management Development*, Post-Conference Report of the Coastal Zone Canada 2000 International Conference, Saint John, New Brunswick, September 17-22 2000. (Dartmouth, Nova Scotia: Coastal Zone Canada Association).

Clinton, W., 1994, *Executive Order 12906, April 13, 1994*, Federal Register, vol. 59, no. 720, pp. 1771–1774. (Washington, D.C.: Government Printing Office).

CRMSW, 2000, *Clean Water Action Plan: Coastal Research and Monitoring Strategy, September 2000*, Coastal Research and Monitoring Strategy Workgroup (Washington, D.C.: US Environmental Protection Agency). http://www.cleanwater.gov/coastalresearch/H2Ofin.pdf (May 19, 2003).

Davis, D. *et al.*, 2002, *Using XML Technology for Data and System Metadata for the MBARI Ocean Observing System (MOOS)*, (Moss Landing, CA, USA: Monterey Bay Aquarium Research Institute). http://www.mbari.org/ (10 October 2003).

DFO, 2001, *Marine User Requirements for Geospatial Data: Summary 2001*, Geospatial Projects Integration Office, (Ottawa, Canada: Department of Fisheries & Oceans), http://www.geoconnections.org/english/MGDI/Key_Docs/ Marine_User_Requirements.pdf (May 20, 2003).

European Commission, 2000, *Directive 2000/60/EC of the European Parliament and of the Council of 23 October 2000 establishing a framework for Community action in the field of water policy*, (Luxembourg: Office of Official Publications of the European Communities).

European Commission, 2002, *Proposal for a European Parliament and Council Directive on the Re-use and Commercial Exploitation of Public Sector Document. June 5, 2002, COM(2002) 207*, (Brussels: European Commission), 20 p.

European Communities, 2002, *Guidance Document on Implementing the GIS Elements of the WFD, EUR 20544 EN*, edited by Jürgen Vogt, (Ispra, Italy: DG Joint Research Centre). http://agrienv.jrc.it/publications/pdfs/GIS-GD.pdf (May 19, 2003).

Evans, J. (editor), 2001, The New Digital Earth Reference Model, NASA Digital Earth Office, http://www.digitalearth.gov/derm/v05/index.html (May 22, 2003).

FGDC, 1998, *Shoreline Metadata Profile of the Content Standards for Digital Geospatial Metadata, FGDC-STD-001.2-2001*, Marine and Coastal Spatial Data Subcommittee, (Reston, VA: Federal Geographic Data Committee), http://www.csc.noaa.gov/metadata/sprofile.pdf (May 20, 2003).

FGDC, 2000a, *National Hydrography Data Content Standard for Inland and Coastal Waterways - Public Review Draft - January 2000*, Marine and Coastal Spatial Data Subcommittee, (Reston, VA: Federal Geographic Data Committee), http://www.fgdc.gov/standards/documents/standards/hydro/HydroStd_pr_draft.pdf (May 19, 2003).

FGDC, 2000b, *Geospatial Positioning Accuracy Standards Part 5: Standards for Nautical Charting Hydrographic Surveys - Public Review Draft - November 2000*, (Reston, VA: Federal Geographic Data Committee), http://www.csc.noaa.gov/cgibin/goodbye.cgi?url=http://www.fgdc.gov/standard/status/sub1_4.html (May 22, 2003).

FGDC, 2002, *Charter for Subcommittee on Marine and Coastal Spatial Data*, NOAA Coastal Services Center, http://www.csc.noaa.gov/fgdc_bsc/overview/charter.htm (May 23, 2003).

GeoConnections, 2003, *Marine Advisory Network Node Terms of Reference*, (Ottawa, Canada: GeoConnections Secretariat), http://www.geoconnections.org/english/MGDI/index.html (May 22, 2003).

Gillespie, R., Butler, M., Anderson, N., Kucera, H. and LeBlanc, C., 2000, MGDI: An Information Infrastructure to Support Integrated Coastal Management in Canada, *GeoCoast*, **1**, pp. 15-24, http://www.theukcoastalzone.com/geocoast/Volume1/Gillespie.pdf (May 22, 2003).

GSDI, 2001, *Developing Spatial Data Infrastructures: The SDI Cookbook version 1.1*, edited by D. Nebert, Gloal Spatial Data Infrastructure Technical Working Group Chair, http://www.gsdi.org/pubs/cookbook/cookbook0515.pdf (May 22,2003)

IHO, 1996, *IHO Transfer Standard for Digital Hydrographic Data Edition 3.0 - March 1996. Special Publication No 57*, (Monaco: International Hydrographic Bureau).

INSPIRE, 2002, *Reference Data and Metadata Position Paper*, RDM Working Group edited by Rase, D., Björnsson, A., Probert, M. and Haupt, M-F.

INSPIRE, 2003, *Contribution to the extended impact assessment of INSPIRE*, INSPIRE FDS Working Group and Dr. M. Craglia, Univ. of Sheffield, (Environment Agency for England and Wales, UK).

IOC, 2003a, *The MarineXML Project portal - inaugural meeting summary report*. http://ioc.unesco.org/marinexml/contents.php?id=9 (13 October 2003).

IOC, 2003b, *Marine XML Web portal*. http://ioc.unesco.org/marinexml/ (May 13, 2003).

ISCGM, 2000, *Global Map, version 1.1 Specifications, 16 March 2000, Capetown*, http://www.iscgm.org/html4/index_c5_s1.html#doc13_3787 (May 20, 2003).

ISCGM, 2001, *Resolutions of the 8th ISCGM Meeting, Cartagena, Columbia, 25 May 2001,* http://www.iscgm.org/more-iscgm/reso-8th-meeting.pdf (May 18, 2003).

ISO, 2002, *ISO/TC 211 Geographical information/Geomatics Scope*, International Standards Organisation, Geneva, http://www.isotc211.org/scope.

htm#scope (May 23, 2003).
Labonte, J., Corey, M., and Evangelatos, T., 1998, Canadian Geospatial Data Infrastructure (CGDI)–Geospatial Information for the Knowledge Economy, *Geomatica*, **52**, pp. 214–222. http://www.geoconnections.org/english/publications/General_information/publications_geomatica_cgdi_e..pdf (May 24, 2003).
Lockwood, M. and Fowler, C., 2000, Significance of Coastal and Marine Data within the Context of the United States National Spatial Data Infrastructure. In *Marine and Coastal Geographical Information Systems* edited by Wright, D.J. and Bartlett, D.J. (London: Taylor & Francis).
Longhorn, R, 2002, Global Spatial Data Sharing Frameworks: The Case of the Intergovernmental Oceanographic Commission (IOC). In *Proceedings of GSDI-6, Budapest, Hungary, 16–19 September 2002*. http://www.gsdi.org/ (April 10, 2003).
Longhorn, R., 2003a, A comparison of spatial information access policies of transnational environmental modelling and global climate change programs. In *Geoinformation for European-wide Integration* edited by Benes, T., (Rotterdam: Millpress), pp. 305–313.
Longhorn, R. 2003b, European CZM and the Global Spatial Data Infrastructure Initiative (GSDI). In *Coastal and Marine Geo-Information Systems* edited by Green, D.J. and King, S.D. (Dordrecht, NL: Kluwer Academic Publishers), pp. 543–554.
NOAA, 2001, *Coastal NSDI*, NOAA Coastal Services Center, http://www.csc. noaa.gov/themes/nsdi/ (May 23, 2003).
NOAA, 2003, *FGDC Marine and Coastal Spatial Data Subcommittee 2003 Work Plan*, http://www.csc.noaa.gov/fgdc_bsc/accomp/2003plan.htm (May 22, 2003).
OGC, 2002, *Open GIS Consortium Vision, Mission and Values*, http://www.opengis.org/info/vm.htm (May 22, 2003)
OMB, 2002, *Circular No. A–16 Revised, August 19, 2002*. (Washington, DC: Office of Management and Budget), http://www.whitehouse.gov/omb/circulars/a016/a016_rev.html (May 27, 2003).
PCGIAP, 1998, *PCGIAP Publication Number 1,* http://www.gsi.go.jp/PCGIAP/tech_paprs/apsdi_pub.htm (May 24, 2003).
PCGIAP, 2001, *Working Group 3 report of Group Meeting at Tsukuba, 25th April 2001*, PCGIAP Working Group 3: Cadastre, http://www.gsi.go.jp/PCGIAP/tsukuba/WG3_back.htm (May 22, 2003).
PCGIAP, 2002, *Report of working group 2 meeting, 18 April 2002*, PCGIAP Working Group 2: Fundamental Data, http://www.gsi.go.jp/PCGIAP/brunei/wg2_report.htm (May 22, 2003).
Sliogeris, P. 2002, *An XML Based Marine Data Management Framework*. (Potts Point, NSW, Australia: Australian Oceanographic Data Centre), http://www.aodc.gov.au/ (20 May 2003).
U.S. Commission on Ocean Policy, 2002, *Developing a National Ocean Policy: Mid-term report, September 2002*, http://www.oceancommission.gov.

CHAPTER TWO

Bridging the Land-Sea Divide Through Digital Technologies

Simon Gomm

2.1 INTRODUCTION

There are many different types of users of coastal zone information, from the casual user who may only want to browse, to the sophisticated user who makes frequent use of mapping and demands continuous improvement. These user communities are diverse in the topics they address, covering such areas as Local and Central Government, environmental and economic analysis, and also increasingly leisure use.

A common mapping framework that bridges the land-sea divide allows users to build applications and decision-making tools necessary to promote the shared use of such data throughout all levels of Government, the private and non-profit sectors and academia. A consistent framework also serves to stimulate growth, potentially resulting in significant savings in data collection, enhanced use of data and assist better decision-making.

As well as a physical division, the land-sea divide has also, for many spatial data suppliers, acted as a limit to their area of responsibility, or formed a data product boundary. As a result users wanting to model the diverse aspect of the coastal zone across this divide have had to identify, obtain and combine separate datasets to provide the data coverage they require. The combination process must resolve integration problems resulting from the differing projections, scale of capture and other specification issues of the source datasets. This process can be time consuming, result in inconsistent data and can cause a hindrance to the management of a particularly sensitive environmental zone.

This chapter will look at the technical issues involved with the integration of data across the land-sea divide and identify means for resolving these. Examples within this chapter have been drawn from the work done by Ordnance Survey of Great Britain, United Kingdom Hydrographic Office and the British Geological Survey on integrated coastal zone mapping project (ICZMap) (Gomm, 2001).

0-41531-972-2/04/$0.00+$1.50

2.2 DATA SPECIFICATIONS

At the commencement of a project the user will need to have assessed the project's spatial and non-spatial requirements, and these will provide the key criteria for defining and selecting suitable data. Typically such criteria will include: physical extent of the project area, data content, attribution, positional accuracy, spatial resolution, currency, projection, datum and transfer format. These criteria are all discussed in more detail below.

2.2.1 Spatial Extent

The extent of the area for which data are required needs to be defined clearly in the coordinate system to be used in the project. At a minimum the extent should be defined by a bounding rectangle using the x and y coordinates of 2 diagonal corners. Ideally the extent should be delineated by a bounding polygon at an appropriate spatial resolution. A bounding polygon will allow for better selection of relevant information and itself provide a tool for analysis within the project. In defining the extent a sufficient margin should be included to allow for inclusion of features which may have an influence on the application.

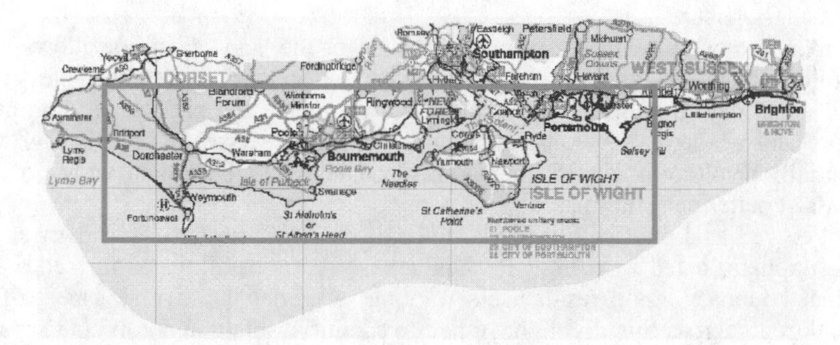

Figure 2.1 Bounding Rectangle. © Crown Copyright 2004. All rights reserved.

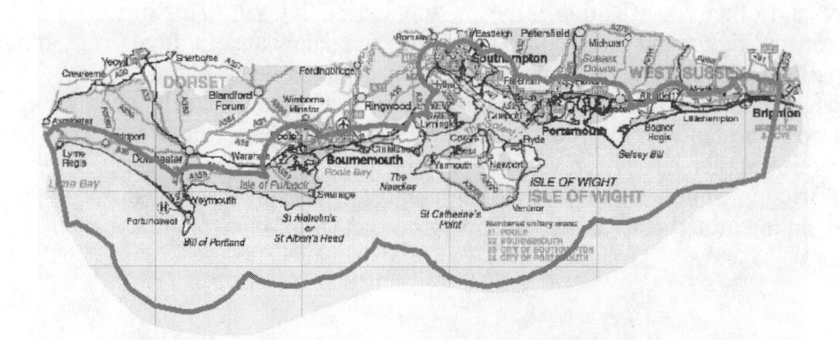

Figure 2.2 Bounding Polygon. © Crown Copyright 2004. All rights reserved.

In Figure 2.1 the extent of the ICZMap coastal zone project is defined as a simple bounding rectangle and in Figure 2.2 as a polygon defined by buffering the coastline 20km offshore and 5km inland. In practice, the latter was more appropriate for the application and datasets of the ICZMap project, with its emphasis on the processes and dynamics of the land-sea interface.

Whilst it is normal to consider defining extent only in two dimensions, it is worth noting that there will be applications where height/depth extents, along with the temporal aspect may need to be specified.

2.2.2 Data Content

The use to which data are to be put will define the features that are required. At the simplest level this may merely be raster imagery for use as backdrop mapping within an application. At a more complex level, where the data are required to form part of the analysis, the user will need to consider what features and object classes are required. In defining these requirements there will be some features that fall uniquely on the land or in the sea, but there will also be others that by their nature bridge this land-sea divide.

2.2.3 Attribution

Attribution of features within a dataset can be at a number of different levels. As a minimum this could take the form of feature coding, allowing selection of required features for analysis and symbolisation. Beyond this, additional attribution concerning the real-world properties of the feature and how it was captured will increase the versatility of the data. These attributes will vary for different features within a dataset depending on the real-world object they represent, and the uses for which the dataset was intended.

2.2.4 Positional Accuracy and Spatial Resolution

An appreciation of the positional accuracy requirements of a dataset is important to ensure that the data are used in an appropriate way with other datasets, and to put results of any analysis in to the correct context. The positional accuracy of spatial data can be expressed both in terms of its absolute accuracy and its relative accuracy.

Absolute accuracy is a measure to which a coordinated position in the dataset corresponds to the true position of the real world feature it represents. Relative accuracy expresses the positional accuracy between points in a dataset, and is a comparison of the distance between features in a dataset with the real world distance. Datasets with a high relative accuracy but low positional accuracy may indicate a systematic shift in the data with respect to the coordinate system.

The spatial resolution represents the coordinate precision to which data are stored in the dataset, and affects the maximum achievable accuracy for a dataset, e.g. if the spatial resolution is only 10m then this could affect the absolute accuracy of a point by up to 7m.

2.2.5 Currency

Currency is sometimes overlooked as an aspect of a dataset's specification. Attribution should ideally allow for the recording of temporal information at feature level. This can include creation date, capture date and modification date (with nature of modification). Occasionally, temporal information will be limited to the date of the dataset's last update. The temporal information is important both from an analysis point of view, but also for the initial selection of data for the project. For example, the ability to select the position of the top and bottom of coastal slopes for a given year can allow predictive analysis of coastal erosion processes to be studied.

2.2.6 Projections and Datum

Coordinates within a dataset will be relative to a given projection and datum. Map projections attempt to represent the curved surface of the Earth on a flat plane. All projections are approximations and some will better represent large areas of the world, albeit with large distortions, while others are more suitable for small specific areas with much smaller distortions.

A geodetic datum or spheroid is a mathematical approximation of the surface of the Earth, which itself is an imperfect sphere. Numerous geodetic datums have been calculated over the years, some suitable for global applications and others calculated to minimise errors on a country-by-country basis.

Height datum represents a base level from which elevations are measured and each information source may have its own. Datasets on land will often share a common height datum (e.g. use of Ordnance Survey Newlyn Datum in Great Britain); however different datums will usually be used for marine datasets. Frequently, with marine datasets having their origin from navigation charts, depth values will be typically based on local lowest astronomic tide values.

For a project it is important to determine what projection and datum are most suitable for the application, and to ensure that you are aware of what the projections and datum of the source datasets are.

2.2.7 Data Transfer Format

Datasets are transferred between media using a chosen transfer format. Whilst there is no single common transfer format there are a number of commercial ones, such as Autodesk-AutoCAD DXF, ESRI Shapefile and MapInfo TAB and MIF/MID formats, which are becoming more widely adopted as *de facto* standards. Such formats can be limited as they are primarily intended for transfer between users of the same software. When read by other software, information may be lost due to data model differences.

Ideally a neutral, non-software-specific, transfer format is needed. Many attempts have been made to try to implement these, one of the most recent examples being GML (Geography Markup Language) promoted by the OpenGIS consortium as a development of the more widely used XML Web document markup language. However, in practice, the software being used for a project may act as a limiting factor in what formats it can and cannot read.

2.2.8 Metadata

Much of the information describing a dataset should be included in its metadata if present. As well as accompanying a dataset, metadata are now frequently being held on readily accessible databases, allowing users to identify datasets suitable to their requirements. The international standard for digital geospatial metadata (ISO 19115) is now being adopted by many national bodies, such as US Federal Geographic Data Committee (FGDC) for its Content Standard for Digital Geospatial Metadata (CSDGM) (http://www.fgdc.gov/metadata/metadata.html/), and the UK's Association for Geographic Information (AGI) with its GIGateway project (http://www.gigateway.org.uk/default.asp/).

2.3 DATA SELECTION

The properties of datasets that have been outlined above cover some of the points that need to be considered in establishing a project and specifying data requirements. Metadata services provide a tool for locating data, but they will not tell you if they are best suited for your purposes. It is unlikely that any dataset will fully match your requirements, but what should be considered is the ability to integrate and modify them to satisfy your needs. Typically, more flexible datasets will have a richness of feature classes and attribution, and good positional accuracy, but these will come at an increase in cost and data volumes.

2.4 DATA INTEGRATION

It is unlikely that a single dataset will satisfy the full needs of a project. This is particularly true when modelling the coastal zone, where there is likely to be one source for the land, another for the sea and potentially other subsidiary datasets straddling both. In these cases there will inevitably be some data integration issues. Typical problems and potential solutions are discussed here.

2.4.1 Differences in scale

Datasets which have been captured from graphic products can be said to have a nominal scale at which they were intended to be used. In these datasets, differences in scale will show themselves in the positional accuracy and spatial resolution, as well as in the features and attributes that are present e.g. field boundaries may be shown at a scale of 1:25,000 but not at 1:50,000. At smaller scales data will also normally have been generalised, simplifying the geometry of the features. These factors are not a problem in themselves, as long as the data meet application requirements, but problems do nonetheless occur when trying to join together data captured at different scales.

The following example illustrates the relevance of this to the coast, using datasets from the Ordnance Survey of Great Britain and from the UK Hydrographic Office respectively.

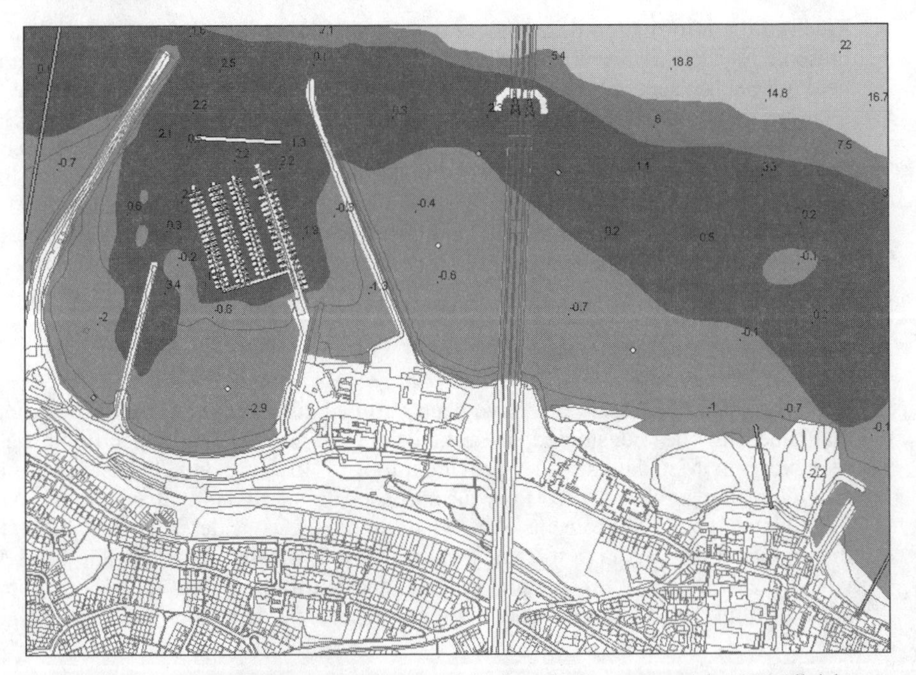

Figure 2.3 Overlay of hydrographic and terrestrial datasets. © Crown Copyright 2004. All rights reserved.

In Figure 2.3 the landward data have been captured at a scale of 1:2,500 and on the seaward side at a scale of 1:25,000 (this being the highest resolution data available for the area). The result of this is a disparity in the features common to both zones, and a greater density of detail on the land compared with the sea. In such situations a choice can sometimes be made as to which feature is most useful to the application, and the software system functionality used to remove or suppress the other. In a similar way features that are superfluous to the application can also be removed or suppressed by the software, provided that features can be distinguished from each other by their attributes.

If the geometry of the large-scale data is too complex, then filtering this using line-simplification or other generalisation algorithms can be considered. Within a project for a user this is probably only appropriate to features where line vertices can be removed whilst keeping the basic shape of the feature.

2.4.2 Projection

Where the datasets to be used are in different projections a choice will have to be made as to which to use, based on the application and the type of output required. Transformation facilities, with parameters to convert between most common map projections, are supplied with mainstream GIS packages. For projections not included these can also be transformed, provided the parameters defining the projection are known.

How the transformation is applied depends on the functionality of the software being used. Frequently this will allow transformation of data coordinates to be done either as a permanent process, storing new coordinates for the features in the new projection, or as a real-time transformation when the dataset is required for display or analysis.

2.4.3 Currency

Where different datasets abut each other or overlap, the same real-world features may be represented in both datasets. In such cases a decision will normally need to be made as to which features to retain. Such selection may be done on the currency of the data i.e. the date of capture of the feature. However, this may not give the best quality in terms of fidelity or resolution. In these situations the final decision as to whether the feature gets included will be based on the specific needs of the application.

2.4.4 Accuracy

As with currency, an appreciation of the relative positional accuracies of datasets that are to be integrated will ensure that data are combined in an appropriate manner. Where features representing the same real-world objects exist in two or more datasets, the positional accuracy will offer one means of selecting which to keep. In some instances it will not be appropriate just to take the data with the highest positional accuracy. For some applications the more accurate data may take up more space, be slower to manipulate, and not enhance the analysis, so ultimately it is up to the user to decide what best fits their needs.

2.4.5 Data Overlap

Many of the issues discussed above cover situations where data in different datasets overlap spatially and represent the same real world objects. It is assumed that in such situations a decision will need to be made as to which features to keep, and that this selection will be driven by the requirements of the application. In the case of a marine and a terrestrial dataset bounded by a coastline, due to the different representations of the coastline in the two datasets, features may overlap (as shown in Figure 2.4).

These overlaps can be treated in a number of ways:

1) Ignore them, accepting that both overlapping features are valid in the context of the source datasets and their specifications. This may not be a real solution when the application requires a single seamless layer of information with no duplication of common features.

2) Spatially intersect overlapping features and retain all the features that result.

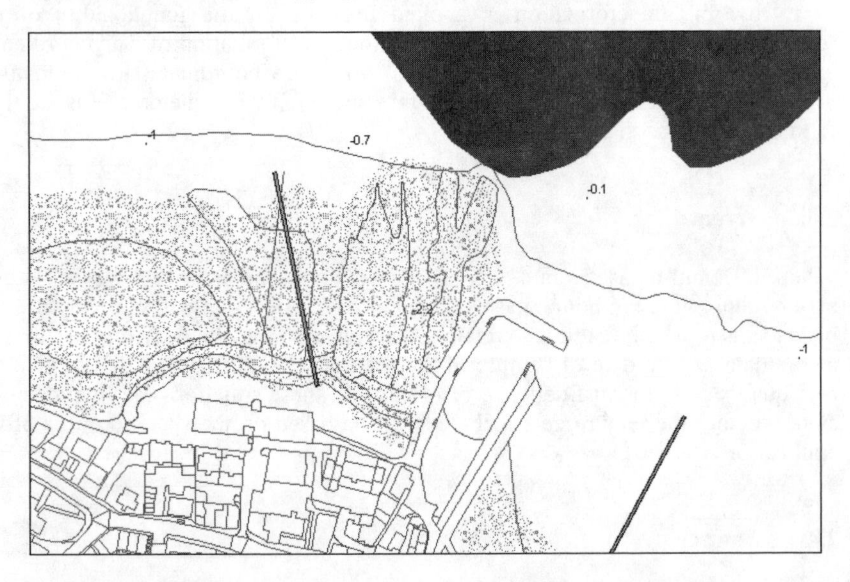

Figure 2.4 Data overlap example. © Crown Copyright 2004. All rights reserved.

Features will be split by others that overlap them and the resulting features will need to retain the attributes of both source datasets to be flexible. This is the most flexible solution but has a potential storage overhead of having to store the two sets of attributes.

3) Spatially intersect overlapping features and retain only one set of features. This will ensure no overlap, but some features will be truncated or split and as a result may present problems when used for analysis.

4) Edit datasets together using a single boundary between them. This can be labour-intensive, but will give the best solution. However, if datasets are updated independently then this work will need to be repeated to maintain currency of the combined dataset.

2.4.6 Height Datum Correction

2D plan data need not necessarily be physically combined into a common dataset to be of use. This is not true in the case of 3-dimensional data. Terrestrial and marine elevation data can be displayed graphically in 2D, but where any use of terrain modelling is required, such as for an analysis of the inter-tidal zone, a combined height model needs to be created. As mentioned previously, the terrestrial and marine elevation data will frequently be relative to different datums. As a result, prior to combining them into a single height model, one or more of the datasets need to be corrected to the other's datum as well as any re-projection of the data onto the other's coordinate system (Milbert and Hess, 2001).

In the simpler cases, correction of a datum can be achieved by a single difference applied to all elevation values in a dataset, but more frequently the difference will vary across an area. In the case of UK Hydrographic Office (UKHO) data, bathymetric values are based on lowest astronomic tides for navigation purposes, as computed for specific locations on each chart. The relationships between these different datums are shown in Figure 2.5.

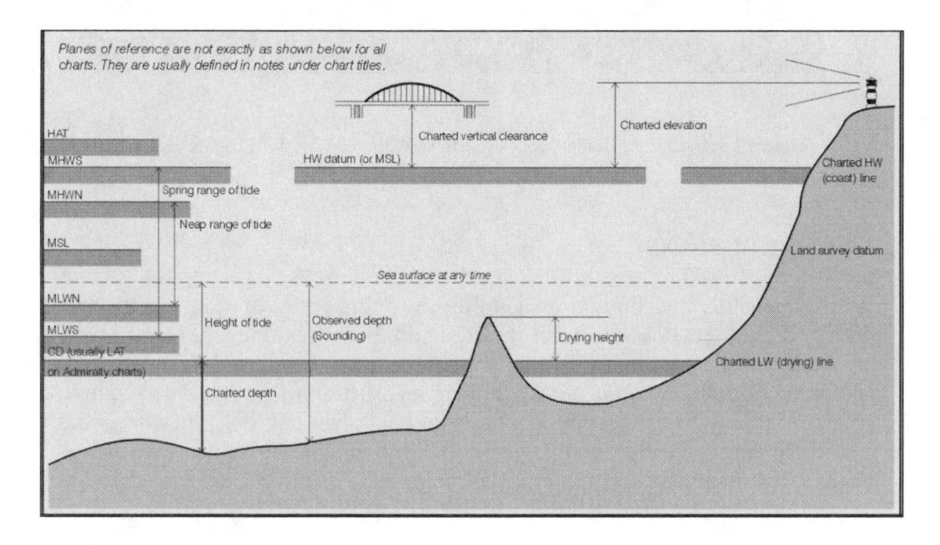

Figure 2.5 Datum Relationships. © Crown Copyright 2004. All rights reserved.

There is a complex and varying relationship between Ordnance Survey of Great Britain's single, standard Newlyn height datum (which applies to all terrestrial mapping by the OS) and the values used in Hydrographic charting, with the latter differing widely along the coast, due to the varying tidal conditions that occur. To convert the UKHO bathymetry values to Newlyn datum, a surface of differences has to be created so that correction value can be calculated for all points (Capstick & Whitfield, 2003). Once a surface of differences has been created this can be used to interpolate values for all bathymetry points. The corrected points can then be used with the terrestrial elevation data to generate an integrated land-sea model for display and analysis.

Figure 2.6 shows a representation of integrated elevation data for the Isle of Wight, UK, with the bathymetry information coming from UKHO sounding values and the terrestrial elevation from Ordnance Survey Land-Form PROFILE contours and spot heights.

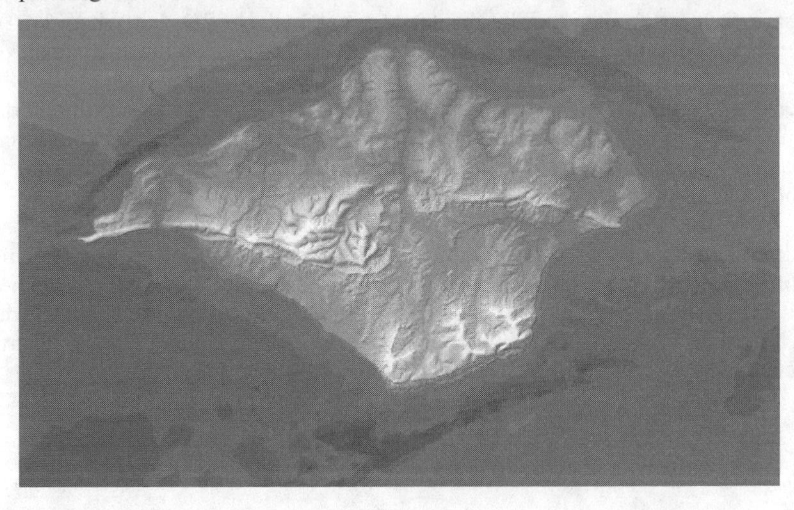

Figure 2.6 Integrated Height Model. © Crown Copyright 2004. All rights reserved.

2.5 CONCLUSIONS

Whilst there may be limited availability of single datasets covering the land-sea divide, digital technologies offer the capabilities to combine and resolve available adjacent datasets to satisfy diverse applications for the coastal zone. In satisfying an application, the user must have a clear specification for their data requirements so as to be able to assess the suitability of candidate datasets, and the issues that are likely to be encountered in their integration. Although not exhaustive, this chapter has attempted to highlight the main areas to be addressed in the selection and integration of spatial datasets for the coastal zone.

2.6 REFERENCES

Capstick, D. and Whitfield, M., 2003, Height integration at the coastal zone. In *Proceedings of the GIS Research UK, 11th Annual Conference 2003*.

Gomm, S., 2001, Integrated mapping of the marine and coastal zone. In *Proceedings of the GIS Research UK, 9th Annual Conference 2001.*

Milbert, D.G. and Hess, K.W., 2001, Combination of topography and bathymetry through application of calibrated vertical datum transformations in the Tamapa Bay region. In *Proceedings of the 2nd Biennial Coastal GeoTools Conference 2001.*

CHAPTER THREE

A Comparative Study of Shoreline Mapping Techniques

Ron Li, Kaichang Di, and Ruijin Ma

3.1 INTRODUCTION

The shoreline is a unique and complex feature and has been recognized as such by the International Geographic Data Committee (IGDC), which designates it as one of 27 key distinct features on the earth's surface. A shoreline is defined as the line of contact between land and a body of water. Although it is easy to define, the shoreline is difficult to capture because of the natural variability of water levels. An accepted method of capturing this line of interface is to determine the *tide-coordinated shoreline*, which is the shoreline extracted from a specific tide water level. NOAA uses Mean Lower Low Water (MLLW) and Mean High Water (MHW) in this way to map shorelines that can be georeferenced. Both the MLLW and MHW are calculated from averages over a period of 18.6 lunar years.

Like the shoreline, shoreline mapping techniques are changing and being improved upon. At present, photogrammetric techniques are employed to map the tide-coordinated shoreline from aerial photographs that are taken when the water level reaches the desired level. Aerial photographs taken at these water levels are more expensive to obtain than remote sensing (RS) imagery. With the development of remote sensing technology, satellites can capture high-resolution imagery with the capability of producing stereo imagery: one such source is the IKONOS satellite images. The question is: is it possible to use RS imagery to map the shoreline so that we can reduce the costs and improve mapping efficiency and accuracy?

During our study we extracted shorelines from aerial photos, simulated and actual IKONOS imagery, and the intersection between a Coastal Terrain Model (CTM) and the water level. We estimated the accuracies of these shorelines and analyzed the potential of these techniques for practical shoreline mapping by comparing the extracted dataset with the shorelines from the United States Geological Survey (USGS) topological maps and National Oceanic and Atmospheric Administration (NOAA) nautical charts.

0-41531-972-2/04/$0.00+$1.50

3.2 SHORELINE EXTRACTION

To evaluate the new shoreline mapping techniques it is necessary to compare the results with a current shoreline product. In our study, the shoreline extracted from the tide-coordinated aerial photos was used as the reference shoreline and the other shorelines were compared with it to assess their relative accuracies. Shorelines from USGS Digital Line Graphs (DLG) and a NOAA Nautical Chart were digitized to facilitate the evaluation of the new results. The extraction methods used are discussed in the following sections.

3.2.1 Shoreline from aerial photos

A shoreline was digitized from tide-coordinated aerial orthophotos. This product was chosen as the baseline in our analysis due to its high accuracy. The aerial photos were commissioned by NOAA in 1997 when the water surface reached the specified level (MLLW). The photo scale is about 1:20,000 and the pixel size is about 0.6 meters on the ground. This shoreline is tide-coordinated to the mean lower low water level. A Global Positioning System (GPS) survey was carried out in 2000 to obtain control points for the bundle adjustment. Eight control points were used in the adjustment. The Root Mean Square (RMS) numbers indicate a good precision in this adjustment: 0.106, 0.107 and 0.073 meters in the X, Y and Z directions, respectively. Based on the bundle adjustment results, Digital Terrain Models (DTMs) and orthophotos were generated in sequence. Then, the orthophoto mosaic was created with a resolution of 1 meter. The shoreline was digitized manually from this orthophoto mosaic. The task was accomplished using ERDAS software.

3.2.2 Shorelines from simulated and actual IKONOS images

IKONOS imagery has two resolutions, 1-meter panchromatic and 4-meter multispectral. We first obtained 4-meter multispectral geo-referenced imagery for our project. The stereo 1-meter images were still pending when we produced the shorelines in this paper. To study the potential of 1-meter stereo imagery in shoreline mapping, we simulated the imaging procedure and obtained the simulated IKONOS 1-meter stereo imagery using rational functions (RFs). The shoreline extracted from aerial orthophotos was used for back-projecting to the simulated IKONOS raw backward- and forward-looking images. The resulting shorelines on IKONOS raw images show that RFs work very well. This indicates that we can use RFs to extract high accuracy 3-D shorelines from real IKONOS 1-meter imagery provided that we can find well-matched conjugate points on raw IKONOS images along the shorelines. To improve the accuracy of 4-meter IKONOS real data, 8 control points were used to perform planar polynomial geo-correction. The RMS in the X and Y directions are 1.048 meters and 1.1028 meters respectively. A shoreline was extracted from the refined IKONOS image.

3.2.3 Shoreline from the intersection between a CTM and the water level

This idea comes from the definition of a shoreline as the contact or intersection line between the water level and the land. We tried to obtain the land model and water surface model (WSM) and then compute the intersection line of these two models. This method is intuitive to people's perception of shoreline (See the profile in Figure 3.1 and the colour insert following page 164). The CTM depicts the elevation and bathymetry in the coastal area. The water surface model is obtained from the Great Lakes Forecasting System at The Ohio State University (OSU). To extract the shoreline, these two models were overlaid and the WSM was subtracted from the CTM to get an output layer. Theoretically, the shoreline should be the line with the zero value in the output layer. In practice, the zero pixels do not form a line due to random errors in these two data models introduced from data acquisition and sampling. We chose a small value range for the study and obtained a shoreline strip. We then took one side edge of the strip as the shoreline for our study. We employed this method because the strip is very thin and it is very small compared to the error of the CTM.

3.2.4 Shorelines from other sources

Besides the above shorelines, four other shorelines were used for the analysis in our study. Among them, two digital shorelines were provided by the Ohio Department of Natural Resources (ODNR), the third one was extracted from a USGS DLG and the fourth one was digitized from a NOAA Nautical Chart.

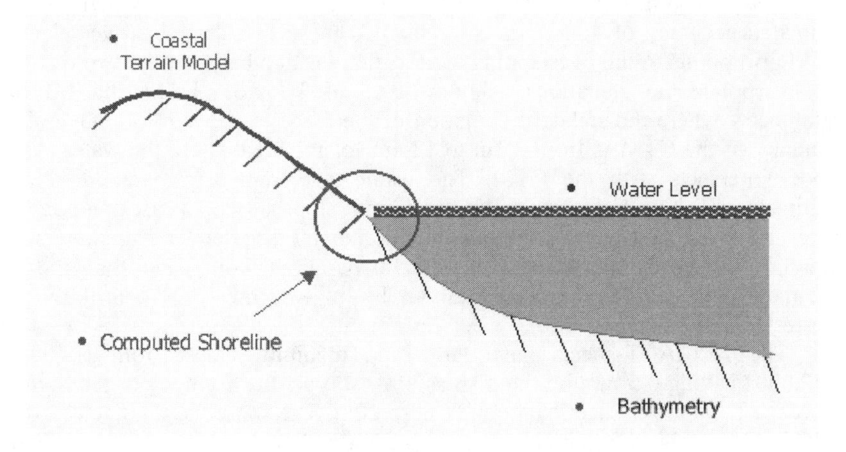

Figure 3.1 Generation of Digital Shorelines

3.3 ESTIMATION OF THE ACCURACY OF THE SHORELINES

The accuracy of these shorelines was analyzed in our study by considering the data sources and the extraction methods. More extensive work will be done in the future to further investigate this important issue.

From the bundle adjustment and conjugate point matching, the DTM standard deviation derived from the aerial photographs is estimated to be 2.1m in X and Y coordinates. Considering the error introduced by resampling from the aerial photos of a 0.6m resolution, the 1-meter orthophoto has an estimated standard deviation of about 2.1m. The identification error of the shoreline from the orthophoto may introduce 1.5 pixels, that is 1.5m, considering those situations where the water and land separation is very hard to define. Therefore, the shoreline from the aerial orthophoto has an estimated standard deviation of about 2.6m if we apply a simple model of error propagation.

The accuracy of the shoreline derived from 1-meter simulated IKONOS imagery should be about 2-4m (Zhou and Li, 2000), considering the fact that the accuracy of 3D ground points reaches 2-3m with GCPs and the accuracy of identifying and locating conjugate shoreline points is about 1.5 pixels (1-2m). The 4-meter IKONOS images have an accuracy of 23.6m (http://www.computamaps.com/ikonos_tech.html). After polynomial georectification using 8 GPS control points, the difference between the GPS and computed coordinates at the control points is about 6m. An optimistic estimation of the shoreline accuracy derived from the 4-meter IKONOS images in this specific case is about 8.5m.

The accuracy of the shoreline derived from a CTM and the water level is affected by the accuracy of the CTM and the accuracy of the water level. The CTM comes from the DTM generated in turn from the aerial photos and the bathymetry. The DTM has an accuracy of 2.1m and the bathymetry has an estimated accuracy of 40m. When we merged these two data sets, we select the DTM grid points in the overlapping area. If there is gap between the two data sets, we interpolate the elevations using weights of 2/3 from DTM and 1/3 from bathymetry where the standard deviation of the CTM is about 13.4m. Overall the accuracy of the CTM is from 2.1m to 13.4m for the area where the water surface model intersects with the CTM. The water level data is accurate to several centimeters. Taking 5 degrees to be the coastal slope in worse cases and 5cm the water level error, we can estimate the horizontal accuracy of the intersected shoreline caused by the water level error to be about 0.6m. Therefore, the final accuracy of the digital shoreline is about 13.4m with the CTM contributing the greatest source of error.

The NOAA T-sheets have large and medium scales: from 1:5,000 to 1:40,000. Taking 0.5 mm on the map as error source, the accuracy of the digitized shoreline from the T-sheet is about 2.5m to 20m. The shoreline digitized from the USGS DLG (1:24,000) should have an accuracy of within 12m. The ODNR map scale is 1:12,000. Again, using 0.5mm as the shoreline digitizing error, the estimated error of the shoreline is about 6m.

Table 3.1 Estimated accuracy of the shorelines derived from various sources

Shoreline	Estimated standard deviation
T-sheet	2.5m to 20m depending on scale
USGS DLG	12m (1:24,000)
ODNR map	6m (1:12,000)
Orthophoto	2.6 meters
CTM and water level	2m-13m dep. on CTM quality
IKONOS 1-meter simulated image	2-4 meters
IKONOS 4-meter image	8.5 meters

3.4 DIFFERENCE AND SHORELINE CHANGE ANALYSIS

The shorelines were acquired at different times and differences can be seen in the results (Plate 3.1). There are two possible interpretations of the shoreline differences. One is that the shoreline indeed changed in the real world. The other possibility is that the differences were introduced as shoreline mapping errors.

From our analysis of the accuracies above, the shoreline from the aerial orthophotos has the highest accuracy, and so we used it as the baseline for the difference analysis. The analysis was performed in raster format because of the efficiency of this method. The first task was to convert the vector shorelines into raster shorelines. The resulting pixel size was 2.5 meters. The procedure is outlined in Figure 3.2.

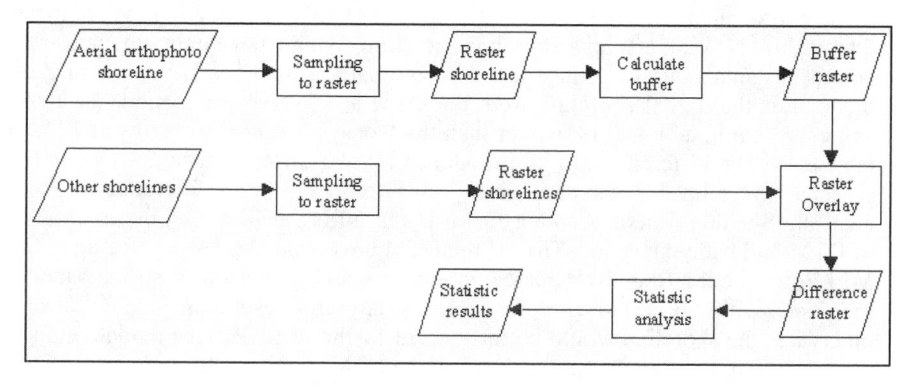

Figure 3.2 Difference Analysis Procedure

This analysis involved comparing every point on the test shorelines with the reference shoreline, which was derived from the aerial orthophoto, in a perpendicular direction toward the reference shoreline. In the buffer zone of the reference shoreline we created pixels whose value represented the distance of each pixel to the reference shoreline. By overlaying other raster shorelines with the buffer image we were able to discern and quantify differences between the shorelines and the reference shoreline. For example, for the shoreline from the simulated IKONOS images, each point on the shoreline has a value of the shortest distance to the reference shoreline. Finally, we calculated some statistic indicators including RMS of the differences for the shorelines derived from the 4-m IKONOS images and the digital models (table 3.2).

Table 3.2 Statistics of the differences between derived shorelines and the reference shoreline

	Points used	**Avg. Dif. (pixel)**	**Avg. Dif. (m)**	**Max. Dif. (pixel)**	**Max. Dif. (m)**	**RMS (pixel)**	**RMS (m)**
Shoreline from 4-m IKONOS images	6652	4.02	10.04	16	40.0	4.91	12.27
Digital shoreline	4590	2.34	5.84	9	22.5	2.74	6.84

The accuracies of the shorelines derived from the 4-meter IKONOS images and the aerial orthophoto are 8.5m and 2.6m respectively; as a result, the estimated accuracy of the difference between them should be 8.9m, which is smaller than the RMS value 12.27m in table 3.2. This may indicate the actual shoreline change during the period from July 1997 when the aerial photos for the reference shoreline were taken to October 2000 when the 4-meter IKONOS images were taken. In addition, the IKONOS shoreline is an instantaneous shoreline, instead of a tide-coordinated shoreline. The water level in October was 0.2-0.3 meters lower than in July creating artificial shoreline accretion. We are developing methods that correct instantaneous shorelines to tide-coordinated shorelines and then compare them. For the digital shoreline from the CTM and WSM intersection, the RMS value 6.842m in table 3.2 is smaller than the worse case shoreline error of 13m in table 3.1. The difference can be introduced by two different factors. One is the seasonal water level difference, which can introduce a large difference in a flat area like Sheldon Marsh. Another factor is the difference between the accuracies of DTM and bathymetry in CTM. If the DTM covers the coastal surface up to the MLLW or better (this is the case in our research), the water surface model intersects with the DTM portion of the CTM and results in smaller errors; otherwise, the shoreline would be intersected by the water surface model, and the bathymetric part or the interpolated part of the CTM produces large errors.

Observing the differences between the reference shoreline and earlier shorelines from the maps, they may represent the shoreline changes (erosion) that occurred in long time periods. The earliest shoreline in our study is the shoreline from 1973 ODNR map, 24 years before the aerial photos for the reference shoreline were taken in 1997.

3.5 CONCLUSIONS

From the analysis above, we can draw several conclusions. First, we determined that the method for generation of digital shorelines from CTM and WSM can be used in mapping instantaneous shorelines. The generated shorelines can meet the accuracy requirement for certain user communities. Additionally, this work shows that high-resolution satellite imagery, such as IKONOS imagery, has the potential to become a tool in shoreline mapping and coastal change detection, and that it may reduce mapping costs. Finally, our study confirms that to detect coastal changes, tide-coordinated shorelines must be derived from the instantaneous shorelines so that the amount of shoreline change can be estimated objectively.

3.6 ACKNOWLEDGEMENTS

The research was supported by a Sea Grant – NOAA Partnership program with the collaboration of CSC, NGS and OCS of NOAA. Special thanks go to Dr. Keith Bedford and Dr. Philips Chu for providing the water level data. We appreciate the assistance and co-operation of Dr. David Schwab of the GLERL of NOAA during Ruijing Ma's PEGS Professional Development Project in Ann Arbor, MI. We also appreciate the continuous support and help from Dr. Scudder Mackey of the Lake Erie Geology Group of the ODNR in Sandusky, OH.

3.7 REFERENCES

Cheng, P. and Toutin, T., 2000, Orthorectification of IKONOS Data Using Rational Function. In *Proceeding of ASPRS Annual Convention*, Washington D.C.

Dowman, I., and Dolloff, J.T., 2000, An Evaluation of Rational Functions for Photogrammetric Restitution. *International Archives of Photogrammetry and Remote Sensing,* **33**(B3), pp. 254–266.

Horstmann, O. and Molkenthin, F., 1996, *Advanced Grid Modeling for Coastal and Nearshore Regions*. The International Conference for Hydroinformatics, Zurich.

Li, R., 1998, Potential of High-Resolution Satellite Imagery for National Mapping Products. *Photogrammetric Engineering and Remote Sensing*, **64**(2), pp. 1165–1169.

Li, R. and Zhou, G., 1998, Coastline Mapping and Change Detection Using One-Meter Resolution Satellite Imagery. Research proposal, Dept. of Civil Engineering and Geodetic Science, the Ohio State University.

Li, R., Zhou, G., Gonzalez, A., Liu, J. K., Ma, F. and Felus, Y., 1998, Coastline Mapping and Change Detection Using One-Meter Resolution Satellite Imagery. Project Report submitted to Sea Grant/NOAA, pp. 88.

Liu, Ernie, 1998, *Developing Geographic Information System Applications in Analysis of Responses to Lake Erie Shoreline Change*, MS. thesis, The Ohio State University.

NOAA, 1997, Shoreline Mapping. http://anchor.ncd.noaa.gov/psn/shoreline.html.

Peuquet, D.J. and Duan, N., 1995, An event-based spatio-temporal data model (ESTDM) for temporal analysis of geographical data. *International Journal of Geographical Information Systems*, **9**(1), pp. 7–24.

Raper, J. and Livingstone, D., 1995, Development of a Geomorphological Spatial Model Using Object-oriented Design. *International Journal of Geographical Information Systems*, **9** (4), pp. 359–384.

Schwab, D. J. and Sellers, D. L., 1980, NOAA Data Report ERL GLERL-16, ftp://ftp.glerl.noaa.gov/publications/tech_reports/glerl-016/dr-016.html.

Tao, C.V. and Hu, Y., 2000, Investigation of the Rational Function Model. In *Proceeding of ASPRS Annual Convention*, Washington, D.C.

Wang, Z., 1990, *Principles of Photogrammetry (with Remote Sensing)*. (Wuhan, China: Wuhan Technical University of Surveying and Mapping and Publishing House of Surveying and Mapping).

Worboys, M. F., 1992, A Model for Spatio-temporal Information. In *Proceedings of the 5th International Symposium on Spatial Data Handling,* pp. 602–611.

Worboys, M.F., 1994, Unifying the Spatial and Temporal Components of Spatial Information. Advances in GIS Research In *Proceedings of the 6th International Symposium on Spatial Data Handling*, Edinburgh, pp. 505–17.

Yang, X., 2000, Accuracy of Rational Function Approximation in Photogrammetry. In *Proceeding of ASPRS Annual Convention*, Washington, D.C.

Zhou, G., and Li, R., 2000, Accuracy Evaluation of Ground Points from IKONOS High-Resolution Satellite Imagery. *Photogrammetric Engineering and Remote Sensing*, **66** (9), pp. 1103–1112.

CHAPTER FOUR

Monitoring Coastal Environments Using Remote Sensing and GIS

Paul S.Y. Pan

4.1 INTRODUCTION

This paper concerns the monitoring of the marine and coastal environment in South Wales using state-of-the-art survey techniques and a geographic information system (GIS). One of the most important natural resources in South Wales is its marine aggregate. This resource is vital to the regional economy in that it provides the building industry with that most essential of raw materials, sand and gravel. However there are growing concerns as to the possible effects of the commercial extraction of aggregate on the coastal and marine environment, and a number of environmental monitoring procedures are in place to detect changes. These range from traditional beach profile surveys to state-of-the-art airborne remote sensing techniques. The National Assembly for Wales has pioneered the use of airborne LiDAR (Light Detection and Ranging) for the acquisition of highly detailed topographical data on beaches, and CASI (Compact Airborne Spectrographic Imager), for the determination of the state of the vegetation along part of the coastline. LiDAR is capable of accurately detecting changes in beach levels. The procedure began in 1998 and will continue until at least 2003, giving an unprecedented insight into coastal changes over time.

Another remote sensing technique has also been deployed. Close-range photogrammetry has been used to determine the degree of retreat of unstable sea-cliffs. All the data collected is used to populate a GIS. Data acquired in this way is compared with that from various monitoring procedures carried out previously (aerial photography and beach profiles). A number of advanced techniques have been developed in parallel to the GIS for the interpretation, analysis and visualization of the data. A number of invaluable lessons have been learned.

Apart from the site-specific monitoring procedures, other strategic data sets such as the macro-fauna community distribution, modelled parameters, etc., have also been acquired from a number of sources. Amongst these parameters, the most important of all is the sediment environment. It defines the fuzzy geographical boundaries in which distinctive hydrodynamic regimes operate. A summary on the resources and constraints is generated for each of the sediment environments. These resources and constraints summaries together with the GIS form the basis of a decision-support system for assisting the formation of policy for the management

0-41531-972-2/04/$0.00+$1.50

of the marine resource. The findings will shape future decisions about the sustainable use of the marine resource in South Wales.

4.2 BACKGROUND

The marine and coastal environment is important to Wales. About 75% of the length of the country is coastal, allowing the waters of the Irish Sea and the Bristol Channel to lap its shores (Figure 4.1 and the colour insert following page 164). South Wales lies adjacent to the Bristol Channel, a body of water largely unaltered in its current dimensions since its beginnings as a marine transgression in the early Holocene. (The Holocene, or post-glacial epoch, covers the period for the end of the Pleistocene about 10,000 years ago to the present day). Here the coastal areas are not only characterized by a number of urban and industrial centres, such as Newport, the capital city Cardiff, Bridgend, Port Talbot and Swansea, but also by many cherished areas of special landscape and nature conservation interest, such as: the Gwent Levels; the Kenfig National Nature Reserve and candidate Special Area of Conservation (cSAC) designated under the European Habitats Directive; the Gower Area of Outstanding Natural Beauty (AONB) – the first to be designated in the United Kingdom; and the Glamorgan Heritage Coast (Figure 4.2 and colour insert). Together, and for different reasons, these coastal areas attract thousands of visitors every year, providing employment opportunities for some of the local population. One of the unseen resources of the Bristol Channel, however, contributes to the local economy in a different way – by yielding high quality building sand for local industry, an essential prerequisite for many forms of economic activity.

Whilst this economic activity brings undoubted prosperity to South Wales and neighbouring regions in mid Wales and South-West England, many people perceive changes to their familiar coastlines, and in particular, to the sandy beaches. There are growing concerns as to the possible effects of the extraction of marine sand from the Bristol Channel, and not everyone thinks the removal of this resource is acceptable or sustainable.

Generally, in Welsh waters, the extraction of sand from the marine environment by dredging is licensed by the Crown Estate. (The Crown Estate Commission is the representative of the Crown, which, in the UK, constitutes the owner of the seabed out to a 12-mile territorial limit). However, the decision on whether a production license should be granted essentially rests with the Environment Minister at the National Assembly for Wales – a devolved and autonomous arm of central Government that came into being in 1999. With its inception has come a desire to increase outside involvement in policy-making and administration, and to increase transparency and accountability in deciding major issues. The challenge for the Assembly's civil servants is to continue to provide objective advice to Ministers in an ever-evolving social, economic, environmental, cultural and political context. This advice must be based on facts and scientific evidence, together with the specialist and professional judgement of the officers involved. All of the dredging licenses granted in recent years have stringent environmental monitoring conditions attached. It is the scientific data from these monitoring procedures that form the basis of sound advice.

This paper discusses the introduction of a Geographic Information System (GIS) into part of the National Assembly for Wales for the analyses of the monitoring data. It highlights the technique's influence on the monitoring procedures and the way it has helped reshape the environmental monitoring requirements in relation to dredging licenses. By way of illustration, a number of the state-of-the-art procedures currently deployed are examined.

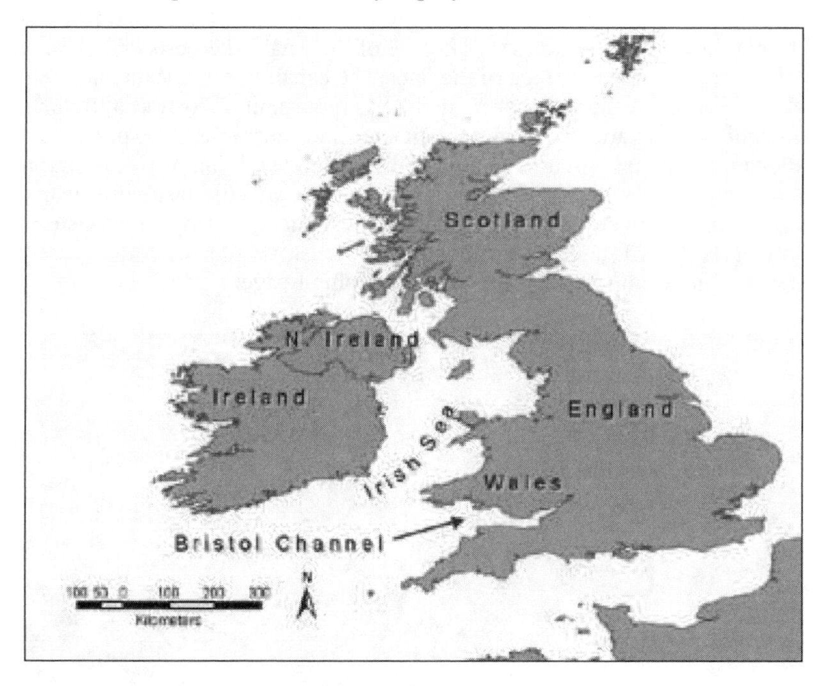

Figure 4.1 Wales and its surrounding areas.

4.3 DATA ANALYSES USING GEOGRAPHIC INFORMATION SYSTEMS

The benefits of modern computerized GIS have been well documented by others, for example, Clark *et al.*, 1991 and Maguire, 1991. GIS was first introduced to the former Welsh Office (now the National Assembly for Wales) in early 1997 for analysis of the monitoring data acquired in respect of the dredging license at Nash Bank. A prototype was developed using ESRI's (Environmental Systems Research Institute Inc.) ArcView GIS. The benefits of the technique over the traditional paper-based reporting were immediately apparent. It:

- provided a stable platform for the integration of disparate data from different sources;
- allowed a large quantity of data to be stored and processed;
- provided a seamless geographical database overcoming the restrictions of traditional map/chart boundaries;

- provided facilities for sophisticated analysis and cross-examination of data; and
- provided advanced facilities for the display and visualization of data to a wider audience.

After a period of evaluation, the technology was adopted for operational use. It would play a key role in re-shaping the environmental monitoring procedures. GIS highlighted the weaknesses of some of the established procedures in terms of both the quality and coverage of the data. The capability of GIS in handling spatial data, in particular, has also presented new opportunities for the introduction and subsequent adaptation of more efficient and cost-effective procedures. The following sections discuss some of these new and innovative environmental monitoring techniques. They are: the assessment of cliff instability using close-range photogrammetry; repeated topographical surveys using Light Detection and Ranging (LiDAR); and habitat mapping of sensitive sites of nature conservation interest using Compact Airborne Spectrographic Imager (CASI).

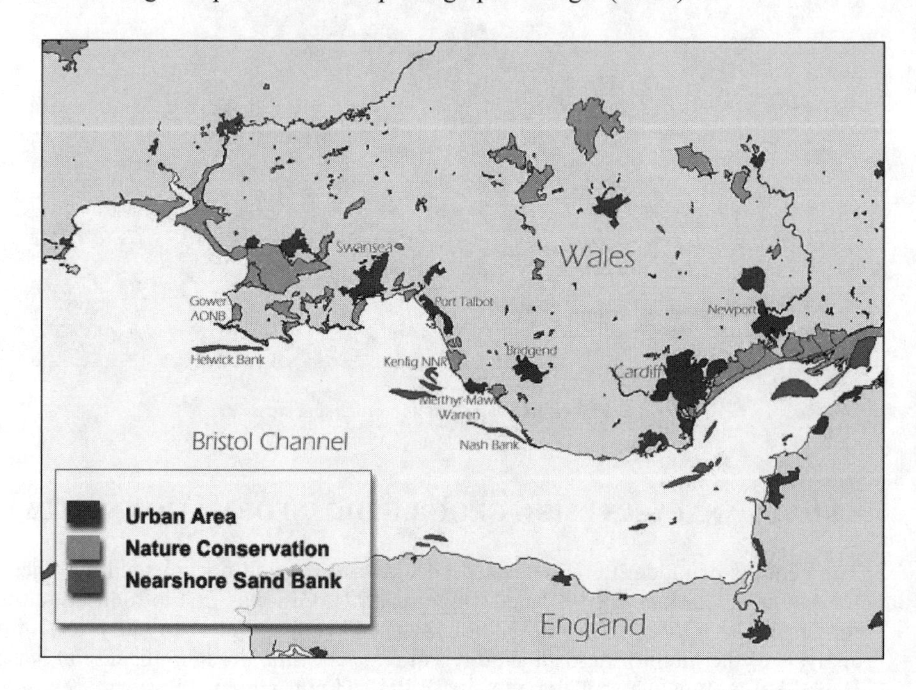

Figure 4.2 South Wales and the Bristol Channel

4.3.1 Assessment of Cliff Instability Using Close-Range Photogrammetry

The Nash Bank lies in the Bristol Channel very close to part of the Glamorgan Heritage Coast. The sandbank contains about 200 million tonnes of material and is a relic feature of the last (or "Devensian") glaciation. The bank is formed ostensibly from pre-existing glacial and glacio-fluvial deposits subsequently moulded by advancing seas. At low water its eastern end is often exposed above the surface of the sea, and it resides, at its nearest point, only 300 metres from the shore. Its physical orientation and proximity to this area of sensitive coastline means that it acts as a "barrier" to incoming waves, and is therefore an important structure in terms of coast protection. Physical processes such as the movement of sediment in, on and around the feature, as well as human activities, are altering the shape of this sandbank. One way of measuring the "vulnerability" of the nearby coastline to such changes is by careful scientific examination of instability in the highly unstable Blue Lias cliffs that predominate. Extremely accurate measurements of the geometry of a representative 800-metre section of this geologically special part of the Glamorgan Heritage Coast, and including part of the Southerndown Coast Site of Special Scientific Interest (SSSI), have been undertaken since 1997 using close-range photogrammetry. Its application in this context is unique in the United Kingdom.

Annual surveys have been undertaken in August of each year between 1997 and 2000. A further one took place in February 2001.

4.3.1.1 Close-Range Photogrammetry: The Technique

Close-range photogrammetry operates on the same principle as aerial photography. It produces highly detailed geometric data of three-dimensional structures. Data capture involves the use of a specialist metric camera oriented horizontally on a tripod, and usually mounted on top of a theodolite. The lenses are calibrated and their distortion characteristics considered for subsequent data processing. The technique is also known as terrestrial photogrammetry. Both aerial and close-range photogrammetry have been described in detail in Wolf (1985).

Close-range photogrammetry involves photographing the features or structures being surveyed using the metric camera in a known orientation from two positions. The camera positions and their orientation can be established by "traditional" surveying methods, such as using visual intersection combined with theodolite measured distances from known control points. Alternatively, a number of control points can be set up in the area being surveyed and subsequently included in the photogrammetric survey. These control points are used to compute the position and orientation of the two camera positions.

Data is derived from the photographic images by simulating the relative orientation of the two camera positions, and processing involves the generation of a three-dimensional stereo model representing the geometry of the structure. It is relatively labour-intensive because the orientation of the two camera positions is not always parallel, and this means additional computational requirements. However, the generation of the 3D geometric model has been helped by advances in automatic image matching techniques in the past decade.

Close-range photogrammetry has been used widely for surveying architectural structures worldwide. Its main advantage is that it can survey the physical dimension of any structure that does not lend itself readily to traditional surveying techniques. It has also been used extensively by traffic accident investigators for collecting data from the scenes of accidents where a very limited period is available for data capture. It is therefore a technique which is characterized by a short set up time, data capture by relatively straightforward photographic means and an ability to survey structures and features which are difficult to examine by any other means.

The intention was twofold: to (a), "test" the technique in the "real-world" of unpredictable environmental change, and (b), establish accurate rates of retreat which could be looked at against historical records. The technique is well suited to the task for several reasons. First, it allows data capture by photographic means without the need to get too close to the area being surveyed – an important consideration given the inherent instability of the Blue Lias cliffs! Also, because the Bristol Channel has the second highest tidal range in the world, only limited periods of time are available for data capture. The technique allows surveyors to record data very quickly around mean low water. Finally, the photographic images captured can be manipulated and rectified to provide a geometrically-correct picture called an orthoimage. An orthoimage is a distortion-free map-like image of an original photographic record. It is produced by rectifying the original photographic image using geometric data derived subsequently. It provides a definitive visual record of the conditions of the cliff at the time of the survey. Figure 4.3 (see colour insert following page 164) is illustrative.

Figure 4.3 The rectified image of the study area for the year 2000.

In addition, the distortion-free orthoimages of the study area can be displayed in the GIS environment where they can be used to help make accurate quantitative measurements, while the map-like orthoimages can be used in conjunction with the 3D geometry derived earlier in the process to produce a realistic representation of the cliffs using advanced computer visualisation techniques such as Virtual Reality.

The combined use of orthoimages and 3D geometric models in a GIS environment has provided a robust platform for analysis of changes to the cliffs. Quantitative contours are interpolated at 0.5 metre and 1-metre intervals using information derived from the geometric data sets. These contours, representing the degrees of change, are displayed on the rectified orthoimage. This innovative visualisation technique allows quantitative measurements of yearly change to be related directly to the cliff face and its geology. This may be vital in guiding interpretative analysis of why change occurs and in attempting to understand whether or not rates of recession are increasing.

An example will help illustrate the techniques. A large rock fall occurred in the winter of 2000-01. Figure 4.4 illustrates the extent and magnitude of the event (in the middle right of the picture) from an oblique angle.

A geometric representation of this event can be displayed and visualised in a couple of interesting ways. Firstly, the interpolated surface can be colour-coded in relation to distance between the viewer and the cliffs. Figure 4.5 shows the data for a 200-metre wide area centred on the fall. Darker shades represent features further away from the observer.

Figure 4.4 The Large rock-fall of winter 2000-01. The cliffs are comprised of Blue Lias rocks and are about 65 metres high. In the foreground is part of a wave-cut platform in the same formation that has been developing for about 7,000 years.

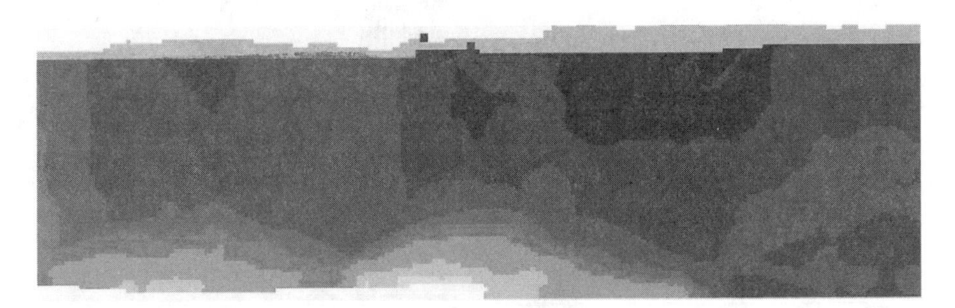

Figure 4.5 Data for a 200-metre stretch of coastline north-west of Cwm Bach

Secondly, the changes brought about by the fall between surveys can be colour-coded to illustrate the changes to the cliffs, and interpolated surfaces may produced using the GIS. Changes at the sampling positions are shaded in reds and blues, depending on the nature of the change recorded. Data collected in February 2001 was compared with that for the summer of 2000. Colour Plate 4.1 (following page 164) shows the results.

The fairly large area of the cliffs represented by deeper reds "failed," causing materials to be deposited as a "classic" debris fan at the base (in deeper blues). The thin area in blue towards the right of the failure represents an "advance." This is, in reality, a rock column which has become detached from the main body of the cliff, and which has been left as a freestanding pillar with a considerable void behind (see Figure 4.4). Future movements will be interesting to monitor!

4.3.2 Topographical Survey Using Light Detection and Ranging

Light Detection and Ranging (LiDAR) is a modern airborne remote sensing technique for surveying topography. It operates on the same basic principle as traditional Radio Detection and Ranging (RADAR), but uses laser as the detection medium. It accurately measures the distance between the instrument and its surrounding environment, and consists of three components: a laser-scanning device, an on-board Differential Global Positioning System (DGPS), and an Inertial Reference System (IRS).

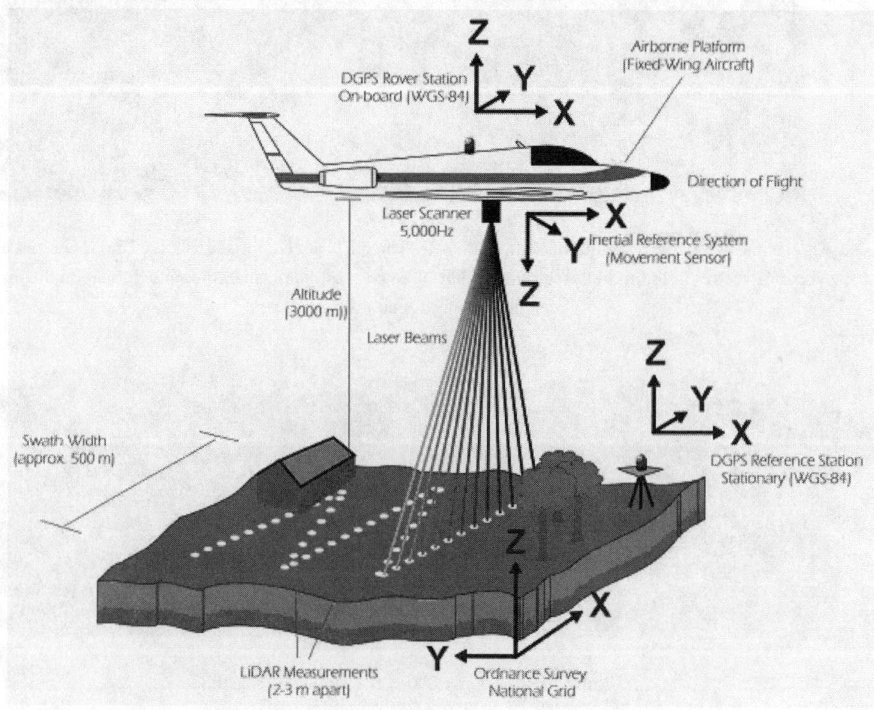

Figure 4.6 A systematic diagram of the three components of airborne LiDAR.

The laser-scanning device is used for the measurement of the distance between the instrument and the topography being surveyed. It emits laser pulses in perpendicular directions to the flight path. By measuring the time required for the laser pulse to return, the distance between the instrument and the topography can be accurately determined. At the same time, an on-board DGPS communicates with navigation satellites and a stationary ground reference station. It gives the positions of the airborne platform, usually a fixed-wing aircraft, in relation to the earth's surface. The last component of the system is the IRS. This is essentially a motion sensor, and detects minute movements of the aircraft in terms of its yaw, pitch and roll (i.e. movements about its three axes). It corrects the orientation of the laser pulses that can become distorted because of aircraft movement. When the three components work in unison, the surveyed topography is recorded very accurately. Routine field validations of the technique by the Environment Agency have indicated an accuracy of better than 15 cm generally. Ground resolution of data is directly dependent on flying height and speed of the surveying aircraft, as well as the scanning rate of the laser pulses. For example, when flying at 70 knots at an altitude of 3,000 feet LiDAR, with a 5,000 Hz scanning rate, can acquire topographical data with a ground resolution of about 2.5 to 3.5 metres. At the time of writing, the Environment Agency has plans to operate a newer version of LiDAR with a scanning rate of 33,000 Hz. The new instrument should produce topographical data with a 0.5 metre ground resolution! Figure 4.6 shows the three main components of airborne LiDAR.

In South Wales, the coastline is diverse, characterized as it is by bays, inlets, estuaries, beaches, rocky-shores, high cliffs, sand dunes, near-shore sand banks and mud flats. Traditional monitoring on some of the beaches using linear profiles has provided some limited data. In addition, accurate identification of ground features from aerial photographs has been difficult and is relatively expensive. LiDAR, however, has provided a high-value and cost-effective solution. Its capability in capturing very detailed topographical data, without the need to physically visit most of the locations being surveyed, makes it well suited for environmental monitoring purposes, especially for inter-tidal zones where it can survey a large area within a short space of time. One of the few restrictions for the deployment of LiDAR is the tidal window.

LiDAR was first available in the UK in early 1998. The National Assembly for Wales has long realized the potential of the technique and has pioneered its use in Wales. As early as March 1998, in relation to dredging at Helwick Bank, the National Assembly for Wales decided to include the technique as one of the environmental monitoring requirements to measure change along the coastline. As such, the procedure would be deployed annually to monitor levels of key beaches in South Gower. This will continue until at least June 2003. This requirement is believed to be the first of its kind in the UK, and highlights the fact that the analyses of the highly accurate topographical data over time is innovative and pioneering.

To complement these activities, in early 1999, a collaborative project called the "Glamorgan Coastal Monitoring Initiative" was launched by the Welsh Office. This was designed to further embrace the potential of this airborne remote sensing technique, and to marry the data to high-resolution habitat information derived from the Compact Airborne Spectrographic Imager (CASI). This is discussed in the following section.

LiDAR produces extensive data sets. In Kenfig National Nature Reserve (NNR), for example, over 30 million data points have been collected under the Glamorgan Coastal Monitoring Initiative. A number of custom-built computer programs have been written to reduce the data files to more manageable sizes. The data sets have been transferred to an ArcView GIS for interpolation and further analysis. Some important derivatives, such as slope and aspect, have also been computed. Plates 4.2 to 4.4 show examples of the interpolated elevation surface, and subsequent derivations of slope and aspect from LiDAR data of the same area in the Kenfig NNR.

4.3.3 Habitat Mapping Using Compact Airborne Spectrographic Imager (CASI)

Compact Airborne Spectrographic Imager (CASI) is a state-of-the-art remote sensing instrument for measuring radiation reflectances in the electro-magnetic spectrum. The classification of CASI data can produce a "map" by positively identifying the radiation reflectances of different "objects." In the case of habitat mapping, CASI can "see" a clear boundary between different types of vegetation as long as they emit different levels of radiation – the so-called spectral signature. Habitat mapping using CASI involves three components: an airborne survey using CASI; a simultaneous ground-truthing exercise; and image classification to produce the habitat map (Plate 4.5).

Table 4.1 The classified values, the classifications and their corresponding display colours in plate 4.6, the classified habitat map of Kenfig NNR.

Category	Classification	Colour
1	water	dirty green
2	bare sand	yellow
3	scrub	dark green
4	*Phragmites*	red
5	*Calluna*	light green
6	*Pteridium*	pale blue
7	successionally-young grassland	blue
8	orchid-rich slack	purple
9	successionally-young slack	light brown
10	tall rank grassland	light blue
11	embryo dune slack	pale blue

The data acquisition stage is characterized by an airborne survey using CASI. The ideal timing for the airborne survey is around midday. This is the time when the sun's angle is at its highest, casting minimum shadow and giving the most accurate representation of radiation reflectance. While the CASI survey is being carried out, a team of specialists undertakes a ground-truthing exercise to determine the ecology of selected locations within the study area. The ecological parameters are measured using traditional methods such as 1m x 1m quadrats with the aid of modern DGPS for accurate determination of position. The exercise provides important information to allow the airborne CASI data to be cross-referenced with known vegetation types. The last stage is the classification of CASI data in conjunction with the ground-truthing information to generate a map of the vegetation types. This classified map is subsequently interpreted to form the basis for a habitat map.

CASI has been used as an integral part of the "Glamorgan Coastal Monitoring Initiative." The main objective has been to provide a benchmark survey for Kenfig NNR which together with Merthyr Mawr Warren makes up the European Union's candidate Special Area of Conservation (cSAC). Both sites are of national and international importance. For example, Kenfig NNR holds over 40% of UK's population of the rare fen orchid (*Liparis loeselii*), whilst Merthyr-Mawr Warren is the second highest mobile sand dune system in Europe.

Kenfig has many other features of conservation interest. The successionally-young stages of dune development, in particular, are species-rich, with more than 40 plant species usually found in a 1m x 1m quadrat. These habitats have been prioritised for management and monitoring purposes under the current European Habitat Directive. However, there are growing concerns that the over-stabilization of the mobile sand dunes at Kenfig NNR is threatening the biodiversity of the cSAC.

For discussion purpose, Kenfig NNR has been divided into two broad areas: a southern and northern section (Figure 4.7). The southern section is mostly dominated by mature dune vegetation, whereas successionally-young seral stages are more prominent in the north of the site. The boundary is situated where one starts to grade into the other.

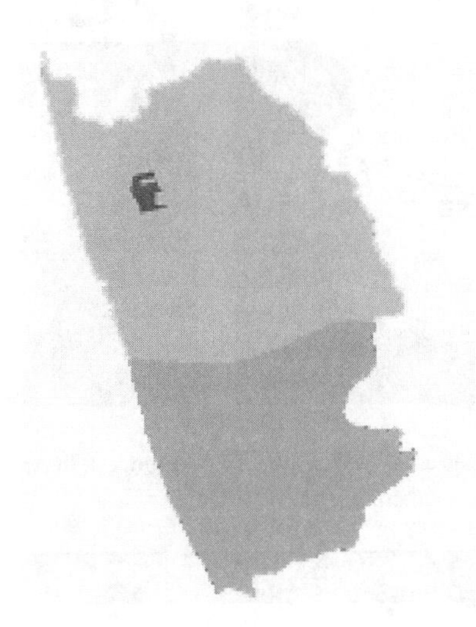

Figure 4.7 Kenfig NNR divided. The section in darker grey represents the ecologically older and more mature southern region of the Reserve. The northern area is depicted in pale grey.

The area Figure for the successionally-young grassland has been calculated by carrying out a pixel count for the habitat type based on a pixel size of 2 x 2m (Table 4.2). The area measurement for the whole site has been calculated in the same way.

Table 4.2 Area measurements for successionally-young grassland at Kenfig NNR, taken from the classified CASI images.

Section	Area of Section (m^2)	Area of successionally-young grassland (m^2)	%
Southern	1,771,276	201,248	11.3
Northern	2,706,684	782,120	28.8
Whole site	4,477,960	983,368	21.9

Preliminary results show that successionally-young grassland not only covers a much greater area in the Northern Section (which might have been expected because it is considerably larger than the Southern Section), but also occupies a much larger percentage of the area than in the South. This evidence confirms observations that the dunes in the southern part of Kenfig are generally in an advanced state of stabilisation, while this is less so in the northern part of the site where successionally-young grassland is relatively well distributed.

Figure 4.8a Areas classified as successionally-young grassland in the southern section

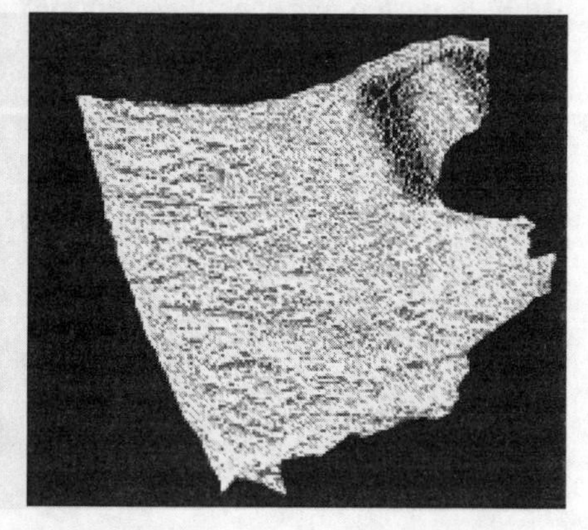

Figure 4.8b Areas with a south-facing aspect

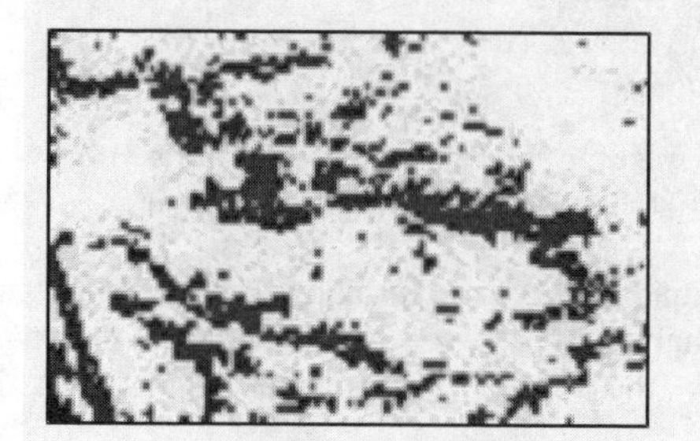

Figure 4.8c Areas classified as successionally-young grassland in the south-western part of the southern section

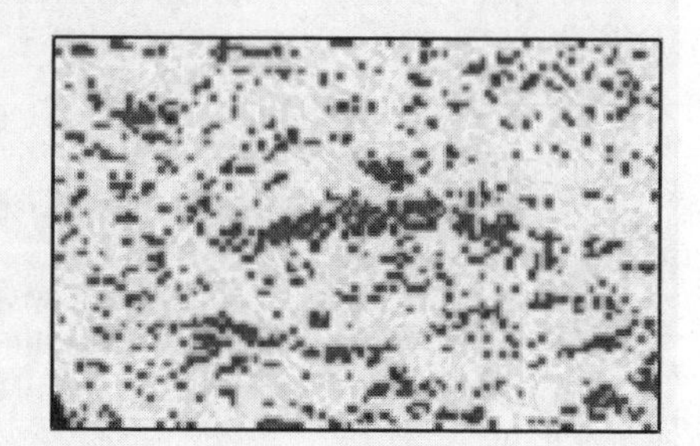

Figure 4.8d Areas with a south-facing aspect

Evidence from the joint analysis of topography and habitat classification in the southern section provides interesting comparison. A high proportion of the areas classified as successionally-young grassland occur close to the top of south-facing slopes. Figures 4.8a and b show the areas classified as successionally-young grassland in the Southern Section, juxtaposed with those showing a south-facing slope. There is also some correlation between the successionally-young grassland and aspect, and this is particularly evident in the south-western part of the section. Figures 4.8c and d are illustrative.

4.4 THE NEXT FRONTIER – MARINE AGGREGATE RESOURCE MANAGEMENT AND PLANNING SYSTEM (MARMPS)

Data collected from these innovative monitoring procedures helps to shed light on some of the plethora of natural processes causing change to the coastal environment. It also helps to expose the implications of some human activities. The data sets are used to populate a GIS. This provides a technical tool to underpin an embryonic decision-support system which will be used for the management and planning of the marine aggregate resource in South Wales to ensure it is only extracted in a sustainable way.

This system, entitled the "Marine Aggregate Resource Management and Planning System (MARMPS)," will operate using data and information on three layers: the site-specific monitoring data sets acquired in relation to the dredging licences, including those relating to bathymetry, 3D cliff analysis and topographical beach data; the regional data from the Glamorgan Coastal Monitoring Initiative, including data from both the LiDAR topographical survey and CASI habitat mapping survey; and the broad-scale and comprehensive strategic data sets derived from the "Bristol Channel Marine Aggregates: Resources and Constraints" project. This research, completed in August 2000, examined the available marine aggregate resources, and the technical, economic, environmental and cultural constraints operating to deter extraction. Two of the many significant outcomes of this work were, firstly, the delivery of an abstractive GIS database with about 100 data themes for the Bristol Channel and, secondly, the derivation of sediment environments representing sub-areas of the physical system that are considered to exhibit similarities across various components of the sediment regime. Together, these will help form the basis for future strategic policy and decision-making.

MARMPS will adopt a multi-criteria approach to aid decision-making at a more local level. In the Bristol Channel, a complicated number of conflicting objectives co-exist, for example, the economic demand for marine aggregate and possible environment impact. MARMPS will define the relationship between these by quantifying them in commensurate terms. In doing so, the conflicting objectives will be summarized and "optimised." The decision-makers will also take part by defining, in their own way, the relative importance of these conflicting objectives, and a weighting mechanism will be applied to take account of these views. Clearly, the weighting may change over time. MARMPS, when completed, will be the first system of its kind in the United Kingdom. It will be operational, designed to tackle real world issues. It will assist decision-makers in reconciling the conflicting interests of different users in the coastal zone.

It is not within the scope of this paper to discuss MARMPS in extended detail. However, the availability of good quality data sets is essential to the success of the development of the system. The National Assembly for Wales will continue to strive to acquire data sets that are robust and meaningful. Continual monitoring of the coastal and marine environment is key to an understanding of the mechanisms of coastal change.

But MARMPS is not just about data collection. There are, for example, a number of identifiable interfaces: the relationship between the coastal and the marine environment; the interaction between various bodies and organizations who have a common interest in the coastal zone; and the need to find lines of communication between science and the general public. The most important interface is, however, the one that brings about an understanding of the strengths and limitations of the natural environment and the formulation of a complementary sustainable policy to ensure wise use of its resources. The outcomes from this interface will, of course, have socio-economic implications.

The development of MARMPS is the latest in an ever-increasing list of new ways of collecting, analysing and using data and information by the newly devolved National Assembly for Wales — the introduction of GIS; routine deployment of LiDAR to monitor coastal change; habitat mapping using CASI; and 3D geometric monitoring of cliff instability. MARMPS will no doubt assist decision-makers at the National Assembly for Wales in making better informed and evidence-based decisions on the future sustainable use of our coastal heritage for many generations to come.

4.5 EPILOGUE

A dissemination seminar for the "Glamorgan Coastal Monitoring Initiative" took place in Porthcawl, South Wales, United Kingdom, on 11 July 2001. (Further details are available at: http://www.swansea.ac.uk/geog/gcmi.) The main objective was to report the results of the Initiative to participating partners, as well as to communicate knowledge about these advanced scientific techniques to the general public.

4.6 ACKNOWLEDGEMENTS

The authors would like to thank the National Assembly for Wales and the University of Wales for their support in the production of this joint work. They would also like to thank Clive Hurford of the Countryside Council for Wales and Graham Thackrah of the University of Wales, Swansea (now at University College London) for their assistance in the habitat mapping of Kenfig cSAC.

The authors are responsible for the contents of this paper, and none are attributable to either the National Assembly for Wales or the University of Wales.

4.7 REFERENCES

Clark, D.M., Hastings, D.A., and Kineman, J.J., 1991, Global databases and their implications for GIS. In *Geographical Information Systems: principles and application, vol. 2* edited by Maguire, D.J., Goodchild, M.F., Rhild, D.W. (London: Longman) pp. 217-231.

Countryside Council for Wales, 2000, *Conservation Report on Kenfig NNR.*

Countryside Council for Wales, 2001, *Habitat Monitoring for Conservation Management and Reporting Volume 1: Case Studies.*

Dargie, T.C.D., 1995, *Sand Dune Vegetation Survey of Great Britain: Part 3 – Wales. JNCC.* Peterborough.

Hurford, C., 1997, *Year 1 Report on the Fen Orchid* Liparis loeselii *Species Recovery Programme at Kenfig NNR, Glamorgan.* Countryside Council for Wales. Unpublished contract report.

Jones, P.S., 1996, *Kenfig National Nature Reserve: A Profile of a British West Coast Dune System, in Studies in European Coastal Management* (Tresaith, Cardigan: Samara Publishing Ltd.) pp. 292.

Maguire, D.J., 1991, An overview and definition of GIS. In *Geographical Information Systems: principles and applications, vol. 1* edited by Maguire D.J., Goodchild M.F., Rhild, D. W. (London: Longman) pp. 9-20.

Pan, P.S.Y. and Morgan, C.G., 1998, Monitoring of the Welsh Coastal Environment, in *Wavelength*, Department of the Environment, Transport and the Regions, Issue 2.

Pan, P.S.Y. and Morgan, C.G., 1999, Glamorgan Coastal Monitoring Initiative 1999, in *The Proceedings of the 25th Annual Technical Conference of the Remote Sensing Society*, Cardiff, September 1999, ISBN 0 946226 27 X.

Pan, P.S.Y., Morgan, C.G., Hurford, C. and Thackrah, G., 2001, Remote Sensing, GIS and the Coastal Environment in South Wales, in *The Proceedings of the GIS Research UK 9th Annual Conference GISRUK 2001*, pp 14-15.

Posford Duvivier & ABP Research & Consultancy, 2000, *Bristol Channel Marine Aggregates: Resources and Constraints*, National Assembly for Wales.

Sanjeevi Shanmugam and M.J. Barnsley, 1996. Monitoring vegetation succession in coastal dunes using remote sensing: a case study from the Kenfig NNR, South Wales, in *The Proceedings of the 22nd Annual Technical Conference of the Remote Sensing Society*, Nottingham, pp 236-243.

CHAPTER FIVE

Spatial Uncertainty in Marine and Coastal GIS

Eleanor Bruce

5.1 INTRODUCTION

The dynamic nature of coastal landscapes and the inherent complexity of the biophysical processes operating in these environments challenge the application of GIS methods. It is well recognised that spatial data models representing static objects are rife with uncertainty (Fisher, 1999; Foody, 2003). However, the mobility of many coastal and marine phenomena and the nebulous nature of boundaries in these environments provide an additional dimension to the problems associated with spatial data uncertainty. In abstracting the infinite complexity of reality into a finite computer based storage structure, multiple levels of uncertainty are introduced. The more encompassing or inclusive a data set, often the more complex the process of abstraction. Users of coastal and marine GIS are faced with both uncertainty in the information derived from spatial data, and uncertainty that inherently exists in the models. The ubiquitous nature of uncertainty in spatial analysis highlights the need to examine the implications for coastal and marine decision-making. This chapter examines the sources of uncertainty, methods for assessing reliability, model uncertainty and the cognitive and practical implications associated with the communication and incorporation of uncertainty in coastal and marine GIS.

5.2 SOURCES OF UNCERTAINTY IN COASTAL AND MARINE GIS

Uncertainties arise in the conceptualisation, measurement and analysis of geographic phenomena (Longley *et al.*, 2001). Distinctions have been made between different types of spatial uncertainty, to assist in identifying the problems and consequences that may emerge. There are a range of terms used to describe uncertainty including accuracy, error, reliability, precision and indeterminacy. Fisher (1999) identified three major types of uncertainty: these include error, vagueness and ambiguity.

0-41531-972-2/04/$0.00+$1.50

Uncertainty is associated with error when both the class of object and the individual represented in the GIS data model are clearly defined (Fisher, 1999). This form of uncertainty is probabilistic in nature. However, some uncertainties are inherent in the geographic phenomena, such as the transition zone between two coastal vegetation communities, and cannot be attributed to randomness (Cheng, 1999; Fisher, 1999; Longley *et al.*, 2001).

Vagueness relates to the difficulty in assigning classes or crisp boundaries to poorly defined objects. For example, the question of whether an area can be described as a saltmarsh requires a threshold value of a measurable parameter, such as the relative or absolute occurrence of saltmarsh species, or recourse to expert opinion (Fisher, 1999; Longley *et al.*, 2001).

Ambiguity occurs when there are differing perceptions of a phenomenon and how it should be classified (Fisher, 1999; Foody, 2003). For example, different standards may be used to define shoreline position, such as Mean High Water (MHW) or Mean Low Water (MLW). These differences in definition will lead to ambiguities in shoreline change analysis (Anders and Byrnes, 1991). Sources of uncertainty frequently encountered in coastal and marine GIS include indeterminate environmental boundaries, the fragmentation of jurisdictional responsibilities over data capture and the scientific uncertainty associated with models.

Despite the prominence and in some cases the requirement for sharply defined boundaries in GIS databases, most objects in geographic space have unclear, fuzzy or non-existent boundaries (Couclelis, 2003; Frank, 1996). This is particularly evident in coastal and marine environments, in which the transition or transient state of many phenomena exacerbates the boundary delineation problem. For example, in defining the extent of a coastal wetland, which is subject to seasonal inundation, there will be temporal variation in boundary location. A range of local interacting environmental factors, including surface hydrology, soil property and groundwater recharge, will influence rate of inundation and consequently the position of the wetland boundary. Potentially compounding this problem, there may be uncertainty in the appropriateness of the definition adopted as the basis for determining the inclusion of an area in the description of wetland. If a wetland has been artificially modified to restrict tidal influences can it still be classified as a coastal wetland? This is described as uncertainty due to indefinite correspondence between concepts and the represented world, where the information may be precise but the classification unclear (Freksa and Barkowsky, 1996). Maps delineating coastal habitats such as wetlands or mangrove communities often form the spatial framework for establishing reserve areas, applying environmental protection legislation or restricting development. Users may assume data certainty based on the neatness of the precise line used to represent an area of habitat on a map while the zone of uncertainty surrounding each boundary remains ignored in the decision process. Management protocols and legislative frameworks often require crisp boundaries for coastal and marine phenomena that are continuous in nature.

Although the legacy of disparate government and private-sector responsibility in coastal and marine management is being addressed in many countries, through the adoption of integrated resource management (Harvey and Caton, 2003), such fragmentation is often still reflected in the capture and custodianship arrangements of spatial data. Bathymetric data is primarily compiled by hydrographic agencies for marine navigation, transport planning and maritime safety (Ward *et al.*, 2000). Terrestrial elevation data is collated by state or federal land mapping agencies operating under different spatial data capture programs. However, the mapping and spatial modeling of coastal phenomena that traverse the water line, such as historical shoreline change, nearshore dynamics or intertidal sediment accretion, requires these datasets to be merged. The merging of these data involves careful consideration of the reference ellipsoid and water level datum (Li, 2000). Paradoxically, data discrepancy, and consequently uncertainty, is often greatest in the zone of highest research significance, the inter-tidal region. GIS based models designed to examine the coastal impacts of sea-level rise scenarios are particularly sensitive to inconsistencies between merged elevation surfaces (Bruce *et al.*, 2003). Recent developments in coastal applications of Airborne Laser Hydrography are providing detailed 'seamless' Coastal Terrain Models (CTMs) (Sinclair *et al.*, 2003). However, if these data are to become available to broader GIS research applications, there will be a need to reconsider the appropriateness of the shoreline as the delineation for the division of spatial data capture responsibilities.

A second consequence of multiple jurisdictional responsibilities is the incompatibility of classification systems used to describe coastal and marine biophysical environments that extend administrative boundaries. This may generate greater uncertainty at state or national boundaries. For example, in Australia the system of Marine Protected Areas (MPAs) is required to meet principles of comprehensiveness, adequacy and representativeness through strategic inclusion of diverse examples of marine life. In achieving this, the mapping of nearshore waters has involved the use of benthic habitats as a surrogate for biodiversity by most jurisdictions (state and territory governments). However, jurisdictional differences in the spatial scale of responsibility, the nature of marine environments (water clarity and depth), and conceptual divergence in definitions of mapping elements have resulted in considerable variation in mapping standards in Australian nearshore waters (ANZECC TFMPA, 2000). An attempt has been made to organise different jurisdictional mapping approaches and construct a single hierarchical ecosystem-based classification (ANZECC TFMPA, 2000). These issues are particularly evident in coastal and marine environments in which biophysical processes transcend jurisdictional boundaries.

It is clear that uncertainty in the conception and measurement of geographic phenomena will lead to uncertainty in analysis results (Longley *et al.*, 2001). However, in many coastal and marine GIS applications uncertainty in the model must also be considered. The exact values of model parameters are rarely known; therefore it is often necessary to estimate values, but there is always an associated error in that estimation (Heuvelink *et al.*, 1989). Increasing concern by researchers and environmental managers for the potential impacts of projected climate change to coastal environments has highlighted the importance of comprehensive model-based approaches for predicting and assessing longer-term change (Capobianco *et al.*, 1999). However, in developing appropriate spatial models there is a need to account for the scientific uncertainty that surrounds predicted sea-level conditions on which these models are based (Cowell and Zeng, 2003). The sensitivity of model results to variation in sea-level predictions can be demonstrated through alternative impact scenarios or coastal risk maps derived by modifying sea-level curves. However, coastal managers are then faced with the challenge of transferring these alternative predictions into response policies.

5.3 METHODS FOR EXAMINING LEVELS OF DATA UNCERTAINTY

The reliability of spatial data sets can be assessed using techniques that extend beyond simple field sampling and verification to provide useful indicators of data uncertainty. Assessment methods provide indicators of data uncertainty that can be used by decision makers to determine a data set's fitness of use for the intended application. These techniques deal with measurement and processing uncertainty in the capture of spatial data. There is a broad range of techniques for assessing the reliability of spatial data that will reflect the nature of the data (categorical or continuous) and the form of uncertainty being examined (position and/or attribute). Two of these include the error matrix and the epsilon boundary model.

An accepted method for reporting accuracies in categorical data derived from remotely sensed classifications is the error matrix (Congalton and Green, 1993; Congalton and Mead, 1983; Veregin, 1999). Error matrixes cross tabulate the classified or mapped data with the referenced or field sampled data (Janssen and van der Wel, 1994). The technique has been applied to provide accuracy measures in the mapping of shallow temperate water habitats including seagrass communities and coral reefs (Bruce *et al.*, 1997; Mumby and Edwards, 2002). This approach allows the appropriateness of the assigned map classification to be assessed. For example, when mapping nearshore coastal habitats using remotely sensed data, light attenuation may distort reflectance signals resulting in confusion in the interpretation of classes such as dense seagrass and reef or low biomass seagrass species and sandy substrate. Direct comparison of field and mapped data in the error matrix provides reliability measures that allow end users to assess the level of misclassification associated with each class providing insight on the appropriateness of the data for its intended use. However, this approach assumes a formal conception or definition of the phenomena being mapped and that there is consistency in the descriptions assigned by the map producers and the field samplers. For example, the rule sets used in the interpretation of satellite imagery to define areas of dense seagrass based on reflectance signatures may not translate in the field for divers measuring the biomass density of sampling quadrats.

A useful technique for the assessment and representation of uncertainty associated with the cartographic representation of mapped boundaries is the epsilon error model (Blakemore, 1984; Chrisman, 1982; Chrisman, 1989). The epsilon error is based on the principle that the true position of a boundary line will occur at a displaced distance from the represented or mapped line (Chrisman, 1989). The epsilon area can be described geometrically as the zone extending either side of the mapped line delineating a probability density function of the boundary's real world location (Chrisman, 1989). True boundary position can be measured in the field through transect surveys to record changes in cover type. Using the line intercept method transect results are then overlaid with the mapped data to determine boundary offset (Skidmore and Turner, 1992). In the marine environment this can be conducted using an underwater video camera towed behind a vessel or detailed dive transects. The width of epsilon bands can be set uniformly according to the standard deviation of the uncertainty of the mapped boundary; in this way the epsilon zone represents a form of mean error (Chrisman, 1989). However, these boundary; error concepts originated in forestry science (de Vries, 1986) and are not easily transferred to the marine and coastal environment. The distribution of error and the corresponding shape of the error band may not be uniform in width (Caspary and Scheuring, 1993; Veregin, 1999). In the nearshore environment factors such as water depth and turbidity may influence the distribution of error around the mapped line. Research conducted in temperate shallow-water environments in southeastern Australia has demonstrated that the depth at which seagrass boundaries can be detected is not an absolute measure but rather a relative measure influenced by the same conditions that control seagrass growth (Mount, 2003).

Quantifying uncertainties in measurement and classification will guides users and decisions makers in assessing the limitations of the data and their decisions. Estimates of measurement error also provide valuable input in examining the implications of uncertainty in analysis results. If spatial data are to be incorporated into coastal and marine models it is important to examine how measurement and processing errors influence model output.

5.4 APPROACHES FOR EXAMINING AND REPRESENTING MODEL UNCERTAINTY

Uncertainty in coastal and marine data and models is inevitable and useful descriptions of the properties of error are provided in the literature (von Meyer *et al.*, 2000). An important consideration for coastal and marine GIS users is how this uncertainty can be managed. Innovative approaches designed to model and represent underlying data uncertainty have emerged from recent spatial uncertainty literature (Morris, 2003). This section briefly describes three of these approaches and presents examples in which they have been applied to modelling coastal and marine environments. These include error propagation, sensitivity analysis and fuzzy set theory. Error propagation and sensitivity analysis measures the impacts of uncertainty in the input data on the analysis or model results. Fuzzy set theory allows an object's inclusion in a class to be assigned along a continuum, offering a definable approach for dealing with inexact concepts (Cowell and Zeng, 2003; Zadeh, 1965).

Analysis of error propagation allows the way in which uncertainties accumulate and affect the end result of the model to be explored (Burrough and McDonnell, 1998). Errors may continue when the output of one spatial operation (such as map algebra or polygon overlay) is used as the input for a subsequent operation (Heuvelink, 1999). The exactness of quantitative results derived from multiple input data layers in spatial models often makes it difficult for the end user to recognise the effects of data and model error. Although uncertainty in both the input data and the model will influence the value of results, it is also critical to examine the way the data and the model interact (Burrough and McDonnell, 1998). For example, uncertainties in elevation will propagate when bathymetric surfaces compiled at different dates are used to compute estimates of sedimentation rates (Van der Wal and Pye, 2003). If sedimentation is measured as the change in elevation the standard error is the root of the squared error deviations of the two bathymetric surfaces (Van der Wal and Pye, 2003). This has significant implications for the quantification of morphological change if standard errors exceed finer scale change estimates. In tracking error behaviour throughout the model comparisons can be made between the contribution of data input and model error (Heuvelink, 1999). This allows developers and those implementing the model to assess the appropriateness of the model in relation to the available spatial data. However, an error propagation analysis can only yield plausible results if the errors associated with each model input have realistic values (Heuvelink, 1999).

Sensitivity analysis is the controlled variation of parameter values in isolation and in combination, and the observed response of model output (Conroy *et al.*, 1995; Lodwick *et al.*, 1990). By imposing variations on the input data sets, measurement of effects on the model results can provide an indication of reliability. An example of this in spatial analysis is the alteration of input data resolution. Identification of the variables with the greatest influence on model results is not only critical to model calibration but also provides direction for future research (Hamby, 1994). The effects of an input parameter on model outcome can be divided into two types. Parameters whose uncertainty contributes to the results are referred to as 'important' parameters, and those that have a significant influence on the results are 'sensitive' parameters (Hamby, 1994). Sensitivity analysis has been used to examine levels of uncertainty in marine habitat modeling (Bruce, 1997). Bayesian probability theory was applied to predict dugong distributions in a semi-enclosed coastal embayment in Western Australia (Bruce, 1997). Sensitivity analysis was applied to examine the effect of species mobility on the model results as dugongs may range over an extensive area within a preferred habitat potentially contributing to uncertainty in sighting location. It was demonstrated that the model was sensitive to positional error in species sighting data at positional offsets of greater than 800 metres (Bruce, 1997). Sensitivity analysis examines uncertainty from output back to input, which is the reverse direction to the error propagation approach that provides a quality test for the input data and the model (Crosetto and Tarantola, 2001). However, these approaches assume certainty in the definition of an object or the conceptual model used to represent a phenomenon. Methods are needed to deal with the polythetic (defined by multiple attributes) and inexact nature of many natural geographic phenomena.

An approach that addresses problems associated with vagueness of definition is fuzzy set theory (Fisher, 1999). In classic set or Boolean theory an object is defined as belonging or not belonging to a class or set. This membership status is coded as values of 1 or 0 respectively (Fisher, 2000; Fisher, 1999; Robinson, 2003). In fuzzy set theory, membership is defined by a real number in a range from 0 (nonmember) to 1 (full member) (Bernhardsen, 2002; Cowell and Zeng, 2003; Fisher, 1999). Fuzzy set theory allows an object membership to a varying degree and to more than one set (Robinson, 2003). For example, in representing the presence of seagrass using classic set theory a seagrass patch may be assigned a value of 1 and an area of bare substrate a value of 0. The same patch represented using fuzzy set theory may have a value reflecting its high degree of membership in the class of seagrass (e.g. 0.85) and a low value indicating a small degree of membership in the class of sand (e.g. 0.1). The assignment of the membership function of a fuzzy set is subjective in nature but the process of assignment is not arbitrary (Robinson, 2003). A range of approaches for assigning grades of membership can be adopted when applying fuzzy sets in GIS (Robinson, 2003).

Fuzzy membership is further complicated when representing natural phenomena that change. Cheng (1999) presents a method to identify coastal features and their dynamics using fuzzy spatial extent (regions) extracted at different epochs. Application of fuzzy set theory for predicting coastal geomorphic hazard associated with sea-level induced recession and storm erosion has also been demonstrated for beach environments in SE Australia (Cowell and Zeng, 2003). Although not a complete solution to the problem of vagueness, an important strength of fuzzy set theory is that it allows partial implementation of the vagueness concept (Fisher, 1999; Robinson, 2003).

5.5 COGNITIVE DIMENSIONS OF UNCERTAINTY

Certainty in the spatial representation of reality has been contested in the broader spatial science community since the earlier stages of GIS (Burrough, 1986; Chrisman, 1982; Chrisman, 1984; Goodchild, 1989). Conceptualization of the nature of uncertainty in GIS has been documented (Fisher, 1999) and methods have been developed for the representation and handling of uncertainty (Agumya and Hunter, 2002; Veregin, 1999). In recent GIS literature there has been recognition of a shift in the perception of uncertainty. Rather than focusing on uncertainty as the consequence of empirical failings Couclelis (2003) argues for accepting uncertainty as an intrinsic property of knowledge. The challenge faced by users of coastal and marine GIS is how these insights on uncertainty translate to the decision-making process.

With the increasing flow of spatial data, facilitated through web-based data clearinghouses, Internet mapping and interoperability, there is potential for unchecked dissemination of associated data uncertainty. Attached metadata documents and national and international spatial data quality standards assist in addressing the unsuspected promulgation of errors and inherent bias. However, in order to accommodate uncertainty coastal and marine managers, planners and communities must convert these data qualifiers into guidelines that will form the basis of meaningful decisions. There is a cognitive dimension to the communication of uncertainty that must extend concepts accepted in the realm of GIS to ensure accessibility to the diverse community of end users. For example, inclusion of an uncertainty measure in the output of a coastal hazard risk model will result in risk being mapped along a continuum rather than an absolute (Zerger *et al.*, 2002). In contrast, emergency managers require clear guidelines for identifying and evacuating risk areas (Zerger *et al.*, 2002).

There is often a requirement from end-users of GIS derived coastal and marine information for unambiguous information that can be incorporated directly into decision structures. These user requirements need to be considered in any definition or representation of uncertainty if end-users are expected to recognise and adopt adequate precautionary approaches. Returning to the wetland boundary example presented earlier, depiction of the 'zone of confusion' surrounding each wetland boundary, although highlighting uncertainty, does not provide a clear basis for administrative decisions. Legislation relating to the environmental protection of a coastal wetland system may require precise definitions of spatial extent. For example, under the New South Wales Coastal Protection Act in Australia definition of the coastal zone includes all coastal rivers to the upstream limit of mangrove plants. Although inclusive of a critical estuarine habitat, this definition does not include saltmarsh communities resulting in the occurrence of these habitats outside the jurisdiction of associated coastal management policies. Many approaches for the representation and handling of spatial uncertainty do not translate neatly into legislative frameworks designed to ensure the protection and ecological sustainability of coastal and marine environments.

Ensuring end-user awareness of spatial uncertainty is not restricted to coastal and marine managers, policy makers and other practitioners. There is a growing role for GIS in encouraging community involvement in environmental decision-making. Recent approaches for integrating local expertise and perceptions into a GIS framework have been described as GIS for participation (GIS-P)(Cinderby, 1999; Weiner *et al.*, 2002). An example includes the adoption of participatory GIS to capture local fisheries knowledge from a Newfoundland fishing community on the northeast coast of Canada (Macnab, 2002). These applications further highlight the cognitive aspect of spatial uncertainty. If results are presented without building the capacity of participating groups to understand limitations associated with the analysis and input data, conflict within communities may be exacerbated by the implementation of GIS-P (Cinderby, 1999). There is value in the use of fuzzy representations over distinct map boundaries in these circumstances. Exact boundaries may infer a level of positional precision that could potentially prompt dispute. Successful development of participatory GIS applications in the coastal and marine environment will require approaches that ensure concepts of uncertainty are communicated in an accessible way to the broader community.

Difficulties associated with incorporating uncertainty measures, fuzzy boundaries or relative values may depend on the temporal scale and the legislative framework of the decision-making process. There is greater capacity for longer-term coastal planning strategies to accommodate fuzzy results and translate these to precautionary decisions based on acceptable risk. However, faced with the immediate responsibility to approve or reject a development proposal under the threat of habitat loss a decision maker may require absolute values on which to base and defend, perhaps in court, their decision.

The exactness of legislative statements relating to the protection of coastal and marine environments is not suited to vague or ambiguous representations and often requires the conversion of fuzzy values into Boolean structures, forced classifications and precise boundaries. This need to simplify the world may reflect an underlying familiarity with the traditional cartographic paradigm of discrete entities. Perhaps there is a need to investigate approaches for accommodating flexible definitions in coastal and marine policies and associated legislative frameworks.

5.6 CONCLUSION

Uncertainty in GIS is inevitable; it is how this uncertainty is perceived and handled which is important. Fragmentation of the landscape into manageable spatial data models is required in many applications of coastal and marine GIS. During this process of abstracting reality the loss of detail potentially introduces levels of uncertainty that may have significant consequence in analysis and modeling results. Compounded by this is the uncertainty that stems from the complexity of coastal and marine systems (Cowell and Zeng, 2003). Techniques exist that allow the quantification, assessment and communication of uncertainty in GIS applications as demonstrated by the range of coastal and marine case studies cited in this chapter. However, if widespread awareness of spatial uncertainty issues is to be achieved amongst the coastal and marine GIS community a greater understanding of users' cognition, perception and incorporation of uncertainty is needed.

5.7 REFERENCES

Agumya, A., and Hunter, G.J., 2002, Responding to the consequences of uncertainty in geographical data. *International Journal of Geographical Information Science*, **16**, pp. 405-417.

Anders, F.J., and Byrnes, M.R., 1991, Accuracy of shoreline change rates as determined from maps and aerial photographs. *Shore and Beach*, **59**, pp. 17-26.

Australian New Zealand Environment Conservation Council, and Task Force Marine Protected Areas. 2000, *Review of Methods for Ecosystem mapping for the National Representative System of Marine Protected Areas*, (Canberra: Environment Australia).

Bartlett D.J. and Bruce, E., 2002. Quality Control in Coastal GIS, In *An Evaluation of Progress in Coastal Policies at the National Level, A Transatlantic and Euro-Mediterranean Perspective*, edited by B. C. Sain, I. Pavlin and S. Belfiore. Dordtrecht, The Netherlands, Kluwer Academic Publishers (NATO Science Series).

Bernhardsen, T. 2002, *Geographical Information Systems: An Introduction*, (New York: John Wiley & Sons Inc.).

Blakemore, M., 1984, Generalisation and error in spatial databases. *Cartographia*, **21**, pp. 131-139.

Bruce, E., Cowell, P.J., and Stolper, D., 2003, Development of a GIS-based estuary sedimentation model. In *Coastal GIS - An Integrated Approach to Australian Coastal Issues* (Wollongong: University of Wollongong). 553pp.

Bruce, E.M., 1997, Application of Spatial Analysis to Coastal and Marine Management in the Shark Bay World Heritage Area, Western Australia. PhD thesis, University of Western Australia, Perth.

Bruce, E.M., Eliot, I.G., and Milton, D.J., 1997, A method for assessing the thematic and positional accuracy of seagrass mapping. *Marine Geodesy*, **20**, pp. 175-193.

Burrough, P.A. 1986, *Principles of Geographical Information Systems for Land Resources Assessment*, (New York: Oxford University Press).

Burrough, P.A., and McDonnell, R.A. 1998, *Principles of Geographical Information Systems*, (New York: Oxford University Press).

Capobianco, M., DeVriend, H.J., Nicholls, R.J., and Stive, M.J.F., 1999, Coastal area impact and vulnerability assessment: The point of view of a morphodynamic modeler. *Journal of Coastal Research*, **15**, pp. 701-716.

Caspary, W., and Scheuring, R., 1993, Positional accuracy in spatial databases. *Computers, Environment and Urban Systems*, **17**, pp. 103-110.

Cheng, T. 1999, *A Process-Oriented Data Model for Fuzzy Spatial Objects*, (The Netherlands: ITC Publications).

Chrisman, N.R., 1982, A theory of cartographic error and its measurement in digital data bases. In *Fifth Symposium on Computer- Assisted Cartography and International Society for Photogrammetry and Remote Sensing Commission*, Virginia, US, (Environmental Assessment and Resource Management) pp.159-168.

Chrisman, N.R., 1984, The role of quality information in the long-term functioning of a Geographic Information System. *Cartographica*, **21**, pp. 79-87.

Chrisman, N.R., 1989, Modeling error in overlaid categorical maps. In *Accuracy of Spatial Databases*, edited by Goodchild, M. F., and Gopal, S., (London: Taylor & Francis), pp. 21-34.

Cinderby, S., 1999, Geographic Information Systems (GIS) for participation: The future of environmental GIS? *International Journal of Environmental Pollution*, **11**, pp. 304-315.

Congalton, R.G., and Green, K., 1993, A practical look at the sources of confusion in error matrix generation. *Photogrammetric Engineering and Remote Sensing*, **59**, pp. 69-74.

Congalton, R.G., and Mead, R.A., 1983, A quantitative method to test for consistency and correctness in photointerpretation. *Photogrammetric Engineering and Remote Sensing*, **49**, pp. 69-74.

Conroy, M.J., Cohen, Y., James, F.C., Matsinos, Y.G., and Maurer, B.A., 1995, Parameter estimation, reliability, and model improvement for spatially explicit models of animal populations. *Ecological Applications*, **5**, pp. 17-19.

Couclelis, H., 2003, The certainty of uncertainty: GIS and the limits of geographic knowledge. *Transactions in GIS*, **7**, pp. 165-175.

Cowell, P.J., and Zeng, T.Q., 2003, Integrating uncertainty theories with GIS for modeling coastal hazards of climate change. *Marine Geodesy*, **26**, pp. 5-18.

Crosetto, M., and Tarantola, S., 2001, Uncertainty and sensitivity analysis: Tools for GIS-based model implementation. *International Journal of Geographical Information Science*, **15**, pp. 415-437.

de Vries, P.G. 1986, *Sampling Theory for Forest Inventory: A Teach-Yourself Course*, (Berlin: Springer-Verlag).

Fisher, P., 2000, Sorites paradox and vague geographies. *Fuzzy Sets and Systems*, **113**, pp. 7-18.

Fisher, P.F., 1999, Models of uncertainty in spatial data. In *Geographical Information Systems: Principles and Technical Issues*, vol. 1, edited by Longley, P. A., Goodchild, M. F., Maguire, D. J., and Rhind, D. W., (New York: John Wiley & Sons Inc.), pp. 191-205.

Foody, G.M., 2003, Uncertainty, knowledge discovery and data mining in GIS. *Progress in Physical Geography*, **27**, pp. 113-121.

Frank, A.U., 1996, The prevalence of objects with sharp boundaries in GIS. In *Geographic Objects With Indeterminate Boundaries*, *GISDATA 2*, edited by Burrough, P. A., and Frank, A. U., (London: Taylor & Francis), pp. 29-40.

Freksa, C., and Barkowsky, T., 1996, On the relations between spatial concepts and geographic objects. In *Geographic Objects With Indeterminate Boundaries*, edited by Burrough, P. A., and Frank, A. U., (London: Taylor & Francis), pp. 109-121.

Goodchild, M.F., 1989, Modeling error in objects and fields. In *Accuracy of Spatial Databases*, edited by Goodchild, M. F., and Gopal, S., (London: Taylor & Francis), pp. 107-113.

Hamby, D.M., 1994, A review of techniques for parameter sensitivity analysis of environmental models. *Environmental Monitoring and Assessment*, **32**, pp. 135-154.

Harvey, N., and Caton, B. 2003, *Coastal Management in Australia*, (Melbourne: Oxford University Press).

Heuvelink, G.B.M., 1999, Propagation of error in spatial modelling with GIS. In *Geographical Information Systems: Principles and Technical Issues*, vol. 1, edited by Longley, P. A., Goodchild, M. F., Maguire, D. J., and Rhind, D. W., (New York: John Wiley & Sons Inc.), pp. 207-217.

Heuvelink, G.B.M., Burrough, P.A., and Stein, A., 1989, Propagation of errors in spatial modelling with GIS. *International Journal of Geographic Information Systems*, **3**, pp. 303-322.

Janssen, L.L.F., and van der Wel, F.J.M., 1994, Accuracy assessment of satellite derived land-cover data: a review. *Photogrammetric Engineering and Remote Sensing*, **60**, pp. 419-426.

Li, R., 2000. Data models for Marine and Coastal Geographic Information Systems. In *Marine and Coastal Geographical Information Systems*, edited by Wright, D., and Bartlett, D., (London: Taylor & Francis), pp.25-36.

Lodwick, W.A., Monson, W., and Svoboda, L., 1990, Attribute error and sensitivity analysis of map operations in geographical information systems: suitability analysis. *International Journal of Geographical Information Systems*, **4**, pp. 413-428.

Longley, P.A., Goodchild, M.F., Maguire, D.J., and Rhind, D.W. 2001, *Geographic Information Systems and Science*, (Chichester, England: John Wiley & Sons Inc.).

Macnab, P., 2002, There must be a catch: Participatory GIS in a Newfoundland Fishing Community. In *Community Participation and Geographic Information Systems*, edited by Craig, W., Harris, T., and Weiner, D., (London: Taylor & Francis), pp. 173-191.

Morris, A., 2003, A framework for modeling uncertainty in spatial databases. *Transactions in GIS*, **7**, pp. 83-101.

Mount, R., 2003, The application of digital aerial photography to shallow water seabed mapping and monitoring - how deep can you see? In *Coastal GIS - An Integrated Approach to Australian Coastal Issues*, (Wollongong: University of Wollongong). 553pp.

Mumby, P.J., and Edwards, A.J., 2002, Mapping marine environments with IKONOS imagery: Enhanced spatial resolution can deliver greater thematic accuracy. *Remote Sensing of Environment*, **82**, pp. 248-257.

Robinson, V.B., 2003, A perspective on the fundamentals of fuzzy sets and their use in geographic information systems. *Transactions in GIS*, **7**, pp. 3-30.

Sinclair, M., Townsend, N., and Barker, R., 2003, Survey operations for coastal zone management in New South Wales using Airborne Laser Hydrography. In *Coastal GIS - An Integrated Approach to Australian Coastal Issues*, (Wollongong: University of Wollongong) 553pp.

Skidmore, A.K., and Turner, B.J., 1992, Map accuracy assessment using line intersect sampling. *Photogrammetric Engineering and Remote Sensing*, **58**, pp. 1453-1457.

Van der Wal, D., and Pye, K., 2003, The use of historical bathymetric charts in a GIS to assess morphological change in estuaries. *Geographical Journal*, **169**, pp. 21-31.

Veregin, H., 1999, Data quality parameters. In *Geographical Information Systems: Principles and Technical Issues*, vol. 1, edited by Longley, P. A., Goodchild, M. F., Maguire, D. J., and Rhind, D. W., (New York: John Wiley & Sons Inc.), pp. 177-189.

Von Meyer, N., Foote, K.E., and Huebner, D.J., 2000, Information quality considerations for coastal data. In *Marine and Coastal Geographic Information Systems*, edited by Wright, D., and Bartlett, D., (London: Taylor & Francis), pp. 295-308.

Ward, R., Roberts, C., and Furness, R., 2000, Electronic chart display and information systems (ECDIS): state-of-the-art in nautical charting. In *Marine and Coastal Geographical Information Systems*, edited by Wright, D., and Bartlett, D., (London: Taylor & Francis), pp. 149-161.

Weiner, D., Harris, T.M., and Craig, W.J., 2002, Community participation and geographic information systems. In *Community Participation and Geographic Information Systems*, edited by Craig, W. J., Harris, T. M., and Weiner, D., (London: Taylor & Francis), pp. 3-16.

Zadeh, L.A., 1965, Fuzzy sets. *Information and Control*, **8**, pp. 338-355.

Zerger, A., Smith, D.I., Hunter, G.J., and Jones, S.D., 2002, Riding the storm: A comparison of uncertainty modelling techniques for storm surge risk management. *Applied Geography*, **22**, pp. 307-330.

CHAPTER SIX

New Directions for Coastal and Marine Monitoring: Web Mapping and Mobile Application Technologies

Sam Ng'ang'a Macharia

6.1 THE VALUE OF COASTAL AND MARINE RESOURCES

Coastal and marine areas are ever increasing in value to the welfare of nations. These areas provide natural, social and economic functions that contribute to increased quality of life. The oceans are instrumental in determining climate that beneficially affects all life on Earth (Payoyo, 1994). Other natural functions include habitat for endangered species, species breeding and resting areas, water treatment, groundwater recharge and flood attenuation. Some social and economic functions include tourism, commercial and recreational fishing, oil and gas development, and construction (Eckert, 1979; Prescott, 1985; Gomes, 1998).

It is clear that coastal and marine areas are of vital importance to human life. Yet human terrestrial and marine activities have proven to have destructive effects on these areas. According to Canada's National Program of Action (CNPA) (2000) the major threats to the health, productivity and bio-diversity of the marine environment result from human activity in the coastal areas and further inland. Approximately 80 percent of marine area contamination results from land-based activities such as municipal, industrial and agricultural waste and run-off, in addition to the deposition of atmospheric contaminants resulting from human industrial activities (CNPA, 2000; Sanger, 1987).

There is a need for a wider dissemination of knowledge relevant to the importance of coastal and marine areas to the world's well-being, and a re-evaluation of societies' attitudes towards these spaces. Good coastal and marine governance (e.g. information dissemination, management, monitoring, etc.) is therefore a key factor in the sustainable use of these environments and will require an integrated, coordinated and equitable approach (Crowe, 2000). If governance is about decision-making and steering, then up-to-date, accurate, complete, usable information (which feeds into the acquisition of knowledge) is indispensable to governance. This is especially critical in the information age of rapid changes, interconnectivity, and globalization that have brought more information to more people making them acutely aware of the unsustainable nature of current social,

0-41531-972-2/04/$0.00+$1.50

economic and political use of marine and coastal spaces (Juillet and Roy, 1999; Rosell, 1999; Miles, 1998).

Where informed decisions have to be made using real-time information there is a need for architecture that quickly disseminates information affecting coastal and marine resources. Accurate, up-to-date, complete and useful spatial information (on many levels) regarding the resources that currently exist, the nature of the environment within which those resources exist, as well as on the users of those resources is always a requirement for effective monitoring of coastal and marine areas. Information on (but not limited to) living and non-living resources, bathymetry, spatial extents (boundaries), shoreline changes, marine contaminants, seabed characteristics, water quality, and property rights all contribute to the sustainable development and good governance of coastal and marine resources (Nichols, Monahan and Sutherland, 2000; Nichols and Monahan, 1999).

6.1.1 Web Mapping and CMM Networks

There are several Coastal and Marine Monitoring (CMM) networks in the world. Most of them are involved in obtaining real-time quantitative indicators that impact on coastal and marine health such as water temperature, water level and meteorological conditions (wind speed and direction, temperature, barometric pressure), together with qualitative indicators such as visual images of the beach and nearshore. Others are involved in collecting quantitative and qualitative indicators of coastal environmental quality. The networks involve government, academic and environmental NGO institutions. They consist of huge repositories containing databases of archival and current material. They vary in scope - from local to regional to national and international networks. Their results are provided in synchronous and asynchronous fashion but increasingly, electronic means of communication are being used to provide information needed by the various interest groups.

The World Wide Web has had a tremendous effect on the way businesses communicate. Large amounts of information can be made available quickly and conveniently to anyone with Internet access and a web browser. The ability to distribute and view spatial information has quickly shifted from a desktop application (fat client) to a browser-based architecture (thin client). This latter architecture is referred to as thin client since the user only needs a web browser to access services and information on the web (Fitzgerald, 2000). With particular reference to spatial information, there has traditionally been the question of accessing the volumes of information, especially if it resides in several different geographical locations. The web lets a data provider make spatial information available to a wider audience. The data provider can therefore provide a virtually centralised repository of resources without having to change the physical location of the data. This prevents any problems that might arise from maintaining or updating duplicate data sources, such as limited space or corrupt data. The web therefore makes it easy to provide the most up to date spatial data (Fitzgerald, 2000).

It is therefore common practice to go to one of several websites dedicated to CMM and download (or interact with) information about a specific geographical location that one is interested in. Whether it is the "Enviromapper for Watersheds" interactive mapping application that provides an index of watershed indicators for US aquatic resources, or the "Scorecard" mapping application, which maps toxic chemicals released from facilities, institutions like the Environmental Protection Agency in the USA are increasingly providing these CMM indicators through their websites (United States, 2003). The US Department of Commerce's National Oceans and Atmospheric Administration (NOAA) and the Canadian Department of Fisheries and Oceans (DFO) provide a variety of information on their websites on oil and chemical spills, tides, marine weather, and fisheries. This information is superimposed on maps to give a geographical context to the information. The superimposition of information on maps – and the provision of such information on the web, together with limited GIS functionality – is what is referred to as web mapping, webGIS or interactive mapping.

6.1.1.1 Web Mapping Technologies

The emergence of web-GIS technologies is providing the catalyst for easier collaboration, integration and cooperation among organizations with a stake in good governance and sustainable development. This is done by providing an environment for data sharing and integration over the Internet, sometimes without organizations having to make any major changes to the structure and formats of the data they maintain. The full range of analytical capabilities available in most contemporary desktop GIS however is not available on the web browser (or WebGIS client) since they are built on the thin-client concept. To include more functionalities at the client end would seem to defeat the concept of the low cost and convenience of utilizing only a web browser to access spatial data.

Traditional mapping issues still pose challenges to web mapping. Some of these new technologies support different data formats (e.g. ESRI shape files, CARIS, MapInfo files etc.), projections, scales, datums, etc., with conversions and visualization being done "on the fly." Certain web-GIS technologies now facilitate the transmission, integration, visualization and analysis of spatial information stored in geographically dispersed locations. A user with permission to access the geographically dispersed data sets need only have access to a web browser in order to view, query, and analyse the data sets. In some instances however, some webGIS can only deal with data residing on one server.

6.1.1.2 A Web Mapping Example - CARIS Spatial Fusion™

Let us briefly describe a web-mapping technology in order to explain the underlying architecture. CARIS Spatial Fusion™ is a "web-mapping" technology that lets users integrate distributed data sources using a web browser. It is an Internet-based technology whose primary function is accessing, visualizing, and analysing heterogeneous, distributed data sources (Fitzgerald, 2000, Webmapper.com, 2000). Spatial Fusion™ combines the speed, convenience and simplicity of the Internet with the ability to read multiple data sources in their native format. Earlier versions of Spatial Fusion™ consisted of a customized Java client and a number of Fusion Data Services. On the server side, Spatial Fusion™ was made up of the following components (CARIS, 1999, Fitzgerald, 2000):

- A Web Server: One must already be running on the network.
- The Orbix™ Runtime needs to be installed on every machine that hosts a Fusion Data Service. The Orbix™ Runtime lets the Spatial Fusion™ applet and the Data Services communicate across the Internet.
- Catalog Service: This service lists all of the available Fusion Data Services.
- Fusion Data Services: These services must be registered with the OrbixWeb™ Implementation Repository. Each service has an accompanying configuration file that contains the name used to register the service with the daemon and the location of the data source.
- Configuration Utilities: CARIS MapSmith™ and CARIS dbMaps™ are provided to help customize the display of CARIS, Oracle 8i Spatial, or Shapefile data.

Recent changes in Spatial Fusion™ Version 3.0 have removed the Orbix runtime component and replaced it with TAO ORB - an open source implementation of an Object Resource Broker using Common Object Request Broker Architecture (CORBA). In Spatial Fusion™, the Object Request Broker (ORB) is the broker for information from the java servlets to various data services. The servlet container is Tomcat developed by the Apache Software foundation to manage and invoke servlets when requested. The normal setup is to have a web server such as Microsoft Internet Information Services (IIS) conFigured to connect to Tomcat. When an action is performed, the Spatial Fusion™ applet sends a request to a servlet, which is handled by the Jakarta Tomcat Servlet container. The request is then sent by the servlet to the data services through the ORB (CARIS, 2003).

As far as the user is concerned, they simply download the Spatial Fusion™ applet from a web server. At that point, a user can easily open data from any fusion service they have access to, providing them with a secure and fully scalable environment (CARIS, 1999, Webmapper.com, 2000). In addition, CARIS Spatial Fusion Developer™ lets users customize the client, so that specialized applets can be rapidly built in a drag and drop environment, giving the ability to tailor applets for specific users (Fitzgerald, 2000).

Thin client web mapping applications such as Spatial Fusion™ have traditionally depended on "wired" infrastructure because large amounts of spatial information have to be sent from the server to the client. The need to have information provided on mobile devices such as PDAs and cell phones is however driving information dissemination towards a wireless infrastructure. But would there be a need for wireless access to CMM networks? Is the general public ready to embrace this technology? In the following section we address these questions.

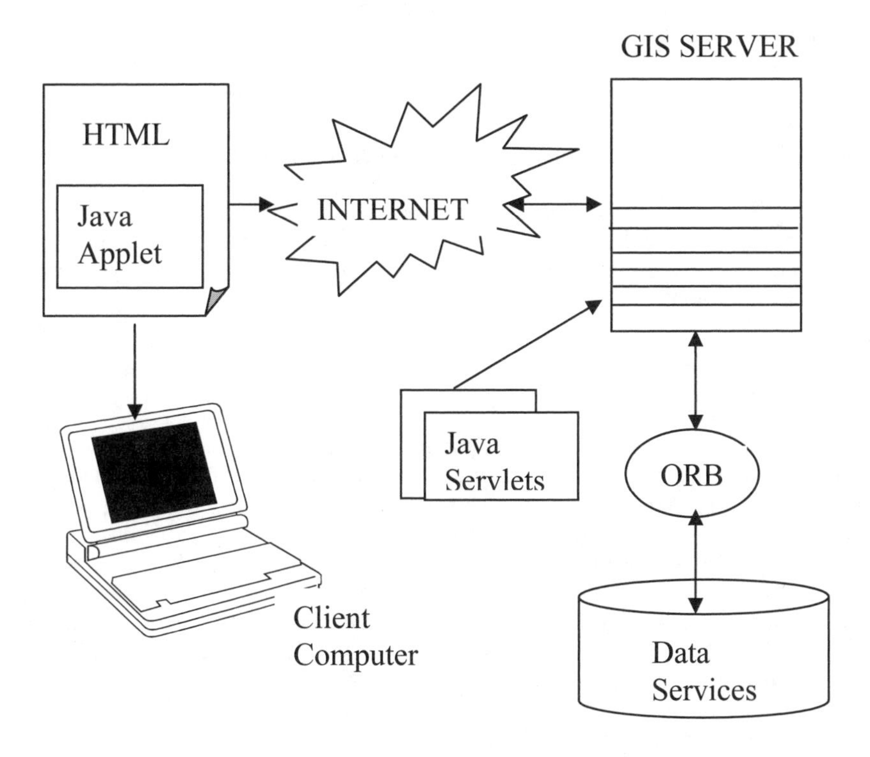

Figure 6.1 CARIS Spatial Fusion Architecture (from the CARIS Spatial Fusion Administrators Guide)

6.2 RE-ENGINEERING MARINE MONITORING NETWORKS

There exists in every jurisdiction a public policy issue associated with information collection (McLaughlin, 1991). This issue is related to the notion of providing ready and efficient access to information about resources, and rights associated with them, to all members of the community. The public is defining itself as an individual property owner – the property being the marine commons and its resources – and is realigning itself to be a primary decision maker. Most people are therefore interested in how resources are being utilized and whether resource use is being policed effectively.

Coastal and marine monitoring networks fulfill the role of providing information effectively. But they are also builders of communities of interest i.e. a forum where participants are able to interact on a specific topic. These online communities are valuable as they are increasingly providing forums for like minded individuals to participate in the decision-making process. With more citizens being involved in advocacy efforts and words like "stewardship of resources" increasingly being used to rally the general public into participation in the decision-making process, provision of information about the space that citizens inhabit is becoming critical.

Sociologists have traditionally viewed this kind of participation as social capital – a relationship that can blur social inequality and enhance transactions that lead to capital accumulation and social growth. We are seeing the public being involved in a culture of participation, through these networks, although this is concentrated on citizen access to information. What is needed is the enhancement of citizen participation, to citizen-based leadership: where the citizens pro-actively seek information (and are able to receive it as it happens) rather than react to information that they come across. This is where wireless access to information becomes important.

6.2.1 Wireless Access to Information

Developments in Internet-enabled spatial data integration and analysis tools are now allowing decision-makers the opportunity to have access in real-time (or near real-time) to spatial information. The public is becoming more technologically "savvy" and is using newly available devices to access various kinds of information. What one finds online is also now available on mobile devices – and with evolving wireless technology we are seeing web-enabled mobile devices being used to integrate and retrieve geographic data as it is collected. Additionally, it makes sense for companies to focus on wireless Internet services given projections for future growth: the Yankee Group predicts that there will be more than one billion mobile devices worldwide by 2003. Some estimate that half of those will be web-enabled (Obeid, 2000).

This is not a new trend. Increasingly, we now have various types of information being sent to us via our mobile devices – allowing us to remain informed about various topics that we are interested in. Take the example of access to information about entertainment or restaurant locations, based on one's geographical location, delivered to a mobile phone. This represents a relatively new trend referred to as Location Based Services (LBS).

6.2.2 Location Based Services – The Driving Force

When this paper was written in 2001, the author had been following the evolution of killer applications in mobile mapping technologies – location based services (LBS). In addition to maps and directions, LBS allows customers to locate branches, service centres, stores or any points of interest near their home. A business can use LBS to locate its customers, manage its supply chain, and to implement its business rules. In fact, recently Microsoft has joined with AT&T Wireless to provide location-based services (Francica, 2002). LBS are now the driving force in wireless mapping applications.

In the coming years, Location Based Services (LBS) are expected to play a defining role in driving not only the mobile Internet but also mobile commerce as well. Location based services find the geographical location of a mobile device and provide services based on that location. LBS applications automatically identify subscribers' locations from the wireless networks and combine this with other information to obtain maps and driving directions, perform proximity searches and find places and points of interest. Although there has been a slow rollout of LBS, Konish (2002) indicates that the slow economy is forcing wireless carriers to look for revenue-bearing services. Inevitably these are being found in making wireless devices location aware and providing location based services. Many LBS applications are already available especially in the field of asset tracking and fleet management (Sweeney, 2002). Other applications include automatic vehicle location and tracking, sale and marketing automation, consumer travel services, wireless call centre tracking, and location-based billing.

6.2.3 Providing Web Mapping to Mobile Devices

LBS can be provided in several different ways – from simple text messages providing a list of options e.g. restaurant names and locations – to a full webcentric interaction with the LBS service – involving GPS locations superimposed over interactive maps. This latter option is accomplished by using a dedicated protocol, the Wireless Application Protocol (WAP), which enables a mobile device to behave like a micro browser (Buckingham, 2000; Mobile Applications Initiative, 2001; Mobile Lifestreams Ltd, 2000). The Wireless Application Protocol (WAP) is an open, global specification that empowers users of wireless devices to easily access and interact with information and services themselves (Buckingham, 2000, Mobile Applications Initiative, 2001). This information is available through a data portal that allows the user to access services in a manner similar to that found on the Internet. The protocol however has certain limitations to it – for example, it only allows the use of sequential menus for users to browse for information. However, WAP is continually evolving to encompass greater functionality (Buckingham, 2000).

The Wireless Application Protocol incorporates a relatively simple micro-browser into the mobile phone. As such, WAP's requirement for only limited resources on the mobile phone makes it suitable for thin clients and early smart phones. WAP is designed to add value-added services by putting the intelligence in the WAP servers whilst adding just a micro-browser to the mobile phones themselves. Microbrowser-based services and applications reside temporarily on servers, not permanently in phones. The Wireless Application Protocol is designed for use with anyone's standard such as Code Division Multiple Access (CDMA), Global System for Mobiles (GSM), or Universal Mobile Telephone System (UMTS) and is aimed at turning a mass-market mobile phone into a "network-based smart phone" (Buckingham, 2000; Mobile Applications Initiative, 2001).

Wireless terminals incorporate a Handheld Device Markup Language (HDML) microbrowser that uses the Handheld Device Transport Protocol (HDTP). HDML is a version of the standard HyperText Markup Language (HTML) Internet protocol that is specifically designed for effective and cost-effective information transfer across mobile networks themselves. This protocol is used to link a mobile device terminal to a server that is connected to the Internet (or Intranet) where the information resides. The content at the site is tagged with

HDML tags that can be easily read and converted by the microbrowser in the handheld device themselves (Buckingham, 2000; Mobile Applications Initiative, 2001).

A user with a WAP-compliant phone uses the in-built micro-browser to make a request for information or service. This request is passed to a WAP Gateway that then retrieves the information from an Internet server either in standard HTML format or preferably directly prepared for wireless terminals using Wireless Markup Language (WML) themselves. In this paper, WML and HDML are used interchangeably. If the content being retrieved is in HTML format, a filter in the Wireless Application Protocol Server may try to translate it into WML. The requested information is then sent from the WAP Gateway to the WAP client, using whatever mobile network bearer service is available (Buckingham, 2000; Mobile Applications Initiative, 2001; Mobile Lifestreams Ltd, 2000).

6.2.4 Extending the Trend to CMM Networks

N'gang'a *et al.* (2001) provided a possible scenario that utilized the technological components mentioned in the preceding sections of this article. A summary of the scenario that had previously been proposed is as follows:

1. A marine event involving coastal or marine environments occurs. This event could be a oil-carrying tanker being involved in an accident, or as diverse as enforcement against banned activities or the enhancement of safety of navigation;
2. The event is detected by remote sensing means or by visual observation. These event sources are important as they establish how the event will enter into the marine event information circle. The marine event information circle is a term that has been coined by the author to represent the flow of information from the event source to the interested group;
3. If the event is detected by visual observation, the observer has the option of using a VHF radio to communicate with someone who has access to a web-mapping server (e.g. a Spatial Fusion™ data server). Alternatively, the user can use other means of event dissemination such as the mobile phone and Internet connectivity. The term "event content" refers to the information that is recorded regarding the event. In this work, this information is the location and the nature of the event;
4. The event is then registered with the event server, which in this case is analogous to a Spatial Fusion™ Data server. In particular, the data needs to be conFigured so that it can be accessed using the fusion services. An event coordinator (or moderator) must exist in the marine environment information circle to serve this purpose;
5. Transmission of information updated with the marine event is then accomplished using electronic means. An email list of a registered interest group can be used to contact the users of the marine environment information circle, or the information can be relayed to a website that the general user can access. Alternatively if there is a WAP server, the information can be relayed (with additional functionality) to members of

the registered interest group who are within geographic proximity of the event. These options allow the event server to provide alternative value-added services.

6. Information dissemination is accomplished using the wireless application protocol and existing GSM networks. In the near future GSM combined with GPRS or the third generation UMTS networks can be used for broadband access to marine events.

Once the information is served up to a user group, decisions can be made regarding implications of the marine event on existing policies, activities, or populations within (or adjacent to) the marine environment. Informed decisions are therefore made with available information about the nature and location of marine events.

6.2.5 What about Bandwidth Issues?

Spatial data involves the transmission of geometric as well as attribute information. It can therefore be argued that a substantial amount of data must be transmitted which implies that a large bandwidth must be available. There is a general consensus in the wireless industry (see Salvo, 2000; Mobile Lifestreams, 2000; Villalobos, 2000) that in the near future bandwidth will no longer be a problem especially when the third generation of mobile phone network becomes available.

We now are connected using the Global System for Mobiles (GSM) wireless network that provides data rates of 9.6 Kbytes/s. An ancillary service on the GSM network is the General Packet Radio Service (GPRS), which provides compression technology that improves data transmission rates to 100 Kbytes/s (Mobile Lifestreams, 2000; Villalobos, 2000). In the near future, the Universal Mobile Telephone System (UMTS) will provide 2 Mbytes/s (Villalobos, 2000). This will definitely change the way we use both wireless and Internet technologies.

Third-generation wireless communication (3G) such as UMTS will definitely play an important role in web based spatial visualisation in the near future. It is forecast that the mobile phone will become a portable terminal where the Internet, video and audio will be supported in a multimedia environment (Pinto, 2001). Therefore, it is expected that the WAP will adapt to the potential offered by this technology. There have been arguments that the WAP protocol is not a protocol for the future[1] due to its graphical limitations and access speed (see Salvo, 2000; Mobile Lifestreams, 2000; Villalobos, 2000); however, the protocol is still evolving and advances in technology and application are expected to eventually address these concerns.

The solution simply requires data connectivity from devices. The amount of data passed over the wireless network for the LBS applications is quite small and doesn't require the higher bandwidth allowed by 2.5G and 3G networks.

---

[1] Note that a competitor to WAP is the imode format, which uses compact HTML (cHTML) as opposed to WAP and is in use in Japan (Archey, 2002; Eurotechnology.com, 2003)

6.2.6 Some Examples

While there are several web mapping software that exist, only recently has the provision of web mapping services begun to move to mainstream mobile devices. Some of these products include (Geoplace.com, 2003;Webmapper.com, 2003)

- AutoDesk's MapGuide and Onsite – Onsite enables viewing of digital design data on Microsoft Windows CE-based handheld computing device.

- ESRI's ArcIMS and ArcPad – ArcPad provides database access, mapping, GIS, and global positioning system (GPS) integration to users out in the field via handheld and mobile devices.

- MapInfo's MapExtend and MapExtreme - *MapInfo® MapXtend®* is a developer tool for creating location-based applications running on wireless handheld devices.

- Microsoft Map Point Webservice – The MapPoint Web Service is the Microsoft® platform for location-based services (LBS) and can be used by mobile applications.

In Canada there are several researchers already working successfully on the kind of approach outlined in this paper. Canadian researchers Nickerson et al. (2003) have already provided a general-purpose architecture for real-time web access to mobile geospatial operations. The research demonstrated that their three-tier architecture, using Wireless Application protocol (WAP), along with eXtensible Markup Language (XML) tags embedded in a Uniform Resource Locator request, provided a sound basis for decision support of real time geospatial operations. Real time field activities were viewed by web users globally as well as stored in (and retrieved from) an online central server and database.

In Ng'ang'a et al., (2000) the authors had drawn attention to some advances in web mapping location-based services such as the automatic vehicle locator (AVL) application used at the city of Fredericton Police department[2]. This was part of the Wireless Public Safety Decision Support System (WPSDSS) for the city of Fredericton (Lunn, 2001). The application uses a cellular digital packet data wireless data network to transmit information about the location and status of police vehicles. The information is overlaid onto the City's GIS mapping layers and orthophoto imagery. While the application does not display information on mobile devices (like cell phones or PDAs), it is just a matter of time before it does.

On the CMM front in Canada, Fisheries and Oceans Canada, Environment Canada, the Defence Department, Transport Canada and the Canadian Space Agency are already collaborating on a program called Integrated Satellite Targeting of Polluters (I-STOP), to use a satellite operated by Radarsat International to identify ocean slicks (Calgary Herald, 2002; CCMC, 2002).

---

[2] Located in the province of New Brunswick, Canada

Canada's Radarsat satellite will be used to provide all weather, day/night images of the Earth's surface and give officials an indication of suspicious activity in a vast area that couldn't easily be covered by any other means, such as aircraft. The technology is only being used on the East Coast for now. If it proves successful, it could also be used on the Pacific coast. The Calgary Herald (2002) reported that the first hit came nine days into the project when the satellite spotted a spill off Newfoundland's south coast, near one of the province's most valued marine bird reserves. A plane was sent to the area off Cape St. Mary's and the ship was ordered back to port, where its operators now face charges under the Migratory Birds Act for allegedly creating a 116-kilometre-long, 200-metre-wide slick. While WAP technology is not being used in this program, one can see that soon enough there will be a situation where individuals interested in coastal and marine monitoring will have information delivered about this fragile ecosystems health to their mobile devices.

6.3 SUMMARY

This paper has begun by discussing the importance of information to coastal and marine governance and then expanded the discussion by outlining existing technologies that can be used to facilitate enhanced decision-making. In particular, the paper has followed developments in wireless application protocol (WAP), and associated technologies, and indicated how this trend can be adopted to disseminate marine information. This paper has then outlined an original scenario that the author had hypothesized, given the technology trend that was being observed in 2001 - and has concluded by giving real world examples of similar scenarios currently in effect.

This paper is not intended to provide an in-depth technological review of the wireless application protocol (WAP) or of advances in mobile applications. It is intended to provide an overview of trends in wireless technology and how they can be used to address the very real issue of disseminating information using commonly available devices to the interested public.

6.4 REFERENCES

Archey, C., 2002, Wireless Browser Standard. *Computer Technology Review*, **22**(10), pp.17.

Buckingham S., 2000, An Introduction to YES 2 WAP, http://www.gsmworld.com/technology/yes2wap.html#1,May.

Calgary Herald, 2002, *Canadian Government using Radarsat to Detect and Track Oil Spills*. October 4th.

Canada's National Programme of Action (CNPA), 2000, http://www.cc.gc.ca/marine/npa-pan/sum_e.htm. Accessed September 2000.

Canadian Centre for Marine Communications (CCMC), 2002, *Marine Information and Communications Technology News*, **13**(29), http://www.ccmc.nf.ca/ diffusion/issues/2002/issue_13,29.htm

CARIS - Universal Systems Ltd, 2003, CARIS Spatial Fusion™ 3.0 Users Guide, CARIS Spatial Components, (Fredericton: CARIS Universal Systems Ltd.).

Crowe, V., 2000, Governance of the oceans, *Short report on the Wilton Park Conference 586*, December 3–5, 1999.

Eckert, R. D., 1979, *The Enclosure of Ocean Resources: Economics and the Law of the Sea*, (Stanford: Hoover Institution Press).

Eurotechnology.com, 2003, *The unofficially independent imode FAQ*, http://www.eurotechnology.com/imode/faq-gen.html

Fitzgerald J., 2000, *CARIS Spatial Fusion: an Internet GIS. White Paper #7*, http://www.spatialcomponents.com/techpapers/fusion.pdf, (Fredericton: CARIS Universal Systems Ltd.).

Francica, J., 2002, SPECIAL REPORT: Microsoft & AT&T Wireless join forces in providing Location-based Services, *Directions Magazine*, July 31, 2002,

Friedheim, R.L., 1999, Ocean governance at the millennium: Where we have been - where should we go. *Ocean & Coastal Management*, **42**, pp. 747–765, http://www.directionsmag.com/article.php?article_id=238.

Geoplace.com, 2003, *Web Mapping Guide – Vendors*, http://www.geoplace.com/gr/webmapping/vendors.asp, Accessed July 2003.

Gomes, G., 1998, *The review of the NOAA National Coastal Zone Management Programs for Estuary and Coastal Wetland Protection*, http://www.oceansconservation.com/iczm/gomes.htm, Accessed September 2000.

Juillet, L. and J. Roy, 1999, Investing in people: the public service in an information age (reflections on the 1999 APEX conference), *Optimum*, **29** (2/3).

Konish, N., 2002, Hope rests on Location-based Services. *Wireless Systems Design*, **7**(10), pp.9.

Lunn, R., 2001, *The Use of Automatic Vehicle Locator (AVL) Technology at the City of Fredericton's Police Department*, Unpublished Report, Department of Geodesy and Geomatics Engineering, University of New Brunswick, Fredericton, New Brunswick, Canada.

Lutz, E. and M. Munasinghe, 1994, Environmental accounting and valuation in the marine sector. In *Ocean Governance: Sustainable development of the Seas*, edited by Payoyo, P.B., (Tokyo, New York, Paris: United Nations University Press).

Manning, E. *et al.*, 1998, Renovating governance: lessons from sustainable development. *Optimum*, **28**(3), pp. 27–35.

Miles, E.L., 1998, *The concept of ocean governance: Evolution towards the 21st century and the principle of sustainable ocean use*. In Proceedings of the SEAPOL workshop on ocean governance and system compliance in the Asia-Pacific context. Rayong, Thailand, 1997.

Mobile Applications Initiative, 2001, *GPRS & 3G Information*. http://www.gprsworld.com, May.

Mobile Lifestreams Ltd, 2000, *Data on 3G - An Introduction to the Third Generation*. http://www.mobileipworld.com/wp/positioning.htm, November.

Ng'ang'a S.M, A. Campos, M. Sutherland and S.E. Nichols, 2001. *Supporting Coastal and Marine Monitoring by Remotely Accessing Data Using Spatial Fusion and WAP*. In Proceedings of the 4th International Symposium on

Computer Mapping and GIS for Coastal Zone Management: CoastGIS Conference, Halifax, Nova Scotia, June 2001.

Nichols, S. and D. Monahan, 1999, Fuzzy Boundaries in a Sea of Uncertainties, The Coastal Cadastre - Onland, Offshore. In *Proceedings of the New Zealand Institute of Surveyors Annual Meeting*, Bay of Islands, NZ, Oct 9-15, pp. 33–43.

Nichols, S., Monahan, D., and Sutherland, M.D., 2000, Good governance of Canada's offshore and coastal zone: Towards an understanding of the maritime boundary issues. *Geomatica*, **54** (4), pp. 415–424.

Nickerson B.G., Shan, Y., and McLellan, J.F., 2003, Web Access to Real Time Wireless Mobile Geospatial Information. *Geomatica,* **57** (1), pp. 13–29.

Obeid, D., 2000, *How WAP Works*, Oracle Publishing Online Article. http://www.oracle.com/oramag/webcolumns/2000/index.html?howwapworks.html

Payoyo, P.B., 1994, Editor's introduction. In *Ocean Governance: Sustainable development of the Seas*, edited by Payoyo, P.B., (Tokyo, New York, Paris: United Nations University Press).

Prescott, J.R.V., 1985, *The Maritime Political Boundaries of the World,* (London, New York: Methuen).

Pinto, N.O., 2001, *UMTS*. Exame Digital nº4, Lisboa, Portugal.

Rosell, S.A., 1999, *Renewing governance: Governing by learning in the information age*, (Oxford: Oxford University Press).

Salvo, R., 2000, *Dossier Telecomonicações*, Fortunas & Negocios nº104, Lisboa, Portugal.

Sanger, C., 1987, Ordering *the Oceans: The making of the law of the sea,* (Toronto: University of Toronto Press).

Sweeney, D, 2002, LBS seeking its day in the sun. *Telecom Asia*, **13**(6), pp.44.

United States Environmental Protection Agency, 2003, *Interactive Web Mapping*. http://www.epa.gov/ceisweb1/ceishome/atlas/learngeog/interactivewebmapping.html

Villalobos, L., 2000, *Como aquecer o primeiro lugar*, Fortunas & Negocios nº104, Lisboa, Portugal.

WAPArch, 1998, WAP Forum 1998, *Wireless Application Protocol Specifications*, http://www.wapforum.org/what/technical.htm

WAP Forum, 2000, Wireless Application Protocol White Paper, *Wireless Internet Today 2000*, June, see http://www.wapforum.org/what/whitepapers.htm

Webmapper.co, 2000, CARIS Spatial Fusion product review. http://www.web-mapper.com/product/CARIS/spatial_fusion/review.html

WebMapper.com, 2003, Web Mapping Products. http://www.web-mapper.com/company/showproducts.cfm, Accessed July 2003.

CHAPTER SEVEN

Exploring the Optimum Spatial Resolution for Satellite Imagery: A Coastal Area Case Study

Chul-sue Hwang and Cha Yong Ku

7.1 INTRODUCTION

Researchers in geography are much interested in issues of scale and resolution, and the strong influence these may have on the results of analyses of spatial patterns (Harvey, 1968; Stone, 1972; Goodchild and Quattrochi, 1997). Issues of scale usually arise from not knowing the optimal unit size for a study area; while MAUP (the Modifiable Area Unit Problem), one of the fundamental research problems in geographical information science, can also be considered to be related to the understanding of the scale for spatial data (Openshaw, 1984; Fotheringham and Wong, 1991).

GIS and remote sensing have shown active interests in both the operational scale and measurement scale of data. Technological developments in GIS and remote sensing have led to the gradual detailing and elaboration of the measurement scale. In the field of remote sensing, for example, the 78m resolution Multi-Spectral Scanner (MSS) data of the 1970s has given way to 30m Thematic Mapper (TM) data in the 1980s, SPOT in the 1990s, and 1m IKONOS imagery in the 21st century. Such improvements in the measurement scale will allow for the new understanding of geographical phenomena that could not be measured at the previous scales.

Changes of the ground resolution of satellite imagery have modified the inherent characteristics of imagery due to the "scale effect" (Goodchild and Quattrochi, 1997). Therefore, researchers have to select the image with the optimum resolution among the satellite imagery at diverse resolutions, at which a specific geographical phenomenon can be understood more effectively. In order to explore the optimum resolution, the objective procedures or indices must be designed for the image characteristics at different resolutions. They should be able to represent the changes in image characteristics that occur at different resolutions.

The aim of the present study is to propose a procedure for identifying the optimum resolution for the study of coastal wetlands. These wetlands, as the point where the ecosystem of the land meets the ecosystem of the ocean, have unique

0-41531-972-2/04/$0.00+$1.50

and diverse topographical features, giving rise to the formation of vegetation colonies, and exhibits diverse spatial distribution due to diverse environmental factors. Due to the dense vegetation and unstable topography of the coastal wetland, it is more effective to use satellite imagery to understand its characteristics, rather than through field investigation (Lyon and McCarthy, 1995). However, it is not necessarily effective to use the satellite imagery of higher ground resolution in the coastal area. For example, in the case of the salt marsh where colonies appear as clusters, the use of low-resolution imagery of the approximate size of the clusters is more effective (McNairn *et al.*, 1993).

In this vein, this study will attempt to find procedures and techniques for the optimal spatial resolution of satellite imagery in the study of coastal wetlands. To achieve this goal, it firstly selected the indices considering the following conditions: (1) the sensitivity of the indices, which change in response to the different resolutions, should be high enough to respond accurately to the changes in image characteristics; (2) the indices should be the ones that can provide the approximate estimate of the final result only with the minimal procedure; (3) the indices should reflect the classification accuracy of the attribute information. We then coarsened the resolution of the satellite data, to produce images at various resolutions, and observed the changes in the textural and attribute information characteristics of these images at various resolutions. As a result, the study was able to decide the operational scale, an optimum resolution in the use of satellite imagery.

7.2 SATELLITE IMAGERY CHARACTERISTICS BY RESOLUTION

7.2.1 Textural Characteristics of Satellite Imagery

The local variance method is frequently used in the search for the optimum resolution. Proposed by Woodcock and Strahler (1987), the local variance method measures the variance by placing a 3 x 3 moving window over the image data, to capture the textural characteristics of the image. However, local variance is limited by the fact that it relies on the overall variance of the image. In other words, the local variance measurements of one image cannot be compared to the results for another image. In addition, using this index makes it impossible to understand the spatial distribution of the variance within the image.

Cao (1992), in an attempt to calculate the fractal dimension of satellite imagery, compared and analysed the Isarithm method (Goodchild, 1980), the Variogram method (Mark and Aronson, 1984) and the Triangular prism method (Clarke, 1986). From the analysis, he found the Isarithm method most useful. In general, the fractal dimension D of satellite imagery appears as a real number between the 2D and 3D. As the complexity of the image's spatial structure increases, the value of the fractal dimension also gets higher. Within this principle, obtaining the fractal dimension gives a numerical representation of the satellite imagery characteristics by resolution.

In order to understand how the distribution characteristics of images change during the coarsening of resolution, the images of variances as well as the images of averages are needed. It is similar logic that we use the variance to make up the weak points in using the arithmetic mean (or average) to represent a data set. The variance of an image set provides a measure of the average squared deviation of a set of values around the mean. While the images of averages represent the image as one pixel, taking the arithmetic mean of the original pixels, the images of variances represent the dispersion characteristics or the variability of the original image during the coarsening of the resolution (Figure 7.1). During the resolution coarsening, if the heterogeneity between the pixels is large, the variance of the generated image will be large. Conversely, if the pixels are of similar value, the variance of the generated image will be small. Low variances are recorded for smooth and wide regions during the resolution coarsening, and high variances are recorded for narrow and complex regions.

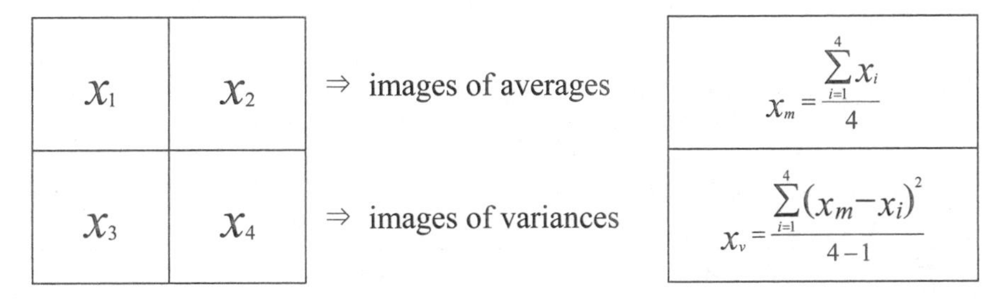

Figure 7.1. Calculating images of averages and images of variances.

Understanding the spatial distribution pattern of the images of variances derived from the process of the resolution transformation leads to the analysis of changes in image characteristics by resolutions. Therefore, the analysis of the spatial distribution of the images of variances can be used to understand the image transformation that follows the resolution coarsening. The spatial autocorrelation of the images of variances is selected to measure the spatial distribution in which the image characteristics transform. The spatial autocorrelation using Moran's I is generally appropriate to measure the distribution pattern for data with a continuous variable such as satellite imagery.

7.2.2 Measuring the Spectral Separability of the Training Set

Divergence is a common measurement to estimate the statistical separability of satellite imagery. Divergence is measured using the average of the statistics by class and the covariance matrix gathered through the supervised classification of the training set. However, since divergence has a tendency to increase abruptly as the distance between the classes increases, it is difficult to generalise divergence (Jensen, 1996).

The Jeffreys-Matusita distance (Matusita, 1966) is a value calculated by the exponential transformation of the Bhattacharyya distance that has been modified from divergence (Kailath, 1967; Wacker, 1971). This J-M distance is more widely used than divergence for calculating the spectral separability of multi-variate classified classes because the J-M distance does not exhibit the abrupt increase in the spectral separability. However, complex calculations and a long processing time are some of the disadvantages of the J-M distance.

Mahalanobis distance (Mahalanobis, 1936) is a value that takes into account both the correlation and distribution features between the data in terms of Euclidean distances. Mahalanobis distance multiplies the Euclidean distance by the covariance matrix between classes. In contrast to the J-M distance, Mahalanobis distance requires simpler calculations and less processing time. In addition, it does not exhibit the abrupt increases in the spectral separability. For these reasons, Mahalanobis distance is the most widely used method for measuring the spectral separability.

7.3 PROCEDURE FOR EXPLORING THE OPTIMUM SPATIAL RESOLUTION

The optimum spatial resolution using satellite imagery is the one that provides good representation of the attribute information, and is referred to as the 'operational scale.' In order to select the image with the optimum resolution from satellite images at different resolutions, the objective procedures or indices must be designed for the image characteristics at different resolutions.

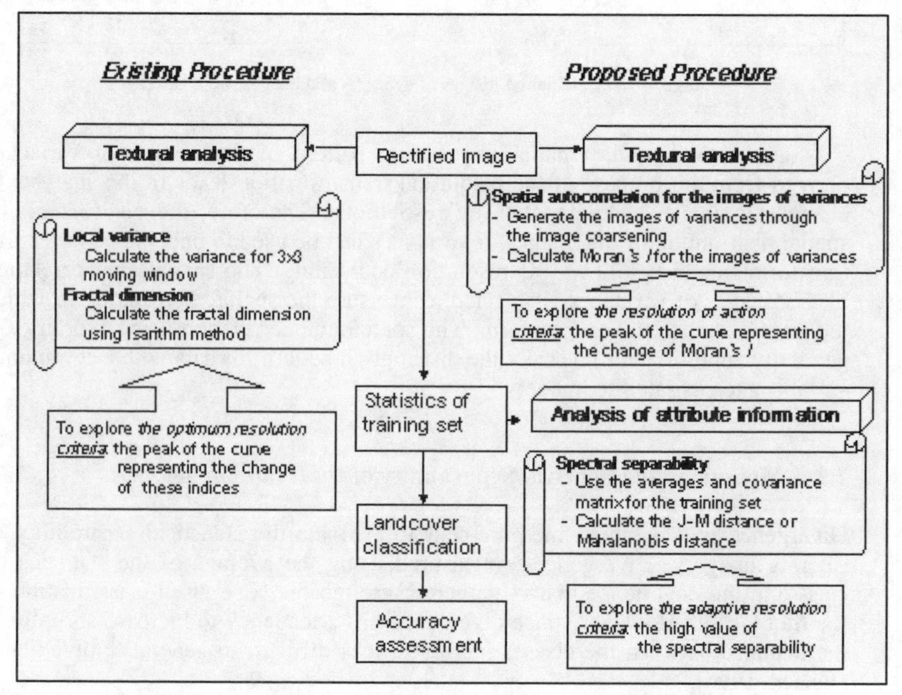

Figure 7.2 The proposed procedure to explore the optimum spatial resolution of satellite imagery.

However, it is unreasonable to rely on a single index to select the optimum spatial resolution. Indices have their own specific objectives to understand the characteristics of the imagery. Therefore, this study will attempt to combine the various indices and present a procedure for exploring the optimum spatial resolution (Figure 7.2).

Until now, existing studies on selecting the optimum resolution have focused only on the textural changes of the images at different resolutions. Local variance and fractal methods are particularly mentioned in such studies. However, because they only reflect the result without illustrating processing procedure of attribute information obtained from the images, they should not be used to understand the changes in attribute information.

This study will attempt to approach the exploration of the optimum resolution using satellite imagery using a two-step procedure. First, through the measurement of the image's textural characteristics, the resolution of action will be selected, and then, through the estimation of the spectral separability, the adaptive resolution for the attribute information will be chosen.

7.3.1 Textural Characteristics: Searching for the Resolution of Action

If the resolution of the original image is gradually scaled down to produce images at low resolutions, the image achieves a structure much like that of a pyramid. This method of the scale transformation is called the pyramid model (Figure 7.3). When producing an image at a low resolution, producing both images of averages and images of variances at the same time is more effective for the analysis of the satellite imagery. The images of variances are the ones that reflect the differences between the original images and the coarsened images during the resolution coarsening.

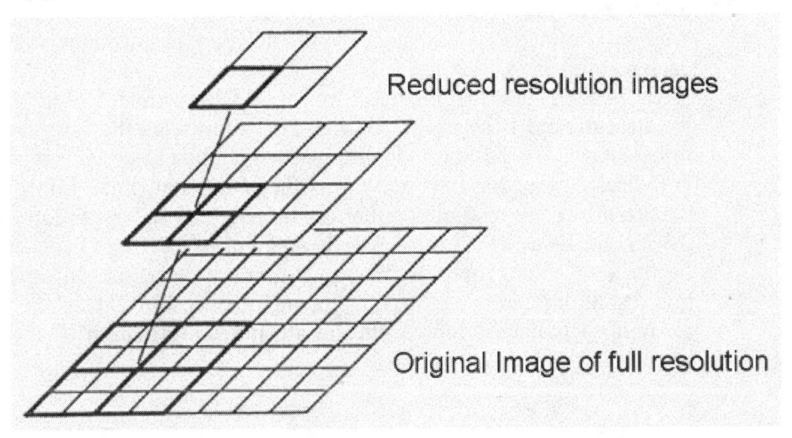

Figure 7.3 Construction of an image pyramid (source: Richards, 1993, p. 33).

- *Step 1*: Moran's *I* is calculated based on the images of variances. It can be said that the higher the Moran's *I* value, the higher the spatial autocorrelation will be and the spatial distribution pattern will then concentrate. The concentrated spatial distribution pattern for the images of variances means that there is a large transformation in some regions following the resolution coarsening.

- *Step 2*: Graph the Moran's *I* values measured at each resolution.
- *Step 3*: Find out the changes in the Moran's *I* value to select the resolution of action. The resolution at the apex of the curve on the graph is the one at which the image information changes. In other words, the image information changes with respect to this resolution (i.e. the resolution at which the satellite image characteristics are changing) and lowers the classification accuracy.

7.3.2 Attribute Information: Searching for the Adaptive Resolution

In searching for the optimum spatial resolution to acquire the proper attribute information, it is effective to apply the measurement of the spectral separability of the training set (Markham and Townshend, 1981). The J-M distance and Mahalanobis distance are the indices that give precise representations of the spectral separability. The resolution where the J-M and Mahalanobis distances are high and the accuracy of the image classification is also high is the one that can best acquire the attribute information of the image. It can be called the *adaptive resolution*.

- *Step 1*: Select the training set in the satellite image.
- *Step 2*: Obtain the statistical calculations (average, standard deviation, minimum and maximum value, covariance between bands, etc.) for the image training set at each resolution and save them as a signature file.
- *Step 3*: Measure the spectral separability, J-M distance and Mahalanobis distance from the signature file.
- *Step 4*: Graph the J-M distance and Mahalanobis distance measured at each resolution.
- *Step 5*: Find out the changes in the J-M distance and Mahalanobis distance to select the resolution that best represents the image's attribute information. The point on the graph at which the indices abruptly increase is the resolution that best represents the image's attribute information. In particular, of the resolutions that are lower than the resolution of action, the resolution at which the spectral separability increases is determined to be the one where the classification accuracy increases. In other words, this resolution can be selected as the adaptive resolution (or the optimum resolution) that best represents the attribute information of the satellite image.

7.4 APPLICATION AND EVALUATION OF THE PROPOSED PROCEDURE

7.4.1 A Case Study Area and Data

The two-stage procedure to explore the optimum spatial resolution proposed in this study was tested by its application to Landsat TM imagery, that covers Suncheon Bay, a typical coastal area in a southern part of Korea (Figure 7.4). Suncheon Bay area has various land cover types such as coastal wetland (i.e., salt marshes, tidal flats, waterways, etc.), urban areas, agricultural areas, and forest.

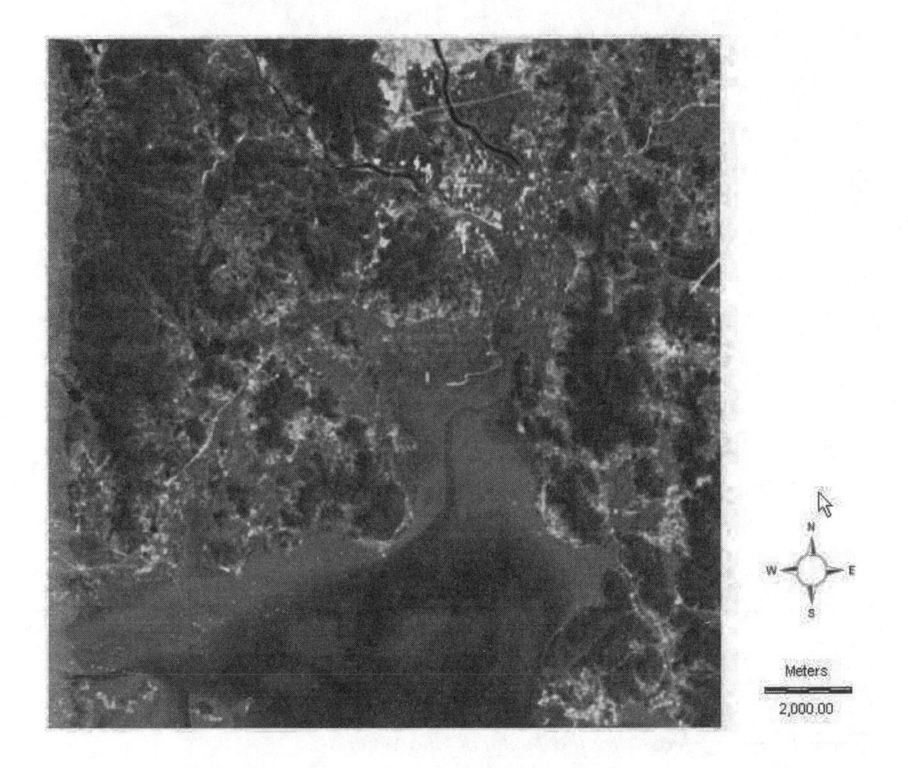

Figure 7.4 Landsat TM image of Suncheon Bay (original in colour).

7.4.2 Applying the Steps Proposed

7.4.2.1 Image Coarsening

To coarsen the resolution, the averages and variances of the adjacent pixels were calculated and then these values were used to produce the images of averages and variances. The images of variances were used to measure textural characteristics through the spatial autocorrelation analysis, and the images of averages were used to measure spectral separability. In this study, the 30m image was coarsened in 30m intervals up to a resolution of 480m, for the production of a total of 16 images (Figure 7.5).

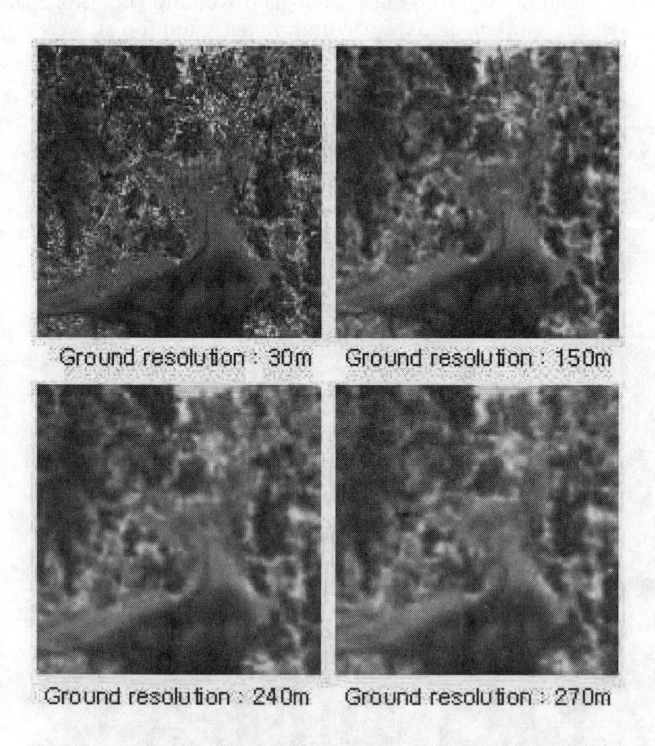

Figure 7.5 Images coarsened into various resolutions.

7.4.2.2 Step 1: Measuring the Textural Characteristics

By measuring the spatial autocorrelation of the images of variances, the following characteristics were discovered (Figure 7.6). In most bands, the Moran's *I* value gradually decreases to a resolution of 120m and then starts to increase at a resolution of 150m. However, this value then begins to decrease again at a resolution of 180m. In other words, the spatial autocorrelation rises abruptly at a resolution of 150m and then falls at later resolutions. Similar phenomena can be seen at resolutions of 300m and 390m. The increase in the Moran's *I* value means that the spatial distribution of the variance is clustered at a certain region. It indicates that there is a large fluctuation in the specific region during the resolution transformation process. As follows, the point at which the Moran's *I* value increases and then decreases can be understood to be the resolution at which the variance drift is changing. In other words, the resolutions of 150m, 300m and 390m are the points at which a change is occurring in the spatial distribution pattern of the images of variances.

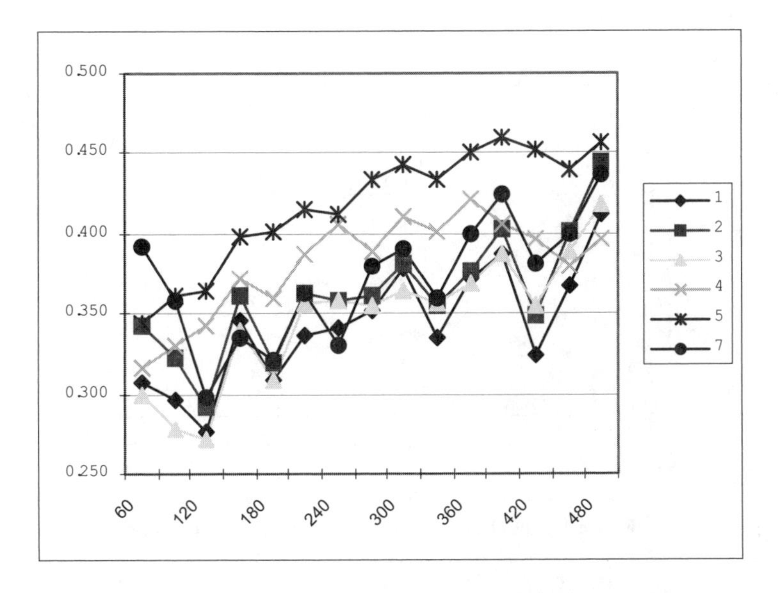

Figure 7.6 The changes of spatial autocorrelation by 7 bands.

In the case of the image of Suncheon Bay area, it can be said that a significant spatial change has occurred in the image as the resolutions changed from 120m to 150m and from 270m to 300m. Therefore, the resolutions of 150m and 300m are the points at which the image's textural characteristics change and can be interpreted as being the resolutions of action. Why does this happen? It is because there exists salt marsh in the study area that has a maximum breadth of 120m with its cluster having the maximum breadth of 280m. Although the image up to a resolution of 120m, which is the maximum size of the salt marsh, can reveal the features of the study area, it is difficult to portray the changes in image characteristics at lower resolutions. At a lower resolution, the salt marsh is portrayed as a cluster, but at a resolution of 300m, which is larger than the size of the cluster, the characteristics can no longer be displayed. It can be said that the characteristics of this study area are reflected in the spatial autocorrelation changes of the images of variances.

7.4.2.3 Step 2: Measuring the Spectral Separability

The land covering types of Suncheon Bay were classified into 6 groups. They were water body, tidal flat, forest, agricultural area, salt marsh and urban area. In addition, two training areas were selected from each land cover type. The information about training sets was saved into the vector formatted file allowing for the same application of the resolution image. The basic statistical information of the training set and the covariance between bands was calculated by every image at resolutions. Based on it, the J-M distance and Mahalanobis distance were measured.

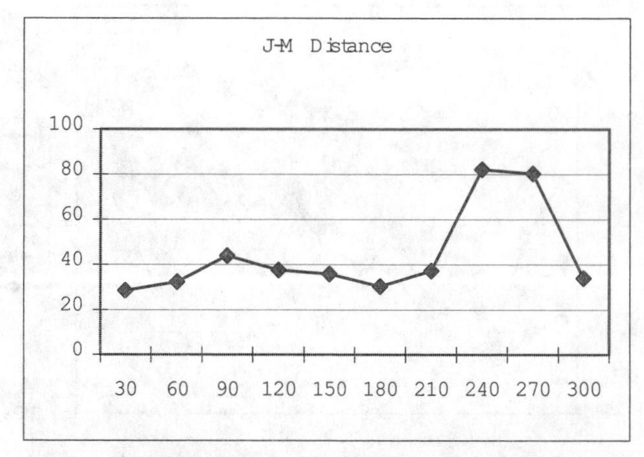

Figure 7.7 The changes of the J-M distance at various resolutions.

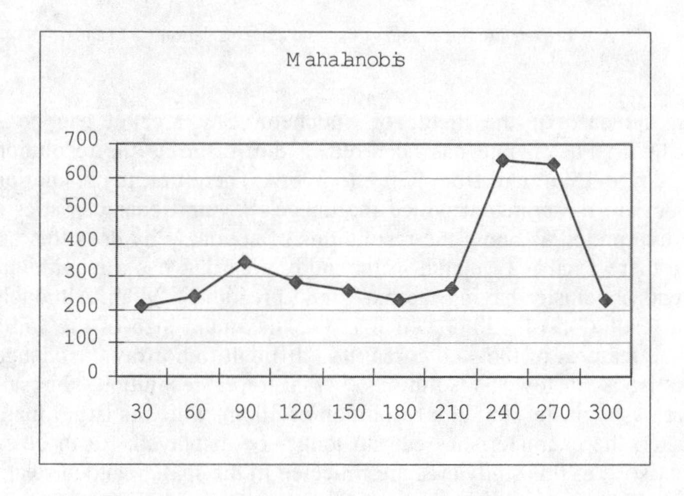

Figure 7.8 The changes of Mahalanobis distance at various resolutions.

Figures 7.7 and 7.8 show the changes in the J-M distance and Mahalanobis distance at each resolution, respectively. It is found that the spectral separability does not change up until a resolution of 150m, which was determined to be the resolution of action in the previous step, but that changes in spectral separability become evident at lower resolutions. In particular, the spectral separability increases at the resolutions of 240m and 270m. At these resolutions, the attribute information is clearly classified making these resolutions the adaptive resolutions (or the optimum resolutions) for representing the attribute changes of the study image.

7.4.3 Evaluation of the Procedure Proposed

7.4.3.1 Comparing Classification Accuracies of Images at Different Resolutions

In order to determine if the resolution of action and the adaptive resolution accurately reflected the image characteristics, the image of the study area was classified by land cover types and the classification results were then evaluated (Figure 7.9).

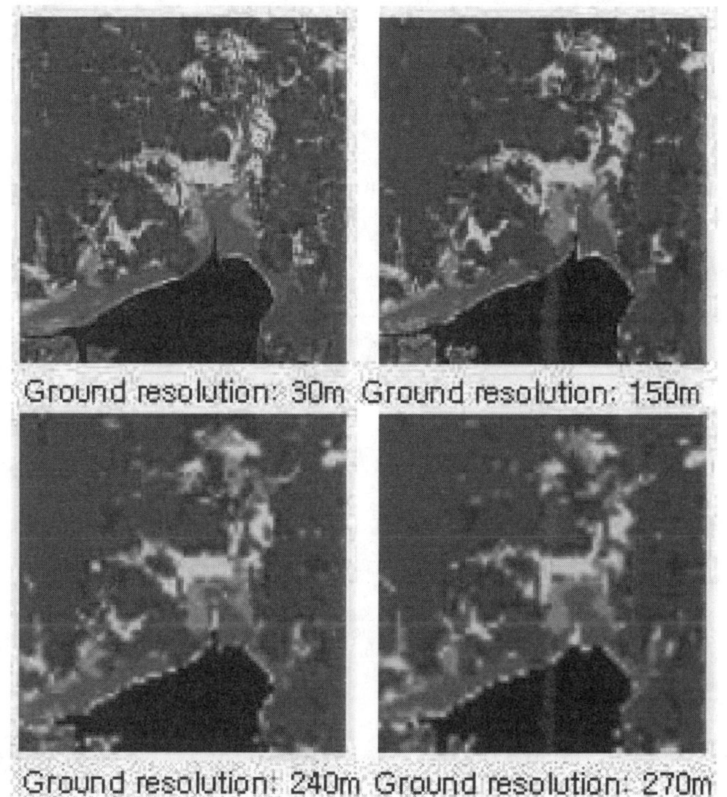

Figure 7.9 The land cover classification at various resolutions.

For the purposes of classification accuracy analysis, the topographic map, forest map and aerial photos of Suncheon Bay area were used to prepare a reference map in Arc/Info vector coverage format, which can be converted into raster data with various resolutions. The reference map had 4 classes, which were water body, tidal flat, agricultural area and forest. In order to measure the image classification accuracy by resolutions, a variety of resolution images were compared to the raster reference data for each resolution. An overlay analysis was conducted on the images at each resolution and the overall accuracies and Kappa Indices were calculated (Figures 7.10 and 7.11).

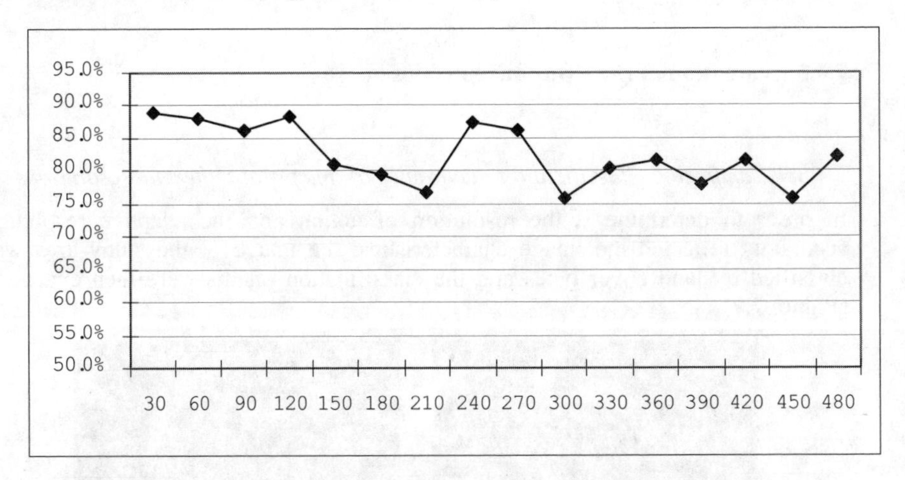

Figure 7.10 Changes of accuracy by spatial resolution.

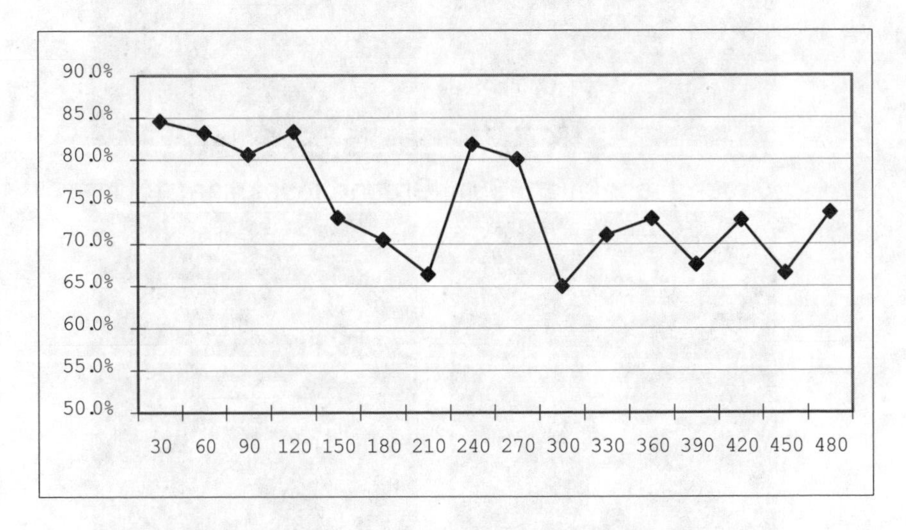

Figure 7.11 Changes of Kappa index by spatial resolution.

The changes of image classification accuracy by resolution were compared to the image characteristics revealed by the procedure for exploring the optimum resolution. The following results were discovered. First of all, there are sharp drops in accuracy at the resolutions of 150m and 300m. These resolutions could be the resolutions of action. They were obtained from the measurement of the spatial autocorrelation of the images of variances (see Figures 7.6 and 7.11). This result is expected to provide the basis for revealing the usefulness of the spatial autocorrelation measurement for the images of variances. Second, the classification accuracy increases at the resolutions of 240m and 270m, the adaptive resolutions to represent attribute information properly, which were selected using the method of the spectral separability measurement (see Figures 7.7, 7.8 and 7.11). In other words, the classification accuracy improves at the adaptive resolutions, which are found at resolutions lower than the resolution of action. In this vein, it can be said that the J-M distance and Mahalanobis distance can be used in the procedure to explore the optimum resolutions.

7.4.3.2 Correlation Analysis Between the Accuracy and the Index

Next, in order to evaluate the indices that reflect the image characteristics by resolutions, the correlation between these indices and the classification accuracies were analysed. Spearman's rank correlation analysis was more suitable than Pearson's product moment correlation analysis in this case because of a small amount of data.

Table 7.1 shows the correlation analysis between the 3 indices, measuring the textural indices of satellite imagery, i.e. local variance, fractal dimension and spatial autocorrelation of images of variances respectively, against the classification accuracy by resolutions.

Table 7.1 Spearman's rank correlation coefficient between the textural indices and the classification accuracies

	Local Variance	Fractal Dimension	Autocorrelation of the images of variances
Correlation	-0.273	-0.200	-0.717

Compared to the low correlations of the classification accuracies with the local variance and fractal dimension, the spatial autocorrelation index exhibits a higher negative correlation. The negative value means that the classification accuracy declines as the spatial autocorrelation index increases. For instance, according to the earlier analysis in this study, the sharp decline in classification accuracy seen at the resolution at the apex of the spatial autocorrelation is a direct reflection of this negative correlation (see Figure 7.6).

Table 7.2 shows the correlation analysis between the 3 methods for measuring the spectral separability, divergence, the J-M distance and Mahalanobis distance respectively, against classification accuracy.

Table 7.2 Spearman's rank correlation coefficient between the indices of the spectral separability and the classification accuracy

	J-M distance	Mahalanobis distance	Divergence
Correlation	0.636	0.588	0.042

Whereas the J-M distance and Mahalanobis distance show a relatively high positive correlation with the classification accuracy, the correlation of the divergence is very low. In other words, with the correlations near 60%, the J-M distance and Mahalanobis distance can be accepted as the indices of the spectral separability. However, for divergence, it does not represent the changes in attribute information by resolutions as an index of the spectral separability.

7.5 CONCLUSION

Analyses of geographical or spatial phenomena can yield different results depending on the spatial data scale selected by a researcher. The scale is also an important variable in research in remote sensing and geographical information sciences, which calls upon quantitative methodologies for the understanding and analysis of spatial processes and patterns. In particular, the spatial resolution determines the scale in such studies that utilise remotely sensed data. This resolution is a decisive factor in the interpretation of the geographical information contained in the satellite imagery.

The aim of the present study was to investigate the proper procedure to select the optimum resolution for satellite imagery. In the first step, the textural characteristics of the original image were analysed and the resolution of action was explored. At this point, Moran's *I* for the spatial autocorrelation of the images of variances was used. The resolution of action, the resolution at which the image characteristics change, was the one where the image's classification accuracy dropped sharply. By exploring the image's resolution of action, the resolution boundaries to maintain the image characteristics were found.

In the second step, the adaptive resolution was explored in order to extract the image's attribute information accurately. In this step, the indices of the spectral separability for the training set were used. It was ascertained that the J-M distance and Mahalanobis distance were the good indices for estimating the accuracy of attribute information. Therefore, the resolution where the J-M distance and Mahalanobis distance were high was the one with the high image classification accuracy.

The procedures for exploring the optimum spatial resolution proposed in this investigation have the following advantages over existing methods: First of all, whereas the procedure to select the optimum resolution in existing methods is vague, the method proposed here searches for the resolution of action first and then offers clearer criteria. Second, it allows the exploration of the resolution that clearly represents the attributes. It can be said that the resolution at which the spectral separability increases is the one where the image's attribute information accuracy is high.

Lastly, this research must be followed by studies of more diverse land covering types, and images with various sensors. Further studies should be able to provide generalised rules that will serve as a clearer foundation for selecting the optimum resolution.

7.6 REFERENCES

Atkinson, P.M., 2001, Geostatistical regularization in remote sensing. In *Modelling Scale in Geographical Information Science*, edited by Tate, J.N. and Atkinson, P.M. (Chichester: John Wiley & Sons), pp. 237-260.

Bailey, A. and Gatrell, A., 1995, *Interactive Spatial Data Analysis*, (Essex: Longman).

Cao, C., 1992, Detecting the scale and resolution effects in remote sensing and GIS, Unpublished Ph.D. dissertation, Louisiana State University.

Cao, C. and Lam, N., 1997, Understanding the scale and resolution effects in remote sensing and GIS. In *Scale in Remote Sensing and GIS*, edited by Quattrochi, D.A. and Goodchild, M.F. (Boca Raton: CRC Press), pp. 57-72.

Clarke, K., 1986, Computation of the fractal dimension of topographic surfaces using the triangular prism surface area method. *Computers and Geosciences*, 12, pp. 713-722.

Cressie, N., 1993, *Statistical Analysis of Spatial Data*, (New York: Wiley).

Duda, R.O. and Hart, P.E., 1973, *Pattern Classification and Scene Analysis*, (New York: Wiley).

Fotheringham, A.S., Brunsdon, C. and Charlton, M., 2001, Scale issues and geographically weighted regression. In *Modelling Scale in Geographical Information Science*, edited by Tate, J.N. and Atkinson, P.M. (Chichester: John Wiley & Sons), pp. 123-140.

Fotheringham, A.S. and Wong, D., 1991, The modifiable area unit problem in multivariate statistical analysis, *Environment and Planning A*, 23, pp. 1025-1044.

Friedl, M.A., 1997, Examining the effects of sensor resolution and sub-pixel heterogeneity on spectral vegetation indices: Implications for biophysical modeling. In *Scale in Remote Sensing and GIS*, edited by Quattrochi, D.A. and Goodchild, M.F. (Boca Raton: CRC Press), pp. 113-140.

Goodchild, M.F., 1980, Fractals and the accuracy of geographical measures. *Mathematical Geology*, 12, pp. 85–98.

Goodchild, M.F. and Quattrochi, D.A., 1997, Introduction: Scale, multiscaling, remote sensing, and GIS. In *Scale in Remote Sensing and GIS*, edited by Quattrochi, D.A. and Goodchild, M.F. (Boca Raton: CRC Press), pp. 1-12.

Griffith, D.A., 1987, *Spatial Autocorrelation: A Primer*, (Washinton, DC: Asscociation of American Geographers).

Harvey, D., 1969, *Explanation in Geography*, (London: Edward Arnold).

Hwang, C.S., 1999, Alternative methods for assessments of DEMs' errors. *Journal of the Korean Society for Geo-Spatial Information System*, 7(2), pp. 23-34. (in Korean).

Hwang, C.S., Hwang, S.Y. and Lee, J.Y., 2000, Spatial data analysis and GIS. *Jirihakchong*, 28, pp. 1-12. (in Korean).

Jeffreys, H., 1948, *Theory of Probability*, (Oxford: Oxford University Press).

Jensen, J.R., 1996, *Introductory Digital Image Processing: A Remote Sensing Perspective*, (Englewood Cliffs: Prentice Hall).

Jupp, D.L.B., Strahler, A.H. and Woodcock, C.E., 1988, Autocorrelation and regularization in digital images: Basic theory. *IEEE Transactions on Geoscience and Remote Sensing*, 26, pp. 463-473.

Kailath, T., 1967, The divergence and Bhattacharyya distance measures in signal selection. *IEEE Transactions on Communication Theory*, COM-15, pp. 52-60.

Lam, N., 1990, Description and measurement of Landsat TM images using fractals. *Photogrammetric Engineering and Remote Sensing*, 56, pp. 187-195.

Lam, N. and De Cola, L., 1993, *Fractals in Geography*, (Englewood Cliffs: Prentice Hall).

Lam, N. and Quattrochi, D.A., 1992, On the issues of scale, resolution, and fractal analysis in the mapping sciences. *Professional Geographer*, 44, pp. 88-98.

Lyon, J.G. and McCarthy, J. (eds.), 1995, *Wetland and Environmental Applications of GIS*, (Boca Raton: CRC Press).

Mahalanobis, P.C., 1936, On the generalised distance in statistics. In *Proceedings of the National Institute of Science in India*, Part A2, pp. 49-55.

Mark, D.M. and Aronson, P.B., 1984, Scale-dependent fractal dimensions of topographic surfaces: An empirical investigation with applications in geomorphology and computer mapping, *Mathematical Geology*, 11, pp. 671-684.

Markham, B.L. and Townshend, J.R.G., 1981, Land cover classification accuracy as a function of sensor spatial resolution. In *Proceedings of the 15th International Symposium on Remote Sensing of Environment*, pp. 1075-1090.

Matusita, K., 1966, A distance and related statistics in multivariate analysis. In *Multivariate Analysis*, edited by Krishnaiah, P.R. (New York: Academic Press), pp. 187-200.

McNairn, H., Protz, R. and Duke, C., 1993, Scale and remotely sensed data for change detection in the James Bay, Ontario, coastal wetlands. *Canadian Journal of Remote Sensing*, 19, pp. 45-49.

Moellering, H. and Tobler, W., 1972, Geographical variances. *Geographical Analysis*, 4, pp. 34-64.

Openshaw, S., 1984, *The Modifiable Areal Unit Problem*, Concepts and Techniques in Modern Geography 38, (Norwich: Geo Books).

Quattrochi, D.A., Emerson, C.W., Lam, N.S. and Qiu, H., 2001, Fractal characterization of multitemporal remote sensing data. In *Modelling Scale in Geographical Information Science*, edited by Tate, J.N. and Atkinson, P.M. (Chichester: John Wiley & Sons), pp. 13-34.

Park, E.J., Seong, J.C. and Hwang, C.S., 2003, A study of drought susceptibility on cropland using Landsat ETM+ imagery. *Korean Journal of Remote Sensing*, 19, pp. 107-115.

Stone, K.H., 1972, A geographer's strength: The multiple-scale approach. *The Journal of Geography*, 71, pp. 354-362.

Wacker, A.G., 1971, The minimum distance approach to classification, Ph.D. dissertation, Purdue University at West Lafayette.

Weigel, S.J., 1996, Scale, resolution and resampling: Representation and analysis of remotely sensed landscapes across scale in geographic information systems, Unpublished Ph.D. dissertation, Louisiana State University.

Woodcock, C.E. and Strahler, A.H., 1987, The factor of scale in remote sensing. *Remote Sensing of Environment*, 21, pp. 311-332.

Xia, Z.G. and Clarke, K.C., 1997, Approaches to scaling of geo-spatial data. In *Scale in Remote Sensing and GIS*, edited by Quattrochi, D.A. and Goodchild, M.F. (Boca Raton: CRC Press), pp. 309-360.

CHAPTER EIGHT

Visualisation for Coastal Zone Management

Simon R. Jude, Andrew P. Jones and Julian E. Andrews

8.1 INTRODUCTION

8.1.1 Coastal Management in the United Kingdom

The United Kingdom has an extensive coastline of over 12,000km in length that is formed by a number of environmental processes and today is subjected to a range of natural and anthropogenic pressures. Coastal management in the UK is underpinned by the development and implementation of Shoreline Management Plans (SMPs). Introduced in 1995 to provide long-term sustainable coastal defence policies and management objectives for sediment cells or sub-cells, SMPs are developed through co-operative discussions between the numerous organisations involved in managing the coastline (Purnell, 1996; Potts, 1999). SMPs can encompass a range of management options including 'do nothing', 'hold the line' of existing defences, 'advance the line' of existing defences or 'retreat the line' (Ash *et al.*, 1996). However, whilst they define the long-term management objectives, individual management schemes remain subject to economic and environmental appraisal as and when they are proposed.

Amongst the key challenges facing coastal zone managers are the need to widen public consultation and strengthen public participation during the selection of management options, and the requirement to improve the information dissemination process once decisions have been made. Shoreline Management Plans are complicated documents for those without prior technical knowledge of coastal processes, and the method in which they are prepared has been criticised for lacking adequate scope for public participation. It has been argued that this has led to suspicion amongst local communities regarding the beneficiaries of the plans (O'Riordan and Ward, 1999). Traditionally, SMPs and environmental and economic appraisals have only been disseminated to a limited number of organisations and interested individuals. The dissemination has generally been paper-based, and two-dimensional maps have been used to illustrate the plans. The

0-41531-972-2/04/$0.00+$1.50

government realises that wider access to information contained within SMPs will be required in the future if they are to gain support for the plans, as the policies outlined in the SMP must be seen to be acceptable to the general public (O'Riordan and Ward, 1999; Potts, 1999). Indeed, a recent government review of the SMP process identified the difficulties associated with facilitating public participation as being very significant (MAFF, 2000). In line with this, the review called for innovative new communication techniques to be developed and incorporated into future SMP documents and dissemination programmes (MAFF, 2000).

The problems found in the United Kingdom are mirrored elsewhere. For example, the European Union Demonstration Programme on Integrated Coastal Zone Management recently noted that stakeholders should be more involved in the development and implementation of coastal management plans (CEC, 2000a). This view is now reflected in recent EU recommendations promoting participatory planning in coastal management and encouragement to develop systems that allow the monitoring and dissemination of coastal zone information (CEC, 2000b).

There is a clear need for the further development of new methodologies that will help enable interested individuals and organisations to be informed of shoreline management decisions in the most inclusive manner possible (Belfiore, 2000; King, 1999). Indeed, King (1999) has specifically called for the use of *electronic* methods to facilitate communication between coastal managers and the public, whilst many others have highlighted the need for research to exploit the potential of GIS in educating, promoting and involving the public in coastal planning and decision-making (Bartlett and Wright, 2000). Certainly, traditional GIS packages are already widely used by organisations involved in coastal management and these systems are frequently cited as one of the tools associated with best practice (e.g. Bartlett, 1994). However, GIS does not provide a universal solution despite its potential for assisting informed decision-making (O'Regan, 1996; Bartlett, 2000). Ultimately a traditional GIS and its output are oriented towards experts with knowledge of complicated terminology, as opposed to the layperson who often has the most to lose from management decisions. These limitations are compounded by the fact that coastal decision-makers are themselves often overwhelmed by the complexity of many GIS applications (Green, 1995). Consequently, the GIS based coastal management systems that have been developed are often simply employed to produce thematic maps of coastal areas for SMPs, and much of the potential of the technology remains unrealised.

8.1.2 VRGIS - a Possible Solution?

One technology with the potential to widen communication in shoreline management planning is Virtual Reality GIS (VRGIS). A VRGIS is in many aspects similar to a traditional GIS, but it encompasses Virtual Reality visualisations as a key output and interaction method. The virtual reality (VR) aspect of VRGIS has evolved mainly as an interface technology within which user interaction issues are of key importance. The more traditional GIS acts as a data storage and manipulation technology.

The important role of visualisation in environmental decision-support has been recorded by a number of authors who have highlighted the need to develop such techniques to assist in the public presentation of complex environmental process models (Bishop, 1994; Bishop and Karadaglis, 1997). The recent development of VRGIS provides an opportunity to further develop public involvement in coastal zone management by providing the functionality to produce realistic virtual reality visualisations of different shoreline management outcomes. These may prove to be a significant advance on traditional methodologies. Using a case study of the north Norfolk coast in eastern England, this article reports on a research project that is developing an integrated VRGIS methodology for the assessment, visualisation and public communication of the environmental impacts of several proposed real-world management schemes.

8.2 METHODOLOGY

8.2.1 The Case Study Area – The North Norfolk Coastline

The north Norfolk coast is a relatively undeveloped low-lying barrier coastline that began to form in its current state around 6,000 to 7,000 years ago (Andrews et al., 2000). Because of its relatively undeveloped nature, the coastline has high scientific, economic and recreational value, reflected by the whole zone being protected by national and international legislation. Management of the coastline is complicated, with numerous statutory and non-statutory bodies involved in overseeing a wide range of sites including a number of nature reserves. The coastline has been studied widely and benefits from an extensive monitoring programme managed by the UK Environment Agency. The development of a first generation SMP began in 1993 and was published in 1996 (Environment Agency et al., 1996).

The SMP covers a very large area. Therefore, a number of smaller project-level study sites were identified through consultation with a range of statutory and non-statutory organisations involved in managing the coastline. In order to illustrate this work, a single scheme at Brancaster West Marshes is described here (Figure 8.1, see colour insert following page 164).

8.2.2 Brancaster West Realignment Scheme

With sea level predicted to rise by up to 88cm by 2100 (Houghton et al., 2001) there is considerable concern regarding the potential for future active management of the coastline because of its vulnerability to North Sea storm surges (Thumerer et al., 2000). A possible option to accommodate future rises includes allowing reclaimed freshwater marshes to revert back to their natural state, a process known as managed retreat, setback, or coastal realignment. Coastal realignment has triggered considerable concern and debate amongst the public (Clayton, 1995), although the European Habitats Directive does require such schemes to offset habitat losses by creating new habitats elsewhere along the coast.

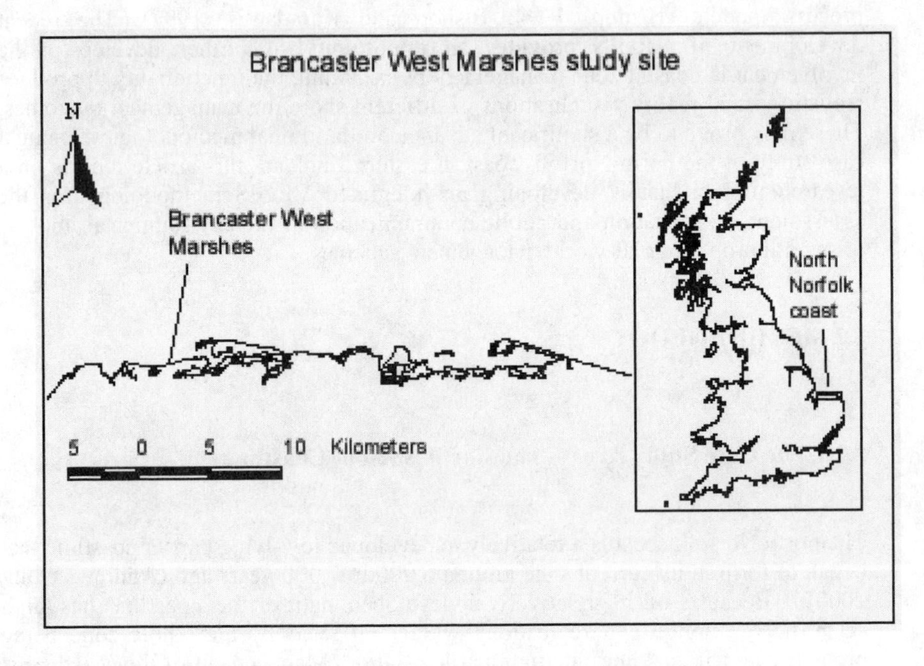

Figure 8.1 The location of the Brancaster West Marshes study site.

Brancaster West Marshes is a site currently under consideration for coastal realignment. The Marshes comprise approximately 40ha of freshwater grazing meadows forming a Site of Special Scientific Interest (SSSI) and a Special Protection Area (SPA) under the European Union Birds Directive (Tyrrell and Dixon, 2000). The site is flanked by earth flood embankments with its frontage protected by defences strengthening the natural dune frontage. The latter were constructed in 1978 to provide protection against storm surges but have degraded to such an extent that the Environment Agency has proposed a managed realignment scheme in which the frontage will be removed, with a new defence constructed 300m inland from the original location (Tyrrell and Dixon, 2000). The freshwater marshes to the north of the new defence will subsequently be allowed to revert to salt marsh.

The scheme has attracted considerable attention because it impacts a site protected under the Birds Directive. Furthermore, it could potentially interfere with the defences and frontage protecting the adjacent Royal Society for the Protection of Birds (RSPB) reserve at Titchwell. Additional complications arise from the privately owned defences belonging to the Royal West Norfolk Golf Club to the east, who plan to construct their own defence to protect their practice ground in response to the scheme.

8.2.3 Database Construction

An extensive GIS database was developed using ArcInfo and ArcView GIS packages (Table 8.1). Data was obtained from organisations involved in managing the coast, and supplemented a number of Ordnance Survey products including Land-Line.Plus, and Land-Form PROFILE. Where management plan data was unavailable in a digital format it was digitised, with permission, from management documents.

Database construction illustrated the difficulties associated with integrating data from a range of different organisations. Firstly, identifying data holdings availability was time-consuming and frustrating, with some organisations wishing to charge for data conversion. Secondly variations in the GIS software used by organisations required further complex conversion into a standard format.

Table 8.1 Sources of data.

Dataset	Supplied by	Provides
Aerial photography	Norfolk County Council and Natural Environment Research Council	Colour aerial photography of the site from 1988 and 2001
CASI	Natural Environment Research Council	5m resolution Compact Airborne Spectrographic Imager image with intertidal zone classification
Geology	British Geological Survey	Solid and drift geology
Landcover Map of Great Britain	Centre for Ecology and Hydrology	25m resolution landcover grid
Land-Form PROFILE	Ordnance Survey	10m DEM grid
Land-Line.Plus	Ordnance Survey	Large-scale vector data
LIDAR	Environment Agency	2m DEM grid
Shoreline Defence Survey	Environment Agency	Flood defence location, design and condition
Shoreline Management System	Environment Agency	Coastal monitoring data including beach profiles

8.2.4 Assessment Methodology

Assessments of future changes at the study site were made using information from a range of sources. Data on the coastline's evolution was provided from results of the Natural Environment Research Council's Land-Ocean Interaction Study (LOIS) together with historical mapping and aerial photography of the site. Furthermore, beach profiles collected by the Environment Agency's Shoreline Management System for the frontage over the last 10 years were incorporated into the GIS. This historical data was complemented with assessments of the potential impacts of future sea level rise on the site. Future sea level rise was calculated using the Model for the Assessment of Greenhouse-gas Induced Climate Change (MAGICC) (Hulme et al., 1995) for each of the latest emissions scenarios from the Intergovernmental Panel on Climate Change (IPCC). The results from MAGICC exhibited considerable uncertainty in predicted rates of future sea level rise. Therefore, for the purpose of the research, the assessments were based on the IPCC's IS92a scenario using the high sea level rise predictions calculated by MAGICC (86cm by 2100). The assessment of sea level rise also accounted for isostatic changes in the land level and used mean isostatic adjustment rates as predicted for eastern England by Shennan (1989).

The estimation of how the combined sea level rise and isostatic adjustment would affect the study site was undertaken in ArcInfo using an Arc Macro Language (AML) script that calculated a new Digital Elevation Model (DEM) for each year between 2000 and 2100. Sources of elevation data employed for this included 2m resolution LIDAR data provided by the Environment Agency and Ordnance Survey Land-Form PROFILE 10m resolution products. This methodology allowed the applicability of different sea level rise scenarios to be quickly assessed before they were used for creating VR visualisations.

8.2.5 Large-scale Data Preparation

Ordnance Survey Land-Line.Plus large-scale digital topological data was used for the basis of the detailed visualisation work. It was supplied in 1km^2 titles at 1:2,500 scale in line-only format, and thus required conversion to a polygon topology to allow attribute data to be incorporated. All editing took place in ArcInfo GIS, where the tiles covering the study site were appended and cleaned to produce a line coverage of the study site. Unfortunately the data was produced in such a way that automated polygon topology generation methods that have been applied for urban areas such as those used by Lake et al. (2000) are unsuitable. This is because not all polygons are completely closed and, although the data is provided with seed points relating to polygons, they fail to cover every land parcel. Hence, where required, polygons were closed and seed points were generated manually.

Once the polygon coverage was created, attribute data was added. Attributes were assigned to each polygon from three separate data sources. For the intertidal zone, 5m resolution Compact Airborne Spectrographic Imagery (CASI) was available from the LOIS Project. This imagery was georeferenced and converted to a grid format using ERDAS Imagine before being used to assign attributes to the polygons. Land-based polygons falling outside the intertidal CASI coverage were classified from aerial photography and landcover data obtained from the Centre for Ecology and Hydrology Landcover Map of Great Britain. In total, 38 different landcover classes were used.

Following construction of the polygon dataset, extensive fieldwork was undertaken to both ground-truth the coverage and to collect digital photographs for use during the production of the visualisations. Plans for the realignment scheme were obtained from the Environment Agency and were digitised to create a new coverage reflecting proposed future state of the site. The Environment Agency dataset was appended to Land-Line.Plus to allow evaluations of the areas of loss to be calculated, and for the individual environmental indicators to be assessed.

8.2.6 Production of Visualisations

Visualisations of the scheme were produced using two techniques. Firstly interactive visualisations were produced using ArcView 3D Analyst Extension to provide 'fly through' Virtual Reality Modelling Language (VRML) experiences, and secondly, static visualisations were produced by exporting the GIS results into World Construction Set, a photorealistic-rendering package. Whilst the virtual fly-through is the most common output of VRGIS, the experience it provides does not necessarily equate to the way members of the public are able to best perceive landscapes. Hence the two methodologies were chosen so as to allow an assessment to be made of their respective roles in widening public understanding of future coastal management schemes. All the visualisations were created using a 1GHz AMD Athlon-based PC with 512MB RAM running Windows 2000.

8.2.6.1 ArcView 3D Analyst

The production of visualisations using ArcView 3D Analyst involved the creation of a Triangulated Irregular Network (TIN) DEM using the Land-Form PROFILE data, over which the Land-Line.Plus polygon dataset was then draped to create a 3D scene. Sea defences and buildings were added as separate coverages, allowing them to be extruded from the DEM to produce 3D surface features. This process involved obtaining sea defence height information from a database of Environment Agency defences. The absence of true height information in Land-Line.Plus meant that, for buildings, a height of 3m per storey was assumed. Sea defence heights were extruded from Ordnance Datum whilst buildings were generated using the DEM as the base elevation. For the production of visualisations of the site following the realignment scheme, the DEM was reprofiled in those areas where change was deemed likely to occur, then it was converted to a TIN. Finally the Land-Line.Plus-derived coverage representing the future site state was used as the drape.

3D Analyst allows the attributes of the resulting visualisation to be queried, and facilities enabling the user to navigate around and zoom in and out of the 3D scene in real-time are provided. For the purpose of this project, static images of the environments were created. Each scene was also exported to a VRML file for use in any Web viewer.

8.2.6.2 World Construction Set

In contrast to ArcView, World Construction Set allows photorealistic visualisations to be generated, and has the advantage for many GIS users that ASCII DEMs and ArcView shapefiles may be imported and used as the basis for these visualisations. However, one drawback, at least for UK users, is that input data must be reprojected to decimal degrees.

The first stage in the production of the visualisations involved the importation of the DEM data into the package. World Construction Set permits terrain features to be generated using Terraffectors. Terraffectors create terrain from two-dimensional vector representations. Their advantage is that a cross-section profile of a study area may be created to represent real world features in which the ability to specify the height or depth of the feature in relation to the underlying DEM is implicit. Area Terraffectors, based on vector polygons, were used to generate representations of the site's sea defences from defence heights and cross-sections as provided by the Environment Agency Shoreline Defence Survey from 1999. They were also employed to create creeks and pools. Despite its poorer vertical resolution, the smoother Land-Form PROFILE data was chosen as the underlying DEM in preference to the Environment Agency LIDAR dataset because Terraffectors were found to work better with that source.

The second stage in visualisation generation involved importing the individual ArcView coverages for which new environment effects were to be created. By splitting the original landcover classification dataset into separate ArcView shapefiles for each landcover type, different features could be assigned different ground effects. For example, grasslands were given a surface texture representing long meadow grass. Later 3D vegetation and trees were included, although the limited selection of pre-generated foliage features and 3D objects in World Construction Set required the creation of new objects from photographs of the study sites. Finally, 3D buildings were added. Their addition involved a multi-stage process where the building polygons were converted from an ArcView shapefile coverage to a DXF file, allowing a package called SoftCAD to be used to create a 3D building. This enabled them to be to converted into a 3D object that was useable in World Construction Set.

During each stage of the visualisation generation process, preview renders were used to test the result of parameter changes and, where necessary, adjustments were made to each feature. Once a satisfactory result was achieved, final rendered images were produced using a number of virtual camera locations that best represented the change that would occur at the site.

8.3 RESULTS

The results of the assessment of the impacts of the scheme revealed that, although there will be a predicted loss of 10.2ha of freshwater grazing marsh following construction, this Figure will most likely be offset by an increase in salt marsh area of 6.3ha and 1.6ha for water channels and creeks. However, what is less certain from this analysis are the impacts on biodiversity that this change may bring about.

The results of the visualisations are provided in Colour Plates 8.1 to 8.4 (following page 164). In each case the same viewing location is used to illustrate the changes at the site following construction of the realignment scheme. The immediate difference between the ArcView 3D Analyst and World Construction Set images is the level of detail, the former being more stylised in comparison to the latter. For example the ArcView 3D Analyst images do not contain the extensive colours, textures and 3D features found in the World Construction Set images. Likewise, the limited VRML functionality in ArcView 3D Analyst results in crude representations of the defences, produced by extruding the defence shapefiles, whilst the Terraffectors available in World Construction Set rendered their detailed cross-sections. This trade-off in detail does, however, have important implications for the rendering times for each method with the ArcView 3D Analyst 3D scenes being continually updated in real-time as the scene was explored, whilst the static rendered images of World Construction Set took approximately 1 minute to render a preview image so it is likely to be some time before explorable World Construction Set environments can be produced.

8.4 DISCUSSION

One of the key criteria for the choice of study area here was that extensive coastal monitoring data was available from a range of organisations. In particular, it was very important that large-scale digital data, plus detailed management plans and information, were available for the chosen study site. Without such information, it would not have been possible to produce assessments and visualisations of future environments that would stand up to public scrutiny. However, even here there were problems in gaining access to some key documents. For example, whilst the Environment Agency Regional Office housed the Shoreline Management System GIS, the project manager for the Brancaster West scheme was based at the local office. This led to difficulties when trying to locate up-to-date plans for the scheme.

The primary drawback associated with the methodology at present is the lack of widespread availability of large-scale digital data in a suitable format. Although Land-Line.Plus represents the best available source, the time-consuming editing required to produce a suitable polygon coverage may prove extremely costly within an organisational context. Despite this limitation, once the data was finally cleaned it lent itself extremely well to use in the visualisation software. Few difficulties were encountered during the GIS analysis stage of the work although problems were identified relating to the interoperability of data between ArcView and World Construction Set. The primary difficulty came from the requirement for ArcView shapefiles to be projected in decimal degrees as opposed to the UK National Grid. Further problems were also typified by the laborious process required to create and import buildings due to the need to convert files to suitable formats. Such findings reflect a general problem relating to the incompatibility of GIS and VR software formats (Williams, 1999).

The two methods presented in this paper were chosen to represent alternative techniques that may be suited to different purposes in coastal management. The 3D Analyst visualisations are more dynamic and interactive, allowing the user to query the results. They may be more suited to those directly involved in the decision-making process and who wish to obtain quantitative information from them. Conversely, World Construction Set provides extremely realistic static images. These may be more suited to inclusion in management documents, such as environmental appraisals, where they could be used as an alternative to photomontages. With calls for SMPs to be distributed electronically (MAFF, 2000), both techniques lend themselves well to dissemination via the Worldwide Web. World Construction Set images could be displayed as scaleable bitmaps, whilst the virtual environments from ArcView 3D Analyst can be easily converted into VRML files, enabling the user to explore the policy impacts for themselves.

The two visualisation techniques do pose a number of questions that reflect the numerous challenges facing the application of VRGIS. The concepts of visualisations and virtual environments are relatively new and as such it has been argued that, until they are more widely used, knowledge of how best to design them will be lacking (Batty et al., 1998). This leads to the question of how the public, as opposed to experts, relate to visualisations and whether simple visualisations like those produced by 3D Analyst are more effective than detailed photorealistic visualisations at conveying complicated information? There are obvious difficulties when trying to compare the applications of visualisation presented here with research using high performance graphics workstations by some other authors (Bishop and Karadaglis, 1997). However, although improvements may be made using customised software on high performance hardware, any practical application in coastal management is likely to be constrained by costs incurred in software, hardware and training. This work has illustrated that effective visualisations may be created using off-the-shelf software running on a high-powered desktop PC.

It is obvious that visualisation techniques on their own will simply produce pretty images that are of little use if not employed properly by coastal managers. VRGIS will clearly only achieve its full potential for coastal management if it is integrated into the planning process (Zube et al., 1987; Lange, 1994). It could potentially be used from the beginning of the SMP or project planning stage, assisting communication between management organisations during the development of alternative options. Later during the decision-making process VRGIS has an obvious use in public consultation and participation. However, such advances will only be achievable if the underlying planning process is opened up to the public to allow their participation, which in the United Kingdom would require major changes to coastal planning legislation. What visualisation techniques should *not* do is, as Lange (1994) points out, be used to sell a particular scheme to the public. This would be a wasted opportunity; visualisations have the potential to *promote* the discussion of alternatives in initiatives to identify the optimal solutions to coastal management problems.

Virtual Reality GIS should not be seen as an immediate and universal solution to coastal management problems because at present there is a lack of research understanding concerning the methods that can be used and the effect of the different contexts in which they may be applied (Zube et al., 1987). As Williams (1999) notes, VRGIS has the potential to enhance conventional GIS but should not be viewed as a replacement. This research is, however, investigating the potential application of the assessment methodology and visualisation techniques in the SMP and planning process. Focus groups and participation seminars have been undertaken to gain feedback both from coastal managers and planners who may adopt these methodologies. Further surveys are being organised with members of the public, who have the most to gain from the use of visualisation techniques as a means of communication and a participatory tool. In this way it is hoped to explore the potential difficulties of applying the methodology in more detail. A second strand of the work will also assess how the visualisations could be used in choice experiments to determine if they may be used in decision-making contexts such as that of extended cost-benefit analyses. It is hoped that this project will illustrate the potential of VRGIS in participatory coastal management.

In conclusion, we believe that this research highlights the potential of GIS and VRGIS as an integrated tool to assess, visualise and potentially value future coastal landscapes through the use of high quality, yet disparate, data related to management of the coastal zone. It also illustrates how VRGIS may communicate, educate, inform and involve the public and stakeholder groups in coastal management decisions by presenting information in a recognisable and understandable format. We firmly believe that the application of VRGIS can stimulate meaningful discussion and dialogue between groups traditionally associated with conflicting opinions. Coastal Zone Management can only benefit from its application.

8.5 ACKNOWLEDGMENTS

The authors would like to thank all of the organisations that provided input into the research and the Environment Agency, Ordnance Survey and Suffolk County Council for providing access to their GIS data. The research was completed whilst the lead author was in receipt of an ESRC/NERC Interdisciplinary PhD Studentship.

8.6 REFERENCES

Andrews, J.E., Funnell, B.M., Bailiff, I., Boomer, I., Bristow, C. and Chroston, N.P., 2000, The last 10,000 years on the north Norfolk coast - a message for the future? In *Geological Society of Norfolk 50th Anniversary Jubilee Volume*, edited by Dixon, R., pp. 76-85.

Ash, J.R.V., Nunn, R. and Lawton, P.A.J., 1996, Shoreline Management Plans: A case study for the North Norfolk coast. In *Coastal Management: Putting Policy into Practice,* edited by Fleming. (Thomas Telford), pp. 318-330.

Bartlett, D., 1994, GIS and the coastal zone: Past, present and future. *AGI Notes* (UK: Association for Geographic Information), pp. 30.

Bartlett, D.J., 2000, Working on the frontiers of science: applying GIS to the coastal zone. In *Marine and Coastal Geographical Information Systems*, edited by Wright, D.J and Bartlett, D.J. (London: Taylor and Francis), pp. 11-24.

Bartlett, D.J. and Wright, D.J., 2000, Epilogue, In *Marine and Coastal Geographical Information Systems*, edited by Wright, D.J. and Bartlett, D.J. (London: Taylor and Francis), pp. 309-315.

Batty, M., Dodge, M., Doyle, S. and Smith, A., 1998, Modelling virtual environments. In *Geocomputation: A Primer,* edited by Longley, P.A., Brooks, S.M., McDonnell, R. and MacMillan, B. (Wiley), pp. 139-161.

Belfiore, S., 2000, Recent developments in coastal management in the European Union. *Ocean and Coastal Management*, **43**, pp. 123-135.

Bishop, I.D., 1994, The role of visual realism in communicating and understanding spatial change and process. In *Visualisation in Geographic Information Systems*, edited by Unwin, D. and Hearnshaw, H. (Wiley), pp. 60-64

Bishop, I.D. and Karadaglis, 1997, Linking modelling and visualisation for natural resources management. *Environment and Planning B: Planning and Design*, **24**, pp. 345-358.

Clayton, K.M., 1993, Adjustment to greenhouse gas induced sea level rise on the Norfolk Coast - a case study. In *Climate and Sea Level Rise: Observations, Projections and Implications*, edited by Warrick, R.A., Barrow, E.M. and Wigley, T.M.L. (Cambridge: Cambridge University Press), pp. 310-321.

Commission of the European Communities, 2000a, *Communication from the Commission to the Council and the European Parliament on Integrated Coastal Zone Management: A Strategy for Europe* (Brussels: CEC).

Commission of the European Communities, 2000b, *Proposal for a European Parliament and Council recommendation concerning the implementation of Integrated Coastal Zone Management in Europe* (Brussels: CEC).

Environment Agency, North Norfolk District Council, King's Lynn and West Norfolk Borough Council, Ministry of Agriculture, Fisheries and Food and English Nature, 1996, *North Norfolk Shoreline Management Plan*: *Sheringham to Snettisham Scalp* (Mouchel Associates).

Green, D.R., 1995, User access to information: A priority for estuary information systems. In *Proceedings of CoastGIS'95*, University College Cork, Ireland, pp. 35-49.

Houghton, J.T., Ding, Y., Griggs, D.J., Noguer, M., van de Linden, P.J. and Xiaosu, D., eds., 2001, *Climate Change 2001: The Scientific Basis*. Contribution of Working Group I to the Third Assessment Report of the Intergovernmental Panel on Climate Change (Cambridge: Cambridge University Press).

Hulme, M., Raper, S.C.B. and Wigley, T.M.L., 1995, An integrated framework to address climate change (ESCAPE) and further developments of the global and regional climate modules (MAGICC). *Energy Policy*, **23**. pp. 347-355.

King, G., 1999, EC Demonstration Programme on ICZM. *Participation in the ICZM processes: Mechanisms and procedures needed* (Hyder Consulting).

Lake, I.R., Lovett, A.A., Bateman, I.J. and Day, B., 2000, Using GIS and large-scale digital data to implement hedonic pricing studies. *International Journal of Geographical Information Systems*, **14** (6), pp. 521-541.

Lange, E., 1994, Integration of computerized visual simulation and visual assessment in environmental planning. *Landscape and Urban Planning*, **30,** pp. 99-112.

Ministry of Agriculture, Fisheries and Food, 2000, *A review of shoreline management plans 1996-1999 - final report March 2000*. A report produced for the Ministry by a consortium led by the Universities of Newcastle and Portsmouth.

O'Regan, P.R., 1999, The use of contemporary information technologies for coastal research and management - a review. *Journal of Coastal Research*, **12** (1), pp. 192-204.

O'Riordan T. and Ward, R., 1997, Building trust in shoreline management: Creating participatory consultation in shoreline management plans. *Land Use Policy*, **14** (4), pp. 257-276.

Potts, J.S., 1999, The non-statutory approach to coastal defence in England and Wales: Coastal Defence Groups and Shoreline Management Plans. *Marine Policy*, **23**(4-5), pp. 479-500.

Purnell, R.G., 1996, Shoreline Management Plans: National objectives and implementation. In *Coastal Management: Putting Policy into Practice* edited by Flemming (London: Thomas Telford), pp. 5-15.

Shennan, I., 1989, Holocene crustal movements and sea level changes in Great Britain. *Journal of Quaternary Science*, **4**, pp. 77-89.

Thumerer T., Jones A.P., and Brown D., 2000a, GIS based coastal management system for climate change associated flood risk assessment on the east coast of England. *International Journal of GIS*, **14** (3), pp. 265-281.

Tyrrell, K. and Dixon, M., 2000, *Brancaster West Marsh Engineers Report.* (Environment Agency).

Williams, N.A., 1999, Four-dimensional virtual reality GIS (4D VRGIS): Research guidelines. In *Innovations in GIS 6*, edited by Gittings, B. (London: Taylor and Francis).

Zube, E.H., Simcox, D.E. and Law, C.S., 1987, Perceptual landscape simulations: History and prospect. *Landscape Journal*, **6** (1), pp. 60-80.

[illegible] (1989) [illegible] coastal management [illegible] 12: [illegible]

O'[illegible] T. and Ward, R. (199[illegible]) 'Building trust in shoreline management: creating participatory [illegible] in shoreline management plans', Land Use Policy, 14(4), pp. 257–276.

[illegible] Coastal Defence [illegible] and Shoreline Management Plans [illegible]

[illegible] R. (199[illegible]) [illegible] (London: [illegible]), pp. [illegible]

[illegible] (1969) [illegible] movements and [illegible]

Thomas [illegible] (2000) [illegible] in [illegible] Journal of [illegible]

[illegible] Agency

[illegible] (1999) [illegible]

[illegible] (199[illegible]) [illegible]

CHAPTER NINE

Application of a Decision Support System in the Development of a Hydrodynamic Model for a Coastal Area

Roberto Mayerle and Fernando Toro

9.1 INTRODUCTION

Recent advances in numerical modelling of physical processes and field survey technology nowadays allow the development of numerical models with extensive data sets. As a consequence, model developers are facing new challenges to handle the increasing amount of data and its analysis.

Furthermore, model development and application concerning coastal areas are heavy time demanding tasks that need tools to assist the researcher. Most of the time they involve analysis of field measured data and its comparison with numerical model outputs. Even though there are good tools for specific tasks, there is a lack of them for integrating measured and modelled data.

An application of a Decision Support System (DSS) in the development of a hydrodynamic model for a coastal area is presented. Description of the DSS components, their interaction, and its GIS capabilities to handle spatial data are also given.

9.2 DEVELOPMENT OF HYDRODYNAMIC MODELS

A hydrodynamic model is the mathematical representation, with the assistance of numerical algorithms, of flow and wave situations to study them more effectively. However to obtain reliable information, development of numerical models must include evaluation of results. This process is time-consuming and involves comparison of numerical model outputs with measured data. Moreover, a unique model development procedure can be restrictive due to the large amount of variables involved in the physical process. The number of variables and their levels of significance depend on the objective of the study. In spite of these difficulties a general trend for the development of hydrodynamic models can be defined and is presented next.

Figure 9.1 shows the phases and steps in the development of a hydrodynamic model. The model development consists of several phases (i.e. sensitivity analysis, calibration, validation and application) in which the modelling experts fine-tune

0-41531-972-2/04/$0.00+$1.50

the numerical and physical parameters to reproduce the field conditions. Each of these phases has several steps that share information with the other phases.

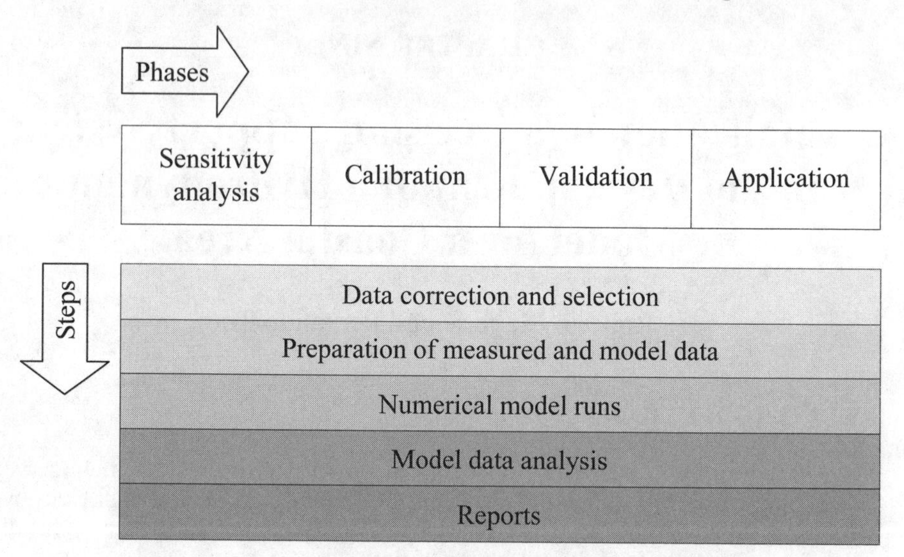

Figure 9.1: Phases and steps in the development of hydrodynamic models

The "*sensitivity analysis*" evaluates the response of the model to changes of the physical and numerical parameters. The numerical parameters are set and the physical parameters having an effect on the results are selected. Physical parameters are adjusted to represent the conditions in the studying area during the "*calibration*". During the "*validation*" the model ability in reproducing the field conditions, within certain level of accuracy, is checked. The model "*application*" is the last phase in the model development and is performed by the end users to make decisions. The validated model is then used in hindcasting, nowcasting and forecasting situations.

Regardless of the model development phase, each of them has several steps as indicated in Figure 9.1. The first one is the "*data correction and selection*" in which the measured data is corrected, selected and stored in a database. This is usually data that needs to be processed to fit the model formats. In situ measurements are in the form of time series at a certain location or recordings at a certain area for a given time. The "*preparation of measured and model data*" includes the creation of the model grid and definition of some default (but logical) numerical and physical parameters. The measured data preparation includes reformatting of the data to be used as initial conditions and boundary conditions for the model and to enhance easy comparison with the model results. "*Numerical model runs*" is the core of each phase and refer to determination of the change in the numerical model outputs caused by changes in input data. "*Model data analysis*" is the step in which the model results are analysed by means of the experience and judgement of the model developer, and compared with other model results or measured data. The data analysis is an iterative process with the model runs and gives feedback to the model developer whether or not to choose new parameters to run the model again. This is a procedure in which software tools are needed to speed up and facilitate model development. Therefore, a "*Decision Support System*" (DSS) that assists the model developers is of vital importance. And finally, the "*reports*" document the work that has been done and summarises the experience and conclusions obtained. These written reports are used as reference in the future phases of the model development.

9.3 A DECISION SUPPORT SYSTEM FOR MODEL DEVELOPMENT

The handling of information for hydrodynamic model development is a difficult task as a consequence of different sources, types and formats of the data and the increasing amount of data involved in such activities nowadays. An efficient way to retrieve, analyse and store information accurately and speedily is crucial for the good performance in the model development process.

A Decision Support System (DSS) for development of hydrodynamic models is an interactive computer-based system intended to help modellers use data from measurements and models to simplify and speed up model creation. It helps to retrieve, summarise and analyse data. It comprises three integrated components: "*interface*", "*database*" and "*models*". Figure 9.2 shows the components of a DSS for development of hydrodynamic models, with the input and output data.

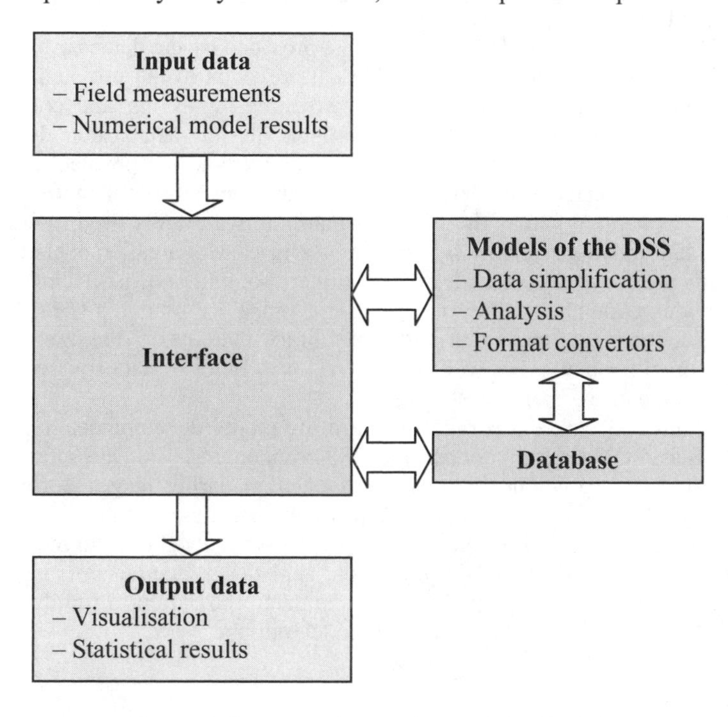

Figure 9.2 Components, input and output data of the DSS

The proposed DSS centres on the model development of depth-integrated (2DH) flow models with field data applied to coastal areas. The DSS has been tailored to handle primarily data such as water levels and current velocities.

The input data to the DSS, from measurements and numerical model outputs, is fed to the system through the interface and is stored in the database component of the system. The GIS application, as a DSS generator, is used to handling spatial data in conjunction with the database. The instructions to the system are given by the decision-maker through the interface that passes the type of analysis to be made by the model components of the DSS. Finally, the output data is presented in the form of statistical tables or visualised in plots and Figures.

The "*interface*" component is the part of the DSS that interacts with the user. It should be user-friendly and usually includes graphical components and it is then called a "*Graphical User Interface*" (GUI). The front end of the system is a graphical interface created customising ArcView 3.2a GIS software (ESRI, 1996).

The "*database*" is the collection of data that is organised so that its contents can easily be accessed, managed and updated. The size and effort spent in the database depend on the amount of available information. It is the data source for the DSS that processes and presents the information. The importance of the database in the system relies on the versatility for the selection of the data for analysis rather than in the storage capacity. In the case of the DSS, its capabilities are implemented using the GIS software generator, with tables connected under the concept of relational database.

The "*models of the DSS*" use the data provided by the database and allow the user a direct opportunity to explore the data relevant to the problem. The model components of the DSS take care of evaluating, analysing and correlating the information from different sources available in the database and reduce the information to make it easy for the user to understand the physical process. The distinction between the models of the DSS and the numerical models should be made clear here. Whereas the latter ones are software tools used to solve partial differential equations for the hydrodynamics in coastal areas, models of the DSS are software tools to link and analyse numeric output and in-situ measurements. The model components are mainly programmed as Matlab (The MathWorks, 2000) functions grouped in a software library that makes the DSS a versatile system open for new types of data analyses, new types of data sources and makes their maintenance easier.

Figure 9.3 makes a parallel between the model development steps in which the DSS assists the researcher and the DSS components. The DSS interface is used in all steps of model development and even in earlier stages during the data acquisition to visualise information. The database component is used all along the development steps and it is queried according to the model development phase that is being analysed. The data simplification, analysis and conversion are performed by the DSS models and is made during the data preparation to run the model and during the evaluation of the numerical model outputs.

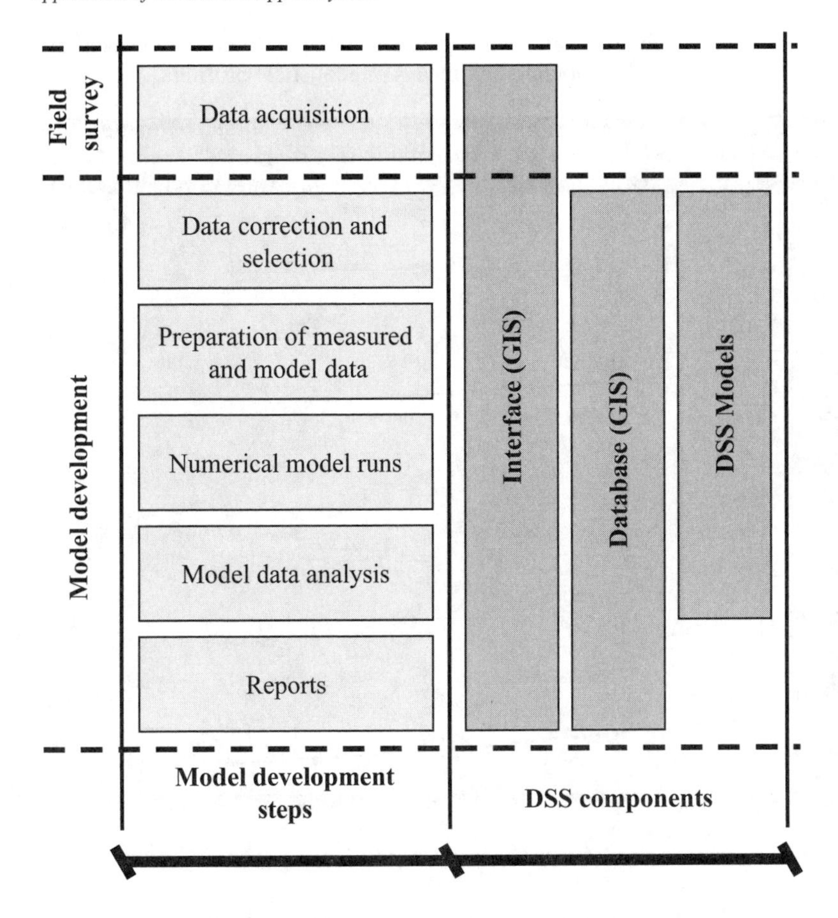

Figure 9.3 Model development phases and DSS components

9.4 APPLICATION

This section presents the application of the DSS during the development of a hydrodynamic model for an area on the German North Sea coast. The multiple possibilities of the DSS for data handling and its versatility to the needs of the user are shown.

The in situ measurements presented herein were collected as part of the project "Predictions of Medium Scale Morphodynamics" (Promorph) (Zielke et al., 1998) and were used for the development of the flow model (Palacio et al., 2003).

The domain of interest is the Meldorf Bight and the adjacent tidal channels covering an area of about 600km^2. Around 50% of the area of investigation is dry during the ebb phase and almost the whole domain is submerged during the flood phase. Figure 9.4 shows the area of investigation, model boundaries and location of in-situ measurements. The tidal channels have a maximum depth of about 20m. Tidal range is around 3.0 to 3.5m. The calibration of the flow model is carried out using measured water levels at several gauge stations and velocity measurements over several cross-sections obtained by means of acoustical profilers.

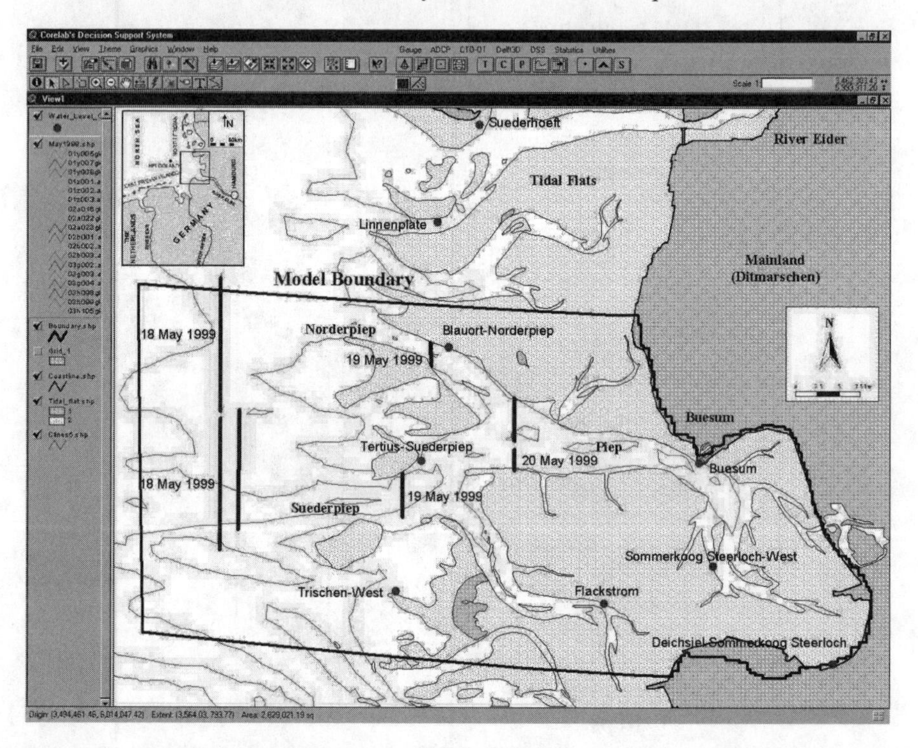

Figure 9.4 Area of investigation, model boundaries and in situ measurements

9.4.1 Data Acquisition

The acquisition of field data describing the model domain is necessary for the model development. Data may be collected from the relevant authorities, research institutes or measured in field campaigns. Here the following data sets were used:

- bathymetry in the tidal channels and tidal flat areas,
- continuously recorded water level measurements from gauge stations,
- atmospheric pressure and wind data,
- current velocity measurements.

Figure 9.4 shows the GUI for the study area with location of the water level gauges and some of the cross-sections selected for current velocity measurements.

An example of measured current velocity along a cross-section is shown in Figure 9.5. This type of plot can be animated when having several cross-section plots in a period of time.

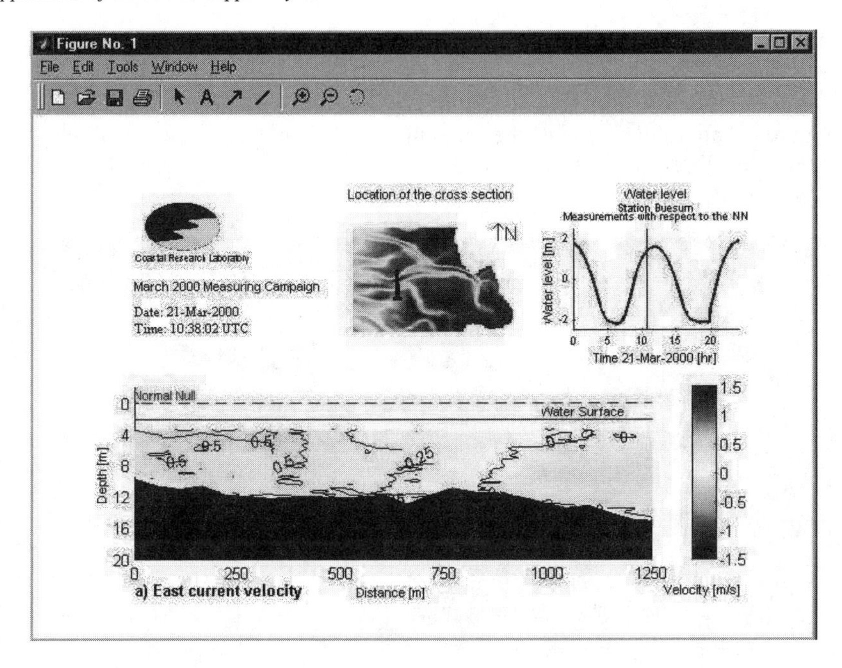

Figure 9.5 Measured current velocity transversal to a cross-section

9.4.2 Data Correction and Selection

Among others, the following corrections are incorporated with the collected measurements:

- the European time used during the recordings is referenced to UTC,
- referenced locations using differential GPS are projected onto the same reference system of the model grid,
- the depth during the measurements is usually referred with respect to the water surface. Due to the changes in the water surface in time accompanying the tide, all data is corrected and referred with respect to a fixed reference level (e.g. mean sea level) to allow comparison of data at different periods of time.

In addition to data correction and projection, some additional data is generated for later use with numerical model outputs; that is for the calculation of the depth-integrated velocity out of the 3D velocity measurements.

In situ measurements are stored in the DSS database and can be accessed by the user. The database tables are used in conjunction with the GIS map layers in such a way that the database can be queried selecting objects in the map views.

Data once entered in the DSS database are available for visualisation and selection for comparison and analysis during model development. The first step in the data selection is to display an inventory of the data showing which in situ measurements and the covered measuring period are available. The time interval considered for model development can be chosen as the time period for which simultaneous measurements of several of the required quantities are available. Or, on the other hand, it gives an idea of the missing field data in the selected interval.

Figure 9.6 shows the in situ measurements available in 2000 in the Meldorf area. The vertical axis shows the available magnitudes in terms of water levels and current velocities and the horizontal axis corresponds to the time axis. Such a Figure enhances detection of the time intervals when data sets are available for the model development. The periods selected for the model development are indicated in the Figure.

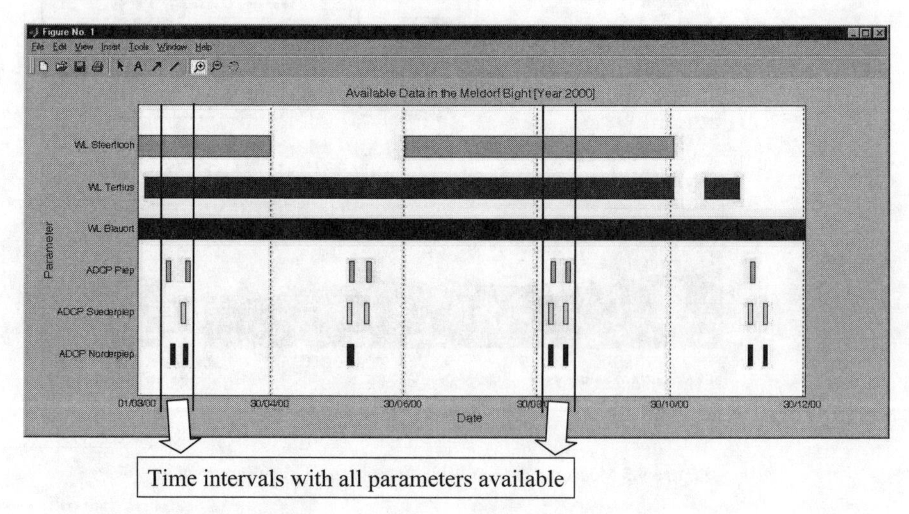

Figure 9.6 Available data in the domain area.

9.4.3 Preparation of Measured and Model Data

The model set-up, grid generation and bathymetry interpolation are done with the software tools of the hydrodynamic model.

A curvilinear grid, adjusted to the bathymetry of the domain, was created, covering an area of 600km^2, with around 35,000 cells and a resolution from 60m (inner part of the model close to the main tidal channels) to 180m (in areas close to the model boundaries). The model has three open boundaries (North, South and West). Along these boundaries the boundary conditions are specified as a function of imposed water levels. Since the bathymetry in the area was obtained from measurements done at different periods, one of the functionalities of the DSS is the selection of the appropriate bathymetry from different data sets. Besides it could also be used in updating the model bathymetry regularly.

The selection of the simulation periods is based on the intervals with available data sets that are going to be used as initial and boundary conditions for the model. These periods can be selected among the intervals shown in Figure 9.6. In the calibration and validation phases additional in situ measurements are needed for the comparison with the model results.

9.4.4 Numerical Model Runs and Model Data Analysis

Subsequently, the DSS assists the modeller in the statistical analysis and comparison of the measured data and/or the numerical model outputs depending on the model development phase being performed, i.e. sensitivity analysis, calibration, validation or application.

9.4.4.1 Sensitivity Analysis

Here the behaviour of the numerical model outputs with respect to the physical and numerical parameters is checked. The aim is to define the numerical parameters and to Figure out the physical parameters to be considered in the calibration of the model. The analysis is done at several locations all over the domain and periods of time that represent the main flow conditions; for instance, in case of tidal dominated areas, tidal phases and tidal cycles as well as periods with calm winds and during storms are considered.

For the Meldorf Bight Model the evaluation was made analysing the computed water levels and current velocities. For the comparison of the modelled water levels, the DSS makes an analysis of the variation of "*amplitude/phase*" with respect to several physical and numerical parameters. The differences in height (amplitude) and time (phase) between the peaks of two time series are computed. This analysis is used for harmonic magnitudes that have a cyclic behaviour, e.g. when evaluating water level data sets. The analysis of the water levels combines data from several locations and periods of time.

Figure 9.7 shows the GUI with a table that summarises the effect in the water levels for different time steps in the model set-up at several locations. The table includes for each location (record) and parameter (field) the number of records, mean, standard deviation and RMS-Error. The aim is to identify an appropriate time step that gives sufficient accuracy in the model results in order to minimise the computational time of the simulation. Different time steps were considered. A period of 30 days was used in the analysis to consider the effect of spring and neap tide. The selected time step is 0.50 min.

For the analysis of the data and the statistical computations the data can be organised in many different data sets. Figure 9.8 shows the DSS interface with the windows, in which the conditions for statistical analysis and selection of the data sets are set.

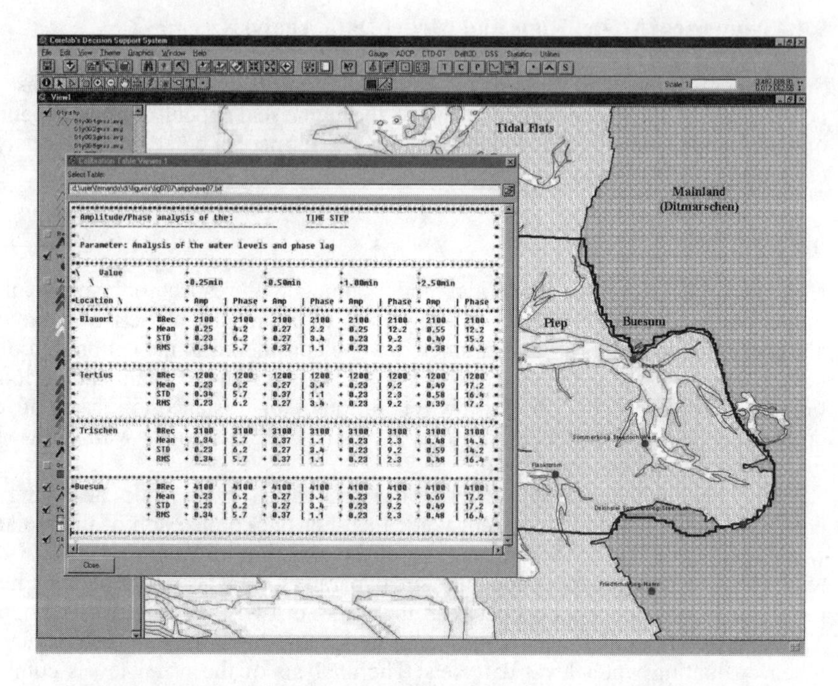

Figure 9.7 Summary of the sensitivity analysis of the time step

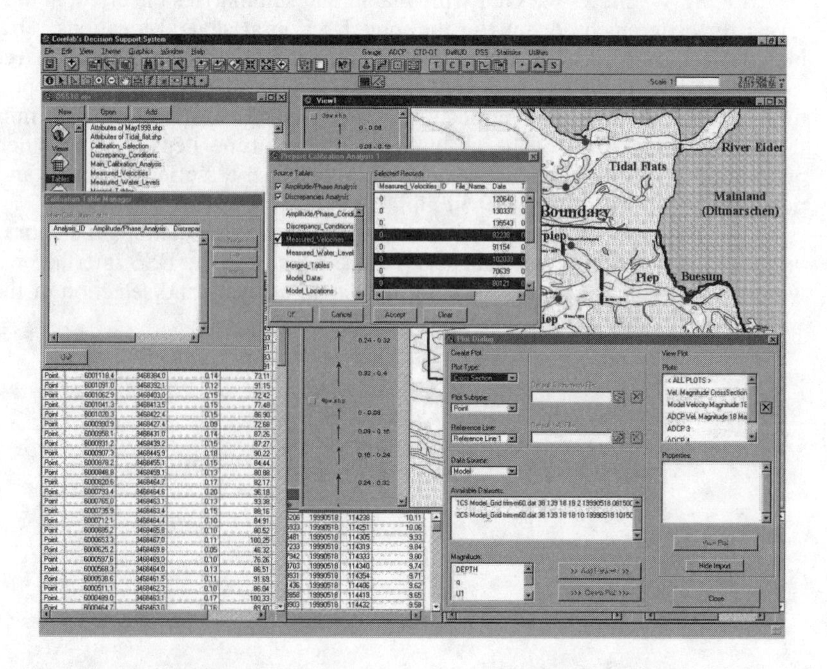

Figure 9.8 DSS interface with input data windows for the statistical analysis

9.4.4.2 Model Calibration

Physical parameters are adjusted to represent the conditions in the studying area during the model calibration. The evaluation was made analysing water levels and flow velocities. Only the calibration of the bottom roughness in terms of Chezy coefficients in the whole domain is presented. Bottom roughness with constant values of Chezy coefficient were considered.

For the comparison of the flow velocities, the DSS makes a "*discrepancies analysis.*" The *discrepancies analysis* computes the differences between the model results and measured data, and performs some statistics with these differences. This analysis can be performed organising the data by locations or velocity intervals (data sets are grouped in several classes according to the velocity magnitude from the minimum to the maximum velocities available in the data sets).

Figure 9.9 summarises the statistics of the analysis of depth integrated velocity for these three types of bottom roughness and several intervals of flow velocity. The table includes some values such as mean, standard deviation and value of the parameters RMS-Error and Relative Mean Absolute Error (RMAE) (van Rijn, 2001) for the flow velocity differences (modelled - measured) used in the analysis. The best option is obtained on the basis of these statistical parameters. In this case there are not significant differences between the three options; however, there are slightly better results with a Chezy coefficient of 60 $m^{0.5}/s$ for the bottom roughness.

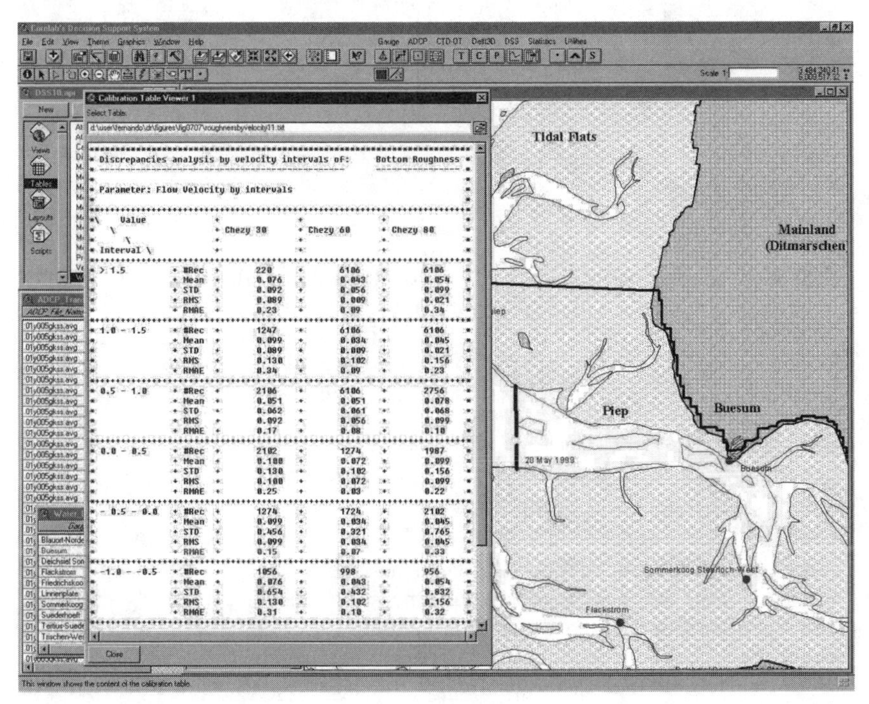

Figure 9.9 Summary of the model calibration of the bottom roughness.

9.4.4.3 Model Validation

During model validation, the model performance to reproduce the field measurements is determined. As in the model calibration, measured and modelled data are compared and therefore the same DSS tools can be applied. The data sets used in the validation are different from those used in the model calibration. The tools presented in section 9.4.2 are employed for selection of additional measurement data.

Figure 9.10 shows a comparison of modelled and measured flow velocity, integrated over the depth, at a cross section within the model domain.

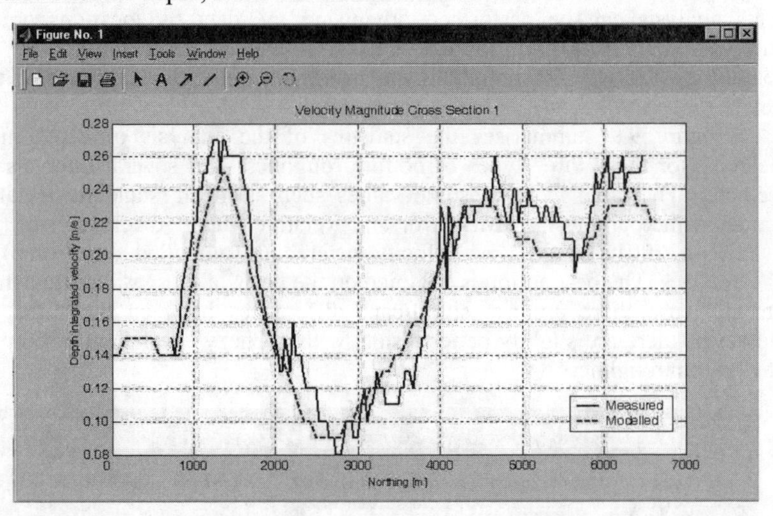

Figure 9.10 Velocities comparison during the model validation.

9.4.4.4 Application

The model, once validated, is used for forecasting flow conditions following several "if" scenarios; this allows to study a problem and possible engineering solutions (e.g. determine water levels for different wind conditions for the design of dikes). In this phase, all the DSS tools already described are available for handling existent and additional data that might be collected after the model validation.

9.4.5 Quasi-Real Time Validation

Another application of the DSS during the data acquisition might be the comparison of the measurements with the model results previously prepared. The steps involved in the quasi-real time validation (Mayerle and Toro, 2002) are as follows:

- prior to the measuring campaign, model simulations are carried out (for different parameters) for the period in question. For this, open sea boundary conditions on the basis of astronomical constituents can be used,

- the model results covering the measuring period are stored and brought to the field,
- during the cross-sectional measurements, the measured current velocities (using acoustical profilers) are transformed into depth-integrated values,
- comparisons between 2DH model results and measurements are done just after the completion of a cross-section,
- after the measuring campaign is achieved (12-13 hours covering one entire tidal cycle), the results taken at all the cross-sections and covering the entire measuring period are analysed and improvements in the model and/or the measuring strategy can be considered on the basis of the statistical analysis. Figure 9.11 shows the DSS in use covering a cross-section along one of the main tidal channels. Figure 9.11a shows the research vessel during the ADCP measurements. Figure 9.11b displays the resulting velocity field from the simulations. In Figure 9.11c a researcher using the DSS during the measurements is shown. Comparisons of modelled and measured current velocities, integrated over the depth, are shown in Figure 9.11d.

9.4.6 Reports

Usually a large number of graphs are created to visualise the model at several locations and certain periods of time. The DSS has been linked to MS Word and MS Powerpoint to summarise these graphs and document the conclusions of the work that has been done.

Figure 9.12 shows several plots of the flow velocities from the Meldorf Model during a tidal cycle already imported into a report. Animation of the behaviour of the velocity field during the tidal cycle can be created with this sequence of Figures.

9.5 CONCLUSIONS

Progress in field measurements and advances in numerical modelling are generating a large amount of data creating new challenges for its handling and analysis. This brings acutely the necessity of software tools to assist the researcher during development of hydrodynamic models.

The use of GIS technology in the model development brings many new capabilities and commodities to the model developer for spatial data handling and easy interaction with a relational database. This chapter shows the application of GIS as a software generator to develop a DSS that assists the modeller in integrating measured data and model results for the development of hydrodynamic models.

The structure of the DSS gives flexibility to the researcher regardless of the methodology of model development and the criteria of analysis for comparing numerical model outputs and field measurement data. It has also flexibility for coupling alternative extensions for new data sources and types of statistical analysis. The DSS also allows the storage and processing of the measured data at the end of a measurement campaign allowing the analysis of the data and to reprogram further measuring campaigns, if necessary.

New technologies, like telemetrics, increase the possibilities of the DSS to use it in quasi-real model validation situations in which the model is prepared *a priori* and is brought to the field to compare the numerical model outputs with the measurements that are being taken. After the completion of a measuring campaign the results taken at all the cross-sections and covering the entire measuring period are analysed and improvements in the model and/or the measuring strategy can be considered on the basis of the statistical analysis.

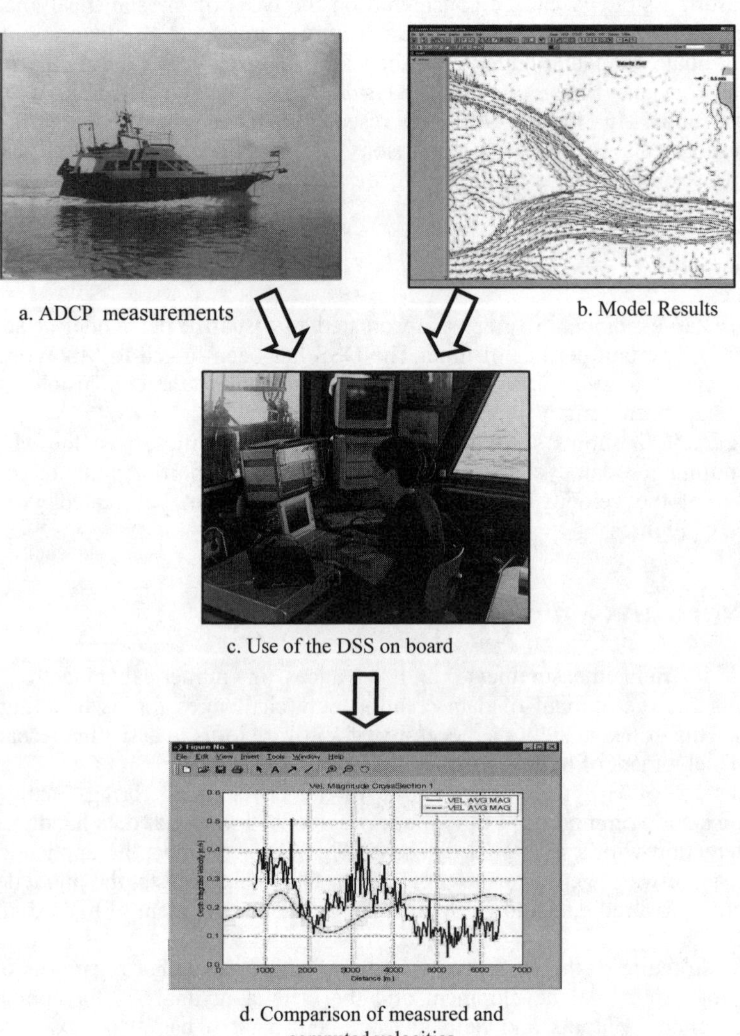

a. ADCP measurements

b. Model Results

c. Use of the DSS on board

d. Comparison of measured and computed velocities

Figure 9.11 Online validation of the numerical model.

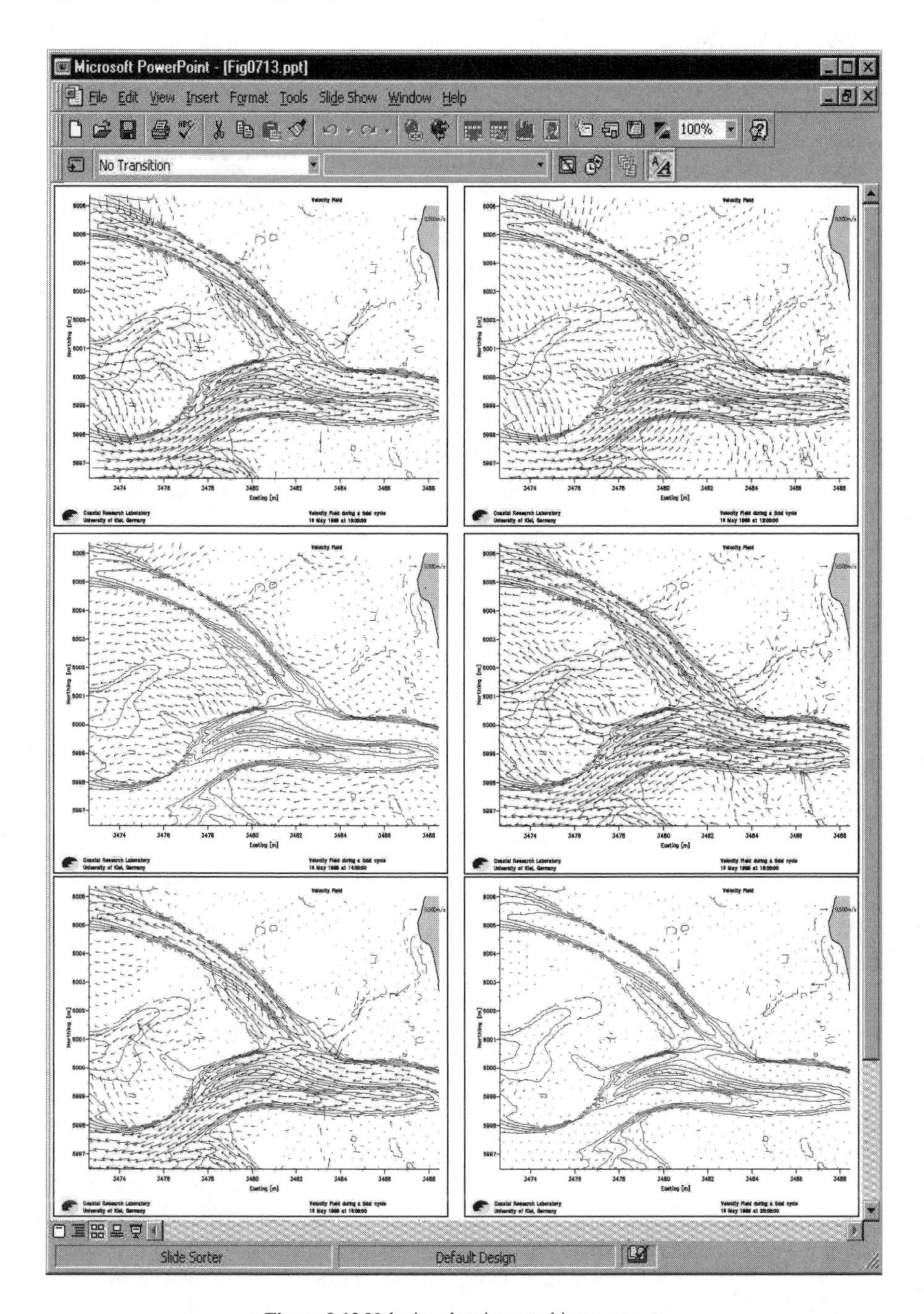

Figure 9.12 Velocity plots imported into a report.

9.6 REFERENCES

ESRI, 1996. Using ArcView GIS, ESRI, Redlands, California, 340pp.

Mayerle, R. and Toro, F., 2002. Effectiveness of a Decision Support System for Quasi-Real Time Model Calibration. Proceedings Hydro 2002, Kiel, Germany, October 2002. pp 497.

Palacio, C.A., Mayerle, R., and Toro, F.M., August 2003. Sensitivity Analysis for the Meldorf Bight Hydrodynamic Model. XXX Congress of the International Association of Hydraulic Engineering and Research (IAHR), Thessaloniki, Greece.

The Mathworks, 2000. Using Matlab, Version 6. The MathWorks, Inc., Natick, MA, USA.

Toro, F. and Mayerle, R., 2001. A Decision Support System for Enhancing Model Development and Application. Proceedings ESRI International User Conference 2001, San Diego, USA, June 2001.

Van Rijn, L.C., 2001, Application of profile and area models to the Egmond site on the time scale of storms and seasons; COAST3D Experiments 1998-1999. Report Z2394, Delft Hydraulics, Delft, The Netherlands.

Zielke, W., Mayerle, R., Gross, G. Eppel, D.P., and Witte, G. 1998. *Predictions of Medium Scale Morphodynamics – Promorph*. German Ministry of Education and Research (contract number 03F0262A/03F0262B), Berlin.

CHAPTER TEN

Decision-Making in the Coastal Zone Using Hydrodynamic Modelling with a GIS Interface

Jacques Populus, Lionel Loubersac, Jean-François Le Roux, Frank Dumas, Valerie Cummins, and Gerry Sutton

10.1 INTRODUCTION AND CONTEXT

There are many kinds of coastal water quality issues. They arise from various sources of inputs to the coastal zone that are either chronic or accidental. Pollution of seawater mostly results from a) point source and diffuse pollution originated from agricultural, industrial and urban activities, or b) pollution from maritime activities, e.g. waste oil as well as all types of toxic substances being dumped into the sea, including radioactive ones (Kershaw, 1997).

Toxic phytoplankton blooms are a new plague. Although they are not the direct result of human behaviour, they are probably linked to human activities, and particularly contaminated ballast water. These harmful disruptions have severe effects on shellfish stocks, entailing long production shutdowns. Problems also currently exist with open sewers discharging into places designated for shellfish production or into recreational waters (Boelens *et al.*, 1999).

After flooding, there is run-off pollution from agriculture and urban areas and nitrate concentrations and bacterial counts increase remarkably. The increase in nitrate concentrations can lead to dramatic eutrophication, whereas pathogenic bacteria can make aquaculture products hazardous to human health (Lees, 2000). Other activities or phenomena such as dredging, deep water sludge disposal or landfill seepage are concerns for water quality and marine living resources (Sullivan, 2001).

While hydrodynamic modelling results and the handling of geo-referenced information are becoming more readily available to coastal management stakeholders within GIS (Geographic Information Systems), there is still a lack of direct interfacing of model results with baseline mapping data (BASIC, 2001). This paper discusses solutions to bridge this gap and illustrates them with two case

0-41531-972-2/04/$0.00+$1.50

studies where effective model outputs are being used for improved environmental management and decision-making.

10.2 HYDRODYNAMIC MODELLING BASICS

Mathematical models (Garreau, 1997) can solve geophysical fluid mechanics equations using a number of simplifying assumptions. In essence, knowledge of the bathymetry, wind tide and currents are used to predict water levels and concentrations of conservative elements over space and time.

The ability of these models to accurately and reliably reproduce the important features of complex systems has improved considerably along with rapid growth in understanding of the underlying physical phenomena, and the ability to quantify them in terms of valid mathematical formulations. The widespread and affordable availability of high-powered computing facilities has also contributed to this development, given the heavy processing loads involved in running all but the simplest models.

Confidence in the predictive capacity of models is governed by the quality of fit between modelled output data and real field measurements. Standard practice in building a typical coastal model entails a number of discrete steps, the first of which involves reconstructing the morphology of the area of interest in the form of a digital terrain model. The next step involves adding water to the system, and then setting it in motion through replication of vertical tidal fluctuations and horizontal wind friction. Together these forcing factors create gradients, which in turn drive horizontal current flows. A range of assumptions is usually made in order to simplify the system, and allow it to run effectively. It is recognised that models do not strive to exactly reproduce in detail all the features of a natural system; however the aim is to arrive at a situation where the model is capable of reliably reproducing the principal features through an iterative process of validation and calibration.

10.3 GIS FOR COASTAL ZONE MANAGEMENT

A Geographic Information System (GIS) is a computer-based information system used to digitally represent and analyse geographic features. It is used to input, store, manipulate, analyse and output spatially referenced data (Burrough and McDonnell, 1998). A GIS can be distinguished from database management systems or from visualisation packages through its specialised capability for spatial analysis. The use of GIS for coastal zone management has expanded rapidly during the past decade and references are numerous (Durand, 1994; Populus, 2000; Wright and Bartlett, 2000). For optimum efficiency, geo-referenced data should be properly stored in geo-databases built on spatial data model design (Prélaz Droux, 1995). Some of the greatest challenges currently faced by those handling coastal zone data are a) the land-sea interface, with different mapping references in both horizontal and vertical modes, b) water dynamics and the related temporal issues, and c) 3D display requirements.

10.4 TOOLS AND DATA

10.4.1 Technical details of regional and local models

MARS-2D is a bi-dimensional model using a finite difference method called ADI (Salomon, 1995). A broad regional model, extending between 40°N and 65° N and from 20° W to 15° E with a 5 km grid, is used as a framework in which to embed further models of smaller extent for areas of interest along French Atlantic and English Channel coasts. Commonly, the embedding system has up to 5 levels allowing phenomena to be examined at resolutions from 1km down to 50m. This type of model is suitable for applications in areas whose waters are typically well mixed (e.g., coastal or mega-tidal areas). The model is designed to solve for tidal and wind driven currents and the transport of dissolved materials.

MARS-3D (Lazure, 1998) is a fully finite difference model, in both vertical and horizontal orientations, which uses a time splitting method based on MARS-2D for the barotropic mode. Good coupling between 2D and 3D modes has largely been achieved through the use of iterative methods. MARS-3D is used at resolutions ranging from 5 km over regional areas of epicontinental seas, down to 100 metres over detailed areas, narrow bays and estuaries. It is currently run operationally at the regional scale (with a 5 km mesh) on all French seaboards. Table 10.1 gives an overview of MARS model features.

Table 10.1 Comparison of MARS-2D and MARS-3D main features

Model	MARS-2D	MARS-3D
Area	English Channel, Bay of Biscay	English Channel, Bay of Biscay
Grid and time step	From 50m up to 10km	~ 5km, 5 – 20 minutes
Period of time	From days to decade	Year to decade
Applications	Tide currents, dissolved matter, salinity under homogenous conditions	Tide, currents, temperature, salinity, transport of dissolved matter
Type	2D Finite difference	3D Finite difference

10.4.2 The modelling system

A modelling chain has been developed at Ifremer over a period of many years. The system has a series of pre-processors and tools for the graphic display of results produced by the MARS computation kernel.

This mathematical model uses some simplified hypotheses to solve the equations that govern how marine currents and sea levels evolve. In order to function, the process requires an input water level along the edge of the area of interest. These boundaries are usually unknown locally, since they are dependent on tidal and weather conditions, which are themselves usually derived from modelling over a much larger area. Thus, the modelling process advances through the generation of a series of sequentially nested models. An initial general model covering a large area of the continental shelf and the Channel is followed by a succession of intermediate models of increasingly smaller scope, but higher resolution. Boundary conditions for the wide-area model are resolved using world tide models, into which modelled meteorological forcing factors have been assimilated. The modelling chain can be summed up as follows:

- A generated link calculates the position, extent and resolution of each sub-model, from the large-area model to the detailed high resolution model based on computational and hydrodynamic design criteria. Computational efficiency is optimised by maintaining a maximum resolution ratio (mesh size ratio) of between four and five between any two consecutively nested models. Hydrodynamic criteria are observed as far as possible in the design through avoidance of islands or zones with violent currents, although at present the system only works when model boundaries are strictly aligned to parallels and meridians. This link of the chain has a user-friendly graphic interface and generates a descriptive file of the entire nesting process.

- The second link in this computation chain is a software program which calculates an interpolated bathymetry for each nested model. The link also has a graphic interface developed in UNIRAS, which restores a depth for each calculation link in a file. The bathymetry used for the large-area model has been validated, and is essentially taken as fixed. However, it is updated on an occasional basis, as new information becomes available. The MARS-2D computation kernel is used in both case studies below, where prevailing tidal currents have a relatively homogenous vertical structure, providing a good approximation of the mean current fields pertaining in the study areas.

- Lastly, a range of graphic tools are used to display the resulting modelled outputs which are written in NETCDF format (Rew and Davis, 1990). NETCDF is a widely used self-documenting format, which also provided a suitable platform upon which the ArcView portal was subsequently developed.

10.4.3 Reference mapping data

Currently, coastal practitioners must refer to common baseline or reference data, i.e., primary data to which secondary (or more application-related) data will be subsequently linked (Allain, 2000). The coastline, bathymetric data, and major administrative boundaries are examples of such baseline data that could be readily provided to a wider public under optimal conditions of accuracy, updating, and scalability to suit various needs.

However, it is noted that the bathymetry of inter-tidal areas, and other near-shore zones, is often less easily available or poorly defined. Such paucity of data is usually associated with the high cost of acquisition and restricted accessibility. Bathymetric data for the two study sites under consideration was available from the French hydrographic service at a scale of 1:50, 000.

ArcView™ was the main GIS platform used for the studies, within which the Spatial Analyst™ extension facilitated a number of operations on raster images such as recoding, resampling, changing extent, computing statistics and the use of algebraic image combination functions. Existing ArcView functionality also facilitated interactions between raster and vector data layers drawing on the attributes of file features attaching to vector data sets. However a major hurdle remains in the efficient handling of large numbers of raster data layers that typically accrue as the output from multiple model runs.

10.4.4 An Integrated GIS/model interface

The primary rationale behind the creation of a GIS/model interface was to allow a) consultation and display of results contained in the output files produced by the MARS hydrodynamic model; and b) extraction of these results to import them to ArcView. The concept was initially tested through the development of ModelView (Loubersac *et al.*, 2000), a prototype interface the functionality of which was successfully demonstrated in the case of hydrobiological contamination in the Bay of Marennes-Oléron, France.

As a further development, in order to broaden the scope of application a platform-independent stand-alone interface, MODELCON, was designed for use in conjunction with a range of GIS packages. In order to ensure optimum compatibility with standard software packages and other GIS (e.g. Excel, MapInfo™ etc.), MODELCONV extracts NETCDF files to a standard ASCII format. The MODELCONV interface was developed in JAVA, with Microsoft's Visual Studio 6 environment and the JAVA library to access NETCDF files (NETCDF JAVA version 2), ensuring maximum portability in anticipation of future use on Unix systems or via the Web.

10.4.5 A Geographic conversion module

An additional processing module was developed in order to address specific geoprocessing requirements beyond those available in the standard version of ArcView. This module operates as an ArcView extension and was implemented in Avenue. It enables the user to perform a range of geodesic processing operations on point, multipoint, polyline and polygon data, as well as 2-D and 3-D related measurements (pointM, multipointM, polylineM, polygonM, pointZ, MultipointZ, PolylineZ, PolygonZ).

This module also allows data to be projected or unprojected, i.e. as geographic co-ordinates (latitude, longitude), or projected Cartesian co-ordinates including 3-D (X, Y, Z). Other operations that are supported include:

- switching between geodesic systems: WGS 84, NTF and Europe 50;
- switching ellipsoids: Clarke 1880, IAG GRS 80 and Hayford 1909;
- switching projections: Lambert (various), Transverse Mercator (UTM).

10.5 CASE STUDY ONE: SHELLFISH PRODUCTION IN THE GOLFE DU MORBIHAN, SOUTHERN BRITTANY

10.5.1 Water quality issues in the Golfe du Morbihan

The Golfe du Morbihan is located in southern Brittany, France. Enclosing an area of 125 km^2, with many islands and extensive intertidal flats, the basin connects to the open ocean via a 1km wide channel. Its perimeter is highly indented, and is traversed by numerous streams and rivers. The main anthropogenic pressures related to two main cities, Auray (11,000 inhabitants) and Vannes (50,000 inhabitants), are augmented by very significant seasonal tourist populations. Important natural shellfisheries (*Venus* spp.) are commercially exploited, and the gulf also supports a valuable oyster farming industry. Both rely on suitable water quality being maintained within narrow sanitary limits. In general, water and shellfish quality criteria are set by the EU Shellfish Hygiene Directive (Lees, 1995). However, at national level slight differences are found in the interpretation of shellfish hygiene *E. coli* guidelines, specifically in stating what proportion of samples must fall under the concentration thresholds and in the treatments required prior to human consumption (BASIC, 2001). In France, these values are defined by the modified decree n° 94-340 of 28 April 1994, as shown in Table 10.2 below. Shellfish farmers must take different measures for depuration with respect to these categories.

Table 10. 2 French shellfish hygiene categories

Categories	Level of contamination in faecal coliforms*			
	300	1,000	6,000	60,000
A	≥ 90%	≤ 10%	0%	
B	≥ 90%		≤ 10%	0%
C	≥ 90%			≤ 10%
D				> 10%

* per 100g shellfish

10.5.2 Microbiological modelling

Faecal coliforms (FC), of which *Escherichia coli* is a major component, are good indicators of bacteriological contamination levels in seawater and shellfish. EU health standards for recreational waters and shellfish harvesting zones are based on organism counts. These bacteria mainly come from rivers and streams which receive waste water from various sources including surface runoff and soil outwash, sewage treatment plant discharges (especially in heavy rainfall situations, when function is impaired by lower residence time) and unauthorised discharges. Other direct sources are storm water drain systems discharging directly into the sea and diffuse contamination in the vicinity of moored boats.

Coliform bacteria do not tolerate exposure to solar UV radiation, and thus have a limited life span in seawater. Their survival time will vary depending on their metabolic state and environmental conditions. Bacterial survival times are positively influenced by the following:

- lower winter temperatures which can slow down bacterial metabolism and extend survival based on slower rates of consumption of energy reserves;

- increased turbidity values, linked to levels of suspended solids. Turbidity has a dual effect, as a potential food source and as protection against solar UV (ultraviolet) radiation. Turbidity is usually higher in winter, due to greater discharge and sediment resuspension.

In general modelling applications, bacterial survival is represented by the term T90 (being time at which 90% of the bacteria will have disappeared), which assumes an exponential rate of decline in numbers. Values for T90 are normally established on an empirical basis, and those used in the current study were based on experience at local sites (Guillaud, 1997).

10.5.3 Simulation descriptions

In order to characterise coliform distributions under a range of environmental conditions, a series of MARS 2-D model simulations were undertaken in which the following parameters were varied:

- Tidal conditions. Simulations were made over a period of three weeks under realistic tidal conditions, i.e. of sufficient duration to capture both spring and neap cycles.

- Seasonal influence. This was investigated by varying discharge volumes, i.e. coliform flux.

- Weather conditions. Three sets of commonly prevailing conditions were selected, in accordance with wind statistics derived from data supplied by the French meteorological office. These were 1) zero wind (baseline condition), 2) westerly $8 m.s^{-1}$ and 3) north easterly $8 ms^{-1}$.

- T90s. A summer value of 10h, with 24h for winter was chosen in accordance with local measurements.
- The impact of major malfunctioning of water treatment plants in heavy rainfall periods. This was modelled by doubling the amounts of bacteria discharged over a 24-hour period, which is typical of what can occur during a summer storm. The objective was to investigate the impact of episodic events on coliform distribution, particularly with respect to existing zonation patterns.

The above scenarios were investigated through a combined series of seven simulations (two seasons under three different conditions, plus one exceptional situation type). In order to assess the validity of the simulations, the resulting coliform concentration distributions were then categorised according to EU shellfish farming criteria (Table 10.2). Model-derived zonation patterns were found to be in broad agreement with existing zonation plans based on both tissue sampling and water quality monitoring network samples.

10.5.4 Simulation results

Having established the overall validity of modelled results, priority was given to locating the zones likely to be subject to the highest contamination. These were found to be in upper reaches of the main rivers and in the bay of Vannes (Figure 10.1). Observations performed during model runs indicated that once steady state conditions are attained, the spring/neap tidal coefficient has little influence on subsequent coliform concentrations. The impact of a constant, moderate wind (8 m/s) on contamination plumes was imperceptible, both in summer and winter. In the summary of results, only the baseline "no wind" situation is referred to.

T90 was found to exert the greatest influence on coliform concentration, and when set to 24hours gave rise to the highest levels of contamination. This is reflected in the final model run for which a 24h T90 linked to winter discharge rates was chosen.

10.5.5 Comparison between the actual classification and a simulation

Figure 10.2 shows the results of zonation categories based on samples collected by the water quality monitoring network (Ifremer, 2003). Zonations are based around 30 such samples per site that are routinely collected over the course of each year at low water during each spring tidal cycle (theoretically least favourable situation). Dots indicate the locations of effluent discharge sources. Figure 10.2 clearly shows the main water body of the gulf officially ranked as category A, whilst the estuaries and their outlets (north and northeast) are in the B or D categories. It is notable that no estuarine areas have been zoned in the C category and have instead been allocated to category D as a precautionary measure in respect of the EU Shellfish Directive. This precautionary designation takes into account the obvious risk of contamination near urban and port areas, as well as the impact of reduced salinity on mariculture products in upper estuarine reaches.

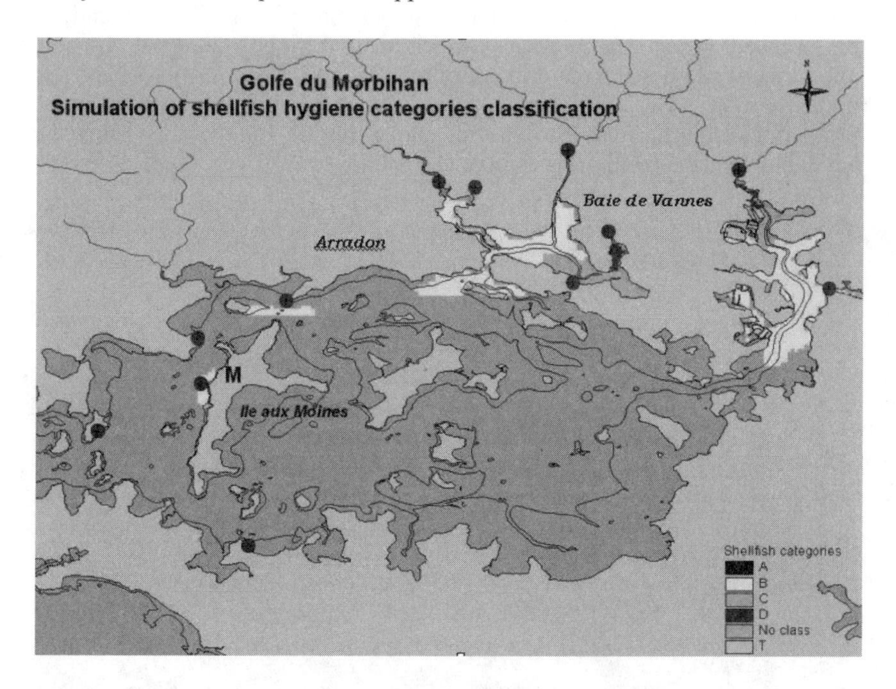

Figure 10.1 Screen capture from GIS showing the Golf du Morbihan. Coloured areas denote EU shellfish classifications based on simulated coliform concentration distributions

Comparing the results highlights the good consistency between simulation results and those obtained from monitoring network observations. Whilst the general configuration of zonation categories (Figure 10.1) based on simulations is broadly consistent with the official classification scheme (Figure 10.2), the former logically appears to give rise to a less conservative regime under which the estuarine areas in the north and northeastern gulf are designated as category C, rather than the precautionary official D designation.

This may be explained by the relative shortness of the period simulated (3 weeks), whereas the official designations are based on an annual monitoring cycle. Other reasons may be that the coliform flows used for the simulation did not include the annual maxima (in the case of water collection and treatment plant malfunctions). Furthermore the actual T90 may exceed the 24h value used in simulations, especially in naturally turbid upper estuarine areas, or as a consequence of winter storm induced sediment resuspension.

By comparing the Figures where the discharge points are identified by a dot symbol, we also note that the model, which minimised coliform concentration levels as seen above, reveals two B category zones in the outer Bay area. These can be seen in Figure 10.1, located to the north and to the west of the Ile au Moines (the island is indicated by an M on the map). Their coincidence with effluent discharge points (dots) suggests the logical cause of reduced water quality in these areas. This result highlights the valuable insights that can be obtained through the use of realistic simulations based on numerical models in identification of localised water quality issues, which may then be addressed through the allocation of additional monitoring resources. In this case, priority was given to the northern most area (associated with effluent discharge Arradon) owing to its proximity to significant shellfish farming areas, resulting in the establishment of an additional hygiene monitoring station.

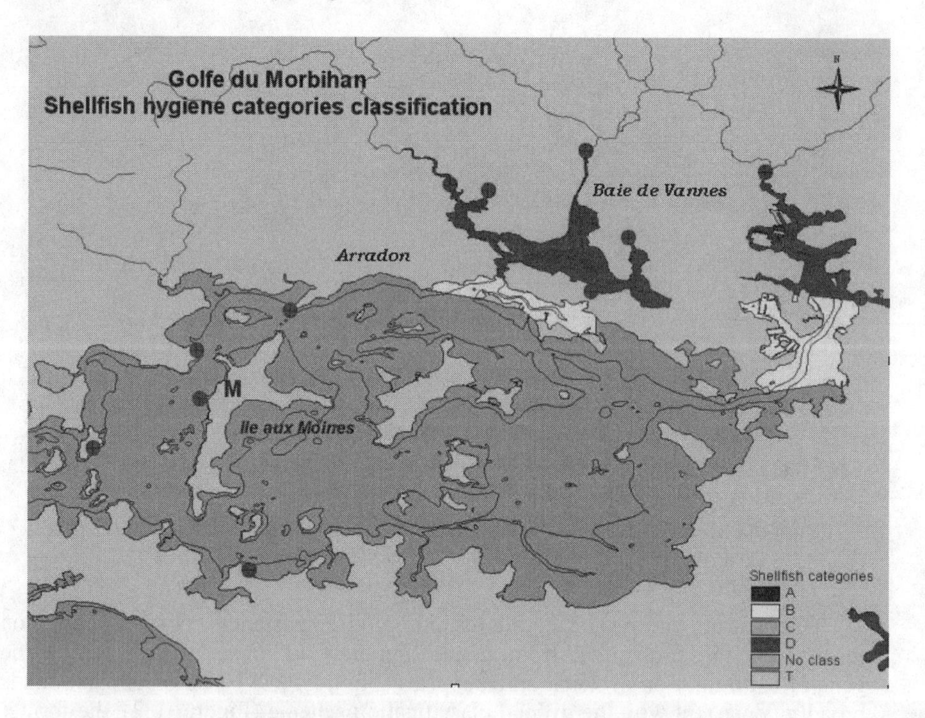

Figure 10.2 Screen capture from GIS showing the Golf du Morbihan. Coloured areas denote EU shellfish classifications based on measured coliform counts.

10.6 CASE STUDY TWO: THE WRECK OF THE *IEVOLI SUN*

10.6.1 Context of the case study

The Italian chemical tanker, *Ievoli Sun,* sailing from Rotterdam to Genoa sank around 9 am local time on 31th October 2000 in the central English Channel, approximately 9 nautical miles north of Les Casquets (Channel Islands) and 20 nautical miles west-north-west of the French Cap de la Hague off the north coast of Normandy (Figures 10.3 and 10.4).

Figures 10. 3 and 10.4 *Ievoli Sun* being towed a few hours before she sank (French Navy sources).

The vessel was carrying three chemical products: 3998 tonnes of styrene (Vinylbenzene), 1027 tonnes of Methyl ethyl ketone (MEK or 2-Butanone) and 996 tonnes of Isopropyl alcohol (Propanol-2). Although the latter two products are considered to be practically non-toxic for aquatic life, styrene is known to be hazardous for the marine environment and for humans. According to the Group of Experts on the Scientific Aspects of Marine Pollution of the International Maritime Organisation (GESAMP, 1989) styrene is bioaccumulable with a propensity to affect edible marine resources. Moreover, it is dangerous for human health when in contact with skin or when inhaled. It can cause major environmental problems due to its persistent irritant toxic odour, which can force the closure of beaches.

French Naval investigations in the aftermath of the sinking showed that the wreck was lying on her port side, and subsequent remotely operated vehicle (ROV) surveys confirmed that the hold was ruptured with a consequent release of styrene into the marine environment.

The wreck was located close to the Channel Islands, in an area of high biodiversity, and in close proximity to shipping lanes, cross-channel traffic, fisheries, aquaculture, and beaches. Neighbouring coastlines are rugged and highly indented, with many sensitive and vulnerable sites, and are under the jurisdiction of several different regional authorities, including those of the Channel Islands.

A rapid response team including experts in pollutant chemistry, ecotoxicology, modelling, fishery science, aquaculture and digital mapping was assembled within two hours of the sinking. The prime objectives of this group were to establish a baseline for environmental contamination, and devise a plan for coastal environmental monitoring in the subsequent hours. This meant being able to characterise the fate of pollutants in the water mass and identify sensitive areas and fish species likely to be affected. Thus the main requirements were (Loubersac *et al.*, 2002):

- to rapidly compile various forms of multi-thematic information from multiple sources;

- to provide simulations of pollutant dispersion at sea for several scenarios in the shortest possible time;

- to standardise these compilations and simulations and make them easy to display, communicate and update for simultaneous use across several decision centres.

10.6.2 Simulation of pollutant behaviour using MARS-2D

The MARS modelling suite (Salomon *et al.* 1995; Garreau 1997) was used to rapidly derive a 500m horizontal mesh hydrodynamic model centred on the wreck position. A 2-D horizontal model was adopted to simulate the local hydrographic regime, where dissolved substances were likely to become well mixed owing to the strength of prevailing tidal currents which are relatively consistent throughout the vertical extent of the water column. Actual wind fields supplied by the French meteorological office were also assimilated into the model.

Simulations were run to reproduce the evolution of instantaneous and residual tidal current fields, tidal trajectories and concentrations of the dissolved fraction of the spilled products. The following three scenarios were considered assuming maximal dissolution of the styrene in seawater: a) a single total discharge of 4,000 tonnes, b) simulation of 4,000 tonnes of the product leaking from the wreck over a period of 3.5 days, and c) simulation of continuous discharge of 20 $l.min^{-1}$ over 25 days (closely matching observed leakage rates).

10.6.3 Setting up a customised GIS

A GIS was established for the wreck site and surrounding maritime region, in which functionality was tailored towards handling and visually representing marine and coastal data. The main steps in this process were:

- Building a reference geographic data set (Allain *et al.*, 2000) in order to ensure stakeholder interoperability;
- Conversion of all data into decimal degree geographic coordinates, and establishing a common vertical reference frame for marine and land-based topographies;
- Adoption of a uniform map template for all data layers optimised for a standard working scale of 1:50,000.

10.6.4 Results

The *Ievoli Sun* sank in an area in which strong tidal currents and an irregular seabed give rise to intense vertical mixing. Under these conditions it is reasonable to assume that styrene escaping from the wreck became rapidly dispersed throughout the water column. A 3-D perspective view showing the spatial extent of dispersed styrene as it appeared after 10 days of simulated advection is illustrated in Figure 10.5. This simulation was based on scenario c) with the following assumptions: *i)* the matter coming from the wreck dissolved completely, *ii)* there was no chemical modification over time, and *iii)* an average current was responsible for the transport of these dissolved substances.

The image presented in Figure 10.5 was produced by merging modelled data with those from the GIS mentioned above, using ArcView 3D Analyst. This image corresponds to the situation on the 9th November at 2 am (10 days after the spill), taking into account the atmospheric forcing data provided by the French meteorological service. The concentration limits indicated on this Figure are 0.30 ppm (parts per million or mg/l) (darkest), 0.15 ppm (lightest), and 0.05 ppm (medium-grey) in the raised area.

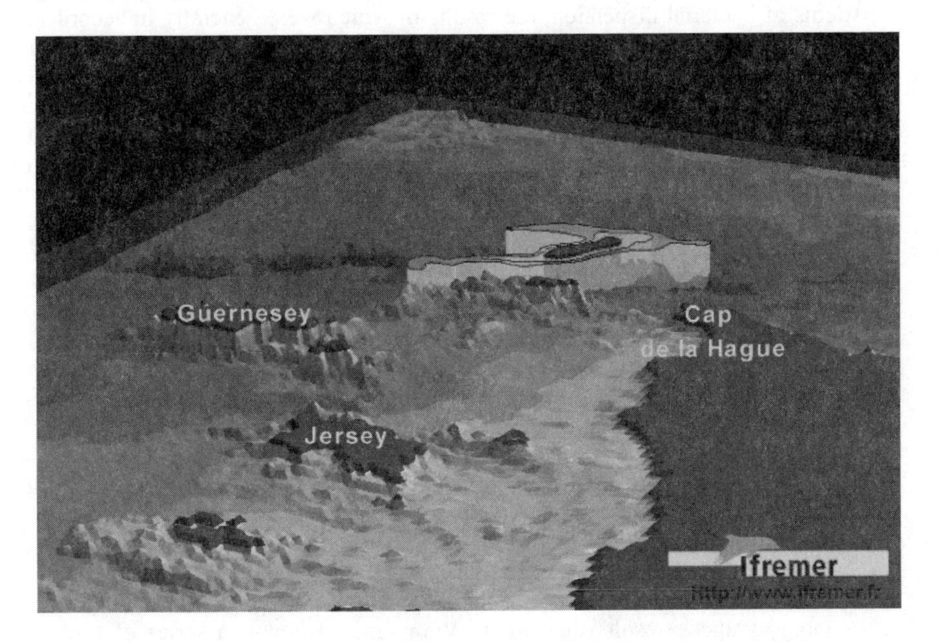

Figure 10.5 3D perspective representation of the plume of dissolved pollutant dispersion after 10 days.

Dispersion of a continuous discharge of 20 $l.min^{-1}$ was also simulated for the period between 30th October and 24th November, incorporating actual wind measurements up to the 19th November and forecasts for the remainder of the simulation period. Plate 10.1 shows the situation at the end of the simulation (24/11/00 at 0600) with contaminant concentration expressed in parts per million (ppm). While the bulk of the dilution patch is located along an axis running East North East – West South West across the wreck site, a distinct branch can be seen extending southward towards the island of Alderney. However, seeing the very low flow simulated, the concentrations of dissolved product in the water column remain extremely low, since the highest concentrations reached in the simulation did not exceed 0.056ppm. However it should be noted that the overall concentrations remained below 0.056ppm throughout the water column; such low values are however consistent with expectations given the low rate of discharge.

10.7 CONCLUSION AND PERSPECTIVES

An operational model of the Golfe du Morbihan was constructed, which successfully reproduced the hydrodynamic regime of the study area to within acceptable tolerances. The model has also been used to effectively simulate patterns of bacterial dispersion, the results of which were generally in accordance with officially established shellfish zonation patterns, based on the sampling regimes of local water quality monitoring networks. Time-series of computed concentrations for specific locations can be extracted from existing model runs, and can be selected to correspond with monitoring sites. The approach thus provides a means of optimising sampling programmes to make best use of available resources, and to improve responsiveness to episodic events with potential consequences for health.

The present arrangement and state of development of these modules allow a user to work effectively with a given number of grids generated as output from a limited number of modelled scenarios. Future developments aimed at automation would reduce the time needed to process and import results, while maintaining the traceability of actions run and all the metadata required to understand and analyse the simulation results. To this end, future efforts will be concerned with improvements in the following areas:

- Ensuring that model-run relevant metadata which is integral in NETCDF files is fully maintained throughout the processing chain;
- Transforming the tables that MODELCONV generates into Arcview grids;
- Implementing batch processing tools, both on the MODELCONV level and to select user-defined time steps, and on operations between variables to generate more synthetic results than in ArcView, and to import a series of files and automatic captioning for them.

The effectiveness of the *Ievoli Sun* pollution response can be attributed to the favourable platform upon which it was based. Pre-existing tools could be activated within 48 hours, facilitating rapid delivery of environmental maps incorporating hourly forecasts of styrene slick position and concentration. Prudent diagnoses of the potential effects on marine life could then be made in the light of various leakage scenarios. Integration was made possible by providing consistent geo-coded information on the land-sea interface, through a GIS interface closely connecting these data and simulations issued from digital models. Integration of consistent geo-coded information on the land sea interface with model simulated outputs within a GIS environment has resulted in the creation of an active interface for pollution response decision makers that incorporates up-to-the-hour scientific assessments.

A further logical extension of this work would be to jointly use hydrodynamic, sedimentological and biological modelling combined with GIS. Links with catchment models to allow regional-scale assessments of human activities (e.g. agriculture) and natural events (e.g. flooding) should also be considered. Planned future work will incorporate sedimentological and biological process modules into the existing hydrodynamic modelling/GIS schema, thereby increasing the diversity of assessments that can be made. Critical coastal management issues including bacterial and viral contamination of bathing waters, occurrence and impact of harmful algal blooms, eutrophication, and definition of coastal water masses (e.g. in relation to the EU Water Framework Directive) could then be addressed.

It is anticipated that schema will ultimately be extended in order to incorporate links with catchment models, thereby allowing regional scale assessments of human impacts (e.g. agriculture) and natural events (e.g. flooding).

10.8 REFERENCES

Allain S., *et al.,* 2000, Données géographiques de référence en domaine littoral. In *"Geomatics and Coastal Environment"*, edited by Populus, J. and Loubersac, L. (Editions Ifremer/Shom), pp. 67-79.

Boelens R.G.V., Walsh A.R., Parsons A.P., and Maloney D.M., 1999, Ireland's Marine and Coastal Areas and Adjacent Seas: An Environmental Assessment. Quality status report. Prepared by the Marine Institute on behalf of the Departments of Environment and Local Government and Marine and Natural Resources, Dublin.

BASIC, 2001, A Scoping Study to Establish a Common Approach to Examining the Impact of Atlantic Arc Water Quality and Dynamics on Coastal Activity and Sensitive Marine Environments, Atlantic area INTERREG-IIC programme. Final report, (Cork, Ireland: CRC).

Durand H., *et al.*, 1994, An example of GIS potentiality for coastal zone management: Pre-selection of submerged oyster culture areas near Marennes Oléron (France). *EARSEL Workshop on Remote Sensing and GIS for Coastal Zone Management*. Delft, The Netherlands, 24 - 26 Oct.

Garreau P., 1997, Caractéristiques hydrodynamiques de la Manche. *Oceanis,* **23**(1) pp. 65-97.

GESAMP [Joint Group of Experts on the Scientific Aspects of Marine Environmental Protection], 1989, The evaluation of the hazards of harmful substances carried by ships: (revision of GESAMP Reports and Studies n° 17). *GESAMP Reports and Studies*, 35: 44p., [11] annexes.

Guillaud J.F., Derrien A., Gourmelon M., and Pommepuy M., 1997, T90 As a tool for engineers: Interest and limits. *Wat. Sci. Tech.*, **35**(11-12), pp. 277-281.

Ifremer, 2003, http://www.ifremer.fr/envlit/surveillance/remi.htm (on-line. Accessed October 2003).

Kershaw, P.J., 1997, Radioactive contamination of the Solway and Cumbria coastal zone: The Solway Firth, *ECSA/JNCC*, pp 43-50.

Lazure P. and Jegou A.M., 1998, 3D modelling of seasonal evolution of Loire and Gironde plumes on Biscay Bay continental shelf. In 5ème Colloque International d'Océanographie du Golfe de Gascogne. *Oceanologica Acta*, **21**(2).

Lees, D.N., Nicholson, M., and Tree, J.A., 1995, The relationship between levels of *E. coli* in shellfish and in seawater with reference to legislative standards.

Lees, D.N., 2000, Viruses and bivalve shellfish. *Int. J. Food Microbiol.* **59**, pp. 81-116.

Loubersac L., Salomon J.C., Breton M., Durand C., and Gaudineau G., 2000, Perspectives offertes par la communication entre un modèle hydrodynamique et un SIG pour l'aide au diagnostic environnemental; caractérisation de la dynamique et la qualité des masses d'eaux . In *"Geomatics and Coastal Environment"*, edited by Populus, J. and Loubersac, L. (Éditions Ifremer/Shom), pp 173-185.

Loubersac *et al.*, 2002, Communication de l'information géographique maritime et côtière pour la gestion d'une crise environnementale. *Revue internationale de Géomatique*. **12**(3), pp. 355-371.

Populus J. and Loubersac L., 2000, (eds), CoastGIS'99: Geomatics and coastal environment. (Brest: Éditions de l'Ifremer).

Prélaz Droux R., 1995, *Système d'Information et Gestion du Territoire*. (Lausanne: Presses Polytechniques et Universitaires Romandes. Coll. META)

Garreau P., 1997, Caractéristiques hydrodynamiques de la Manche. *Oceanis,* **23**(1) pp. 65-97.

Rew, R. K. and G. P. Davis, 1990, NetCDF: An interface for scientific data access. *Computer Graphics and Applications, IEEE*, pp. 76-82, July 1990.

Salomon J.C., Breton M., and Gueguenìat P., 1995, A 2D long term advection-dispersion model for the Channel and Southern North Sea. Special Issue MAST 52, part B and C. *Journal of Marine Systems,* **6**(5-6), pp. 495-528.

Sullivan, N., 2001, The relationship between the disposal site and the SAC in Falmouth bay, south-west England. *Unpublished Dissertation, Msc in Estuarine and Coastal Zone Management*.

Valuing our environment, 1999, *National Trust (UK)*.

Wright D. and Bartlett D., 2000, *Marine and Coastal Geographical Information Systems*. (London: Taylor and Francis).

CHAPTER ELEVEN

Towards an Institutional GIS for the Iroise Sea (France)

Françoise Gourmelon and Iwan Le Berre

11.1 INTRODUCTION

As far as coastal zones are concerned, many challenges faced by the scientific community and policy makers, planners and managers justify adoption of a transdisciplinary approach based on bio-chemical, geophysical and socio-economical factors (Burbridge and Humphrey, 1999). GIS are well-established computer-based systems for storing, retrieving, analysing, modelling and visualising the vast amounts of spatial data that may be collected by several providers (Fabbri, 1998). Nevertheless, the implementation of a coastal database is a complex process which requires institutional support to guarantee the multidisciplinary approach, the sustainability of the project in terms of raising funds and human resources, and to promote relationships with other institutions working on the same area (De Sède and Thiérault, 1996).

The major pollution caused by the *Erika* disaster, coupled with the catastrophic storms which reached the Atlantic coastal zone at the end of 1999, has led the French authorities to propose creation of a national coastal GIS. The implementation of such a project is complex, witnessed by the lack of any decision about which reference geographical data for the coastal zone to adopt (Allain *et al.*, 2000). In fact, coastal data are scattered among many organisations. They are produced for lots of purposes and therefore are available at various scales, typologies and formats. In spite of these difficult conditions, there are a number of smaller, independent GIS projects at work on the French coastal zone (ENR/OELM, 2000; Guillaumont and Durand, 2000).

This contribution describes the GIS implemented by Géosystèmes laboratory (CNRS, European Institute for Marine Studies) on the coastal zone of Finistère (western Brittany, see Figure 11.1) during the last ten years. In the beginning, the GIS was conceived as a support tool for monitoring and managing the Biosphere Reserve of Iroise, but today, it also serves as a powerful tool for carrying on integrated research on the coastal zone of the Iroise Sea, especially within the framework of the European Institute for Marine Studies (Institut Universitaire Européen de la Mer, IUEM). This latter institute brings together a number of

0-41531-972-2/04/$0.00+$1.50

separate marine research teams from the University of Western Brittany, Brest, dealing respectively with oceanography, geology, biology, chemistry, geography, economy and law. The development from a GIS dedicated to scientific applications towards an institutional support is described below.

Figure 11.1 Study area

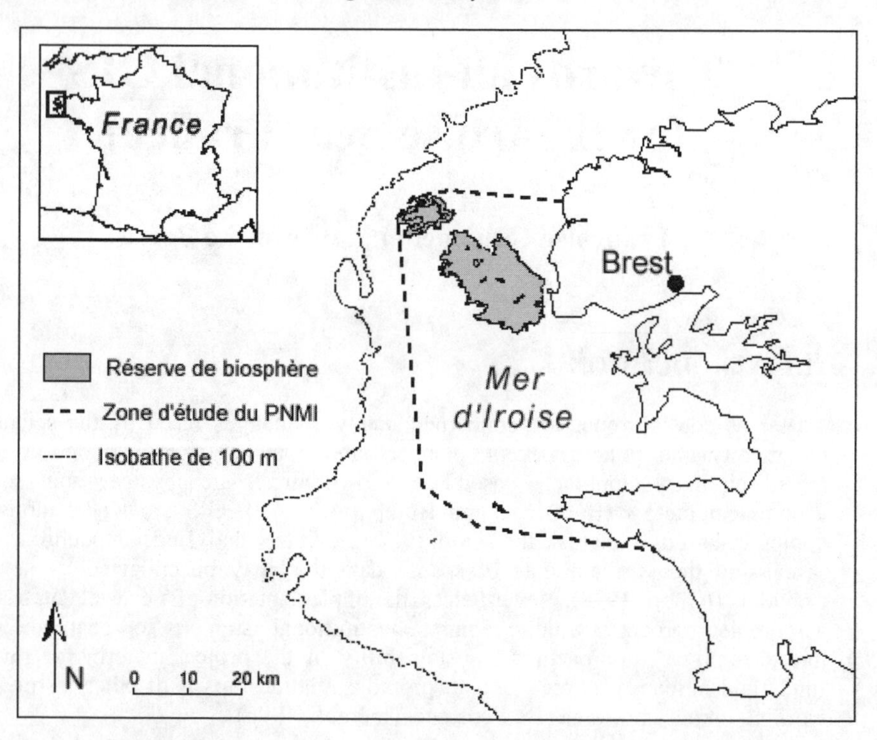

11.2 THE DATABASES

Nowadays the GIS supports two complementary databases, called *SIGouessant* and *BIGIroise*, that were originally developed with separate objectives and data (see Figure 11.2).

The *SIGouessant* database deals with the Mer d'Iroise Biosphere Reserve at a local scale. The biosphere reserve label is allocated to representative terrestrial and coastal areas by the UNESCO Man and Biosphere (MAB) programme. The conception of such an area is based on three main functions: conservation of biodiversity and landscapes; sustainable use in regional units; and logistic support for research, monitoring, education, information and involvement of the local population. In this context, many thematic studies have been performed on the Mer d'Iroise Biosphere Reserve to increase knowledge of dynamic terrestrial and marine processes, and to study the aftermath of human activity such as tourism, management and fishing on biodiversity. The objectives are to provide a basis for land management recommendations as well as for wildlife management programmes. The *SIGouessant* database has been developed since 1990 as a support to the long term ecosystemic approach, and to provide data for scientific and management investigations (Gourmelon *et al.*, 1995). The geographic reference data is provided by orthophotographs produced by the French National Geographical Institute (IGN - Institut Géographique National). The thematic layers describe physical, natural and social parameters with the same classification systems and scales, collected at various dates thanks to continued monitoring.

Figure 11.2 Database organisation

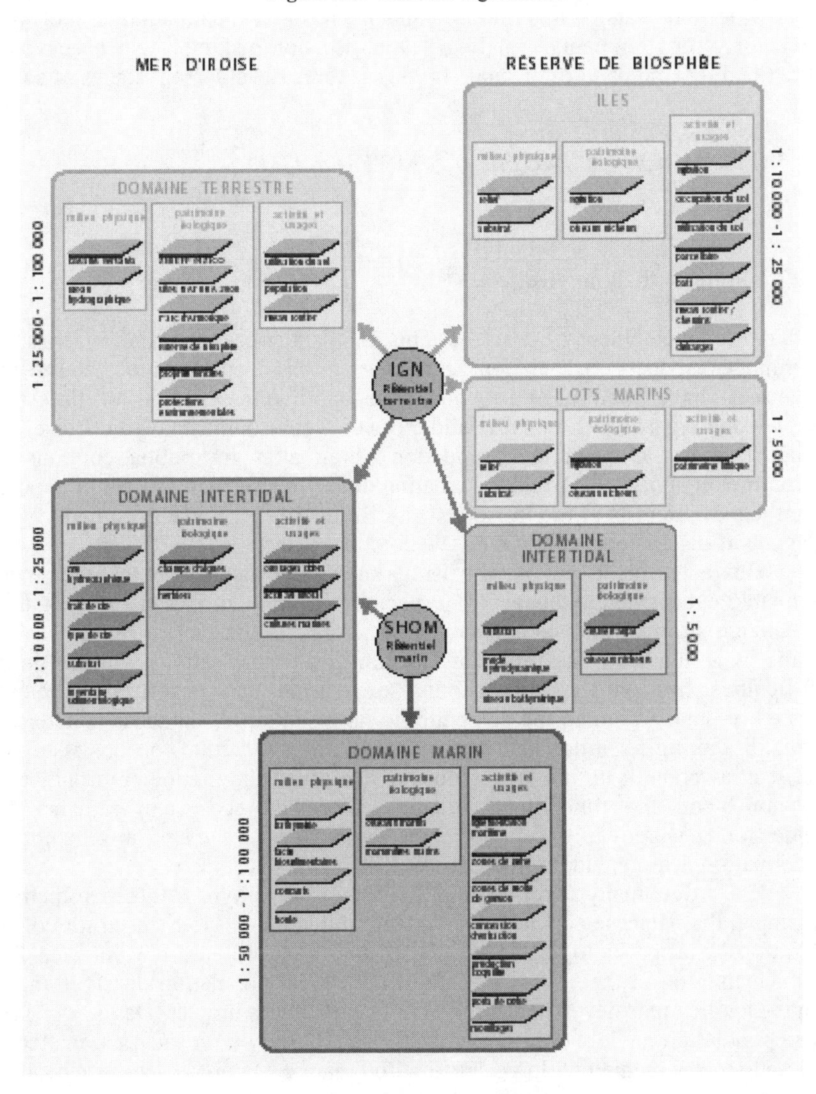

The *BIGIroise* database is an extension of *SIGouessant* to the whole coastal area of the Iroise Sea. Its initial aims were to produce a synthetic environmental mapping through combination of multiple data sets available for the coastal zone (Le Berre *et al.*, 2000). The database uses marine geographical information produced by the French hydrographic service (SHOM – Service Hydrographique et Océanographique de la Marine), and is built up according to the environmental planning methods developed and proposed by the Unesco MAB committee (Journaux, 1985). After the collection of existing digital data, and the digitisation of other data sources (atlases, maps, etc.), 34 thematic layers, concerning physical, biological and socio-economical parameters for marine and terrestrial areas of the Iroise Sea have been integrated in the GIS (see Figure 11.2).

After ten years of functioning, this GIS has become an adopted tool for data inventory, environmental analysis and decision-making, which have been described by Crain and Mc Donald (1984) as three development stages of a GIS.

11.3 OVERVIEW OF IROISE GIS APPLICATIONS

11.3.1 State of the knowledge

After the development of a set of tools for data processing and analysis, using techniques such as association, combination and generalisation, the resulting databases have been used to compile a baseline assessment of the state of knowledge regarding the marine and terrestrial environments of the Iroise coastal area. This has allowed the production of an atlas for public communication (Gourmelon *et al.*, 1995), and the creation of a synthetic map showing the potential conflicts of interests in the Iroise Sea (Le Berre, 1997). These documents are used to support discussions among the stakeholders.

However, this inventory of existing knowledge also shows that the quality of available data is heterogeneous — especially in terms of exhaustiveness and age — though strong efforts to collect and structure data on the Iroise Sea have been made by a number of scientific and institutional organisations. One of the main difficulties lies on the compatibility of marine and terrestrial geographical reference data. Although the compatibility is essential for database coherence, we are still waiting for a national consensus (Alain *et al.*, 2000) on this issue. Other difficulties come from the lack of knowledge about the marine environment. The structure and the functioning of the Iroise Sea ecosystem remains barely understood. Moreover, except for some legal aspects, scarcely any geographical information dealing with human activities is available.

GIS offers many functions that may be used to bypass these problems: For example, the database can help define sampling strategies to be used for collecting additional marine biological data; and it may also provide relevant data for the implementation of theoretical models of ecological population distribution, based on the applications developed for the US Gap Analysis project (Davis *et al.*, 1990). The population models used for Iroise Sea are based on physical parameters such as bathymetry, sedimentology, and hydrodynamic features (marine currents and waves) and have been tested successfully on seaweeds and on the bottlenose dolphin (*Tursiops truncatus*, see the next section, below). Furthermore, integration with remote sensing allows the production of synoptic and multi-temporal data that are particularly useful for marine studies, because of the variety of the sensors used for recording parameters such as water colour, sea surface temperature, shore morphology, etc. (Van Zuidam *et al.*, 1998).

11.3.2 Environmental issues

GIS is now widely used for applications relating to terrestrial studies and management issues. In landscape ecology especially, the capabilities of GIS are successfully used to perform spatial and statistical analysis of many environmental components, and for modelling real-world processes (Haines-Young *et al.*, 1993). In the islands and the islets of Iroise Sea, many similar GIS applications are concerned with exploring relationships between vegetation and land-cover changes, fauna (rabbits and nesting birds) and human activities. For example, land-cover and land-use changes of Ushant island over a 150-year timeframe have been studied, using field studies, aerial photographs and the 19th Century land registry (Gourmelon *et al.*, 2001) as primary sources of data. Over this period, the island underwent a drastic transformation from rural landscape to extensive shrubland. Only traditional extensive sheep breeding is actually maintained. Within a GIS analysis, the relationships between sheep grazing and land-cover have been established, and scenarios of land-cover potential related to changes in the intensity of sheep grazing produced. The scientific results have provided an objective framework for further assessment of fallow land management.

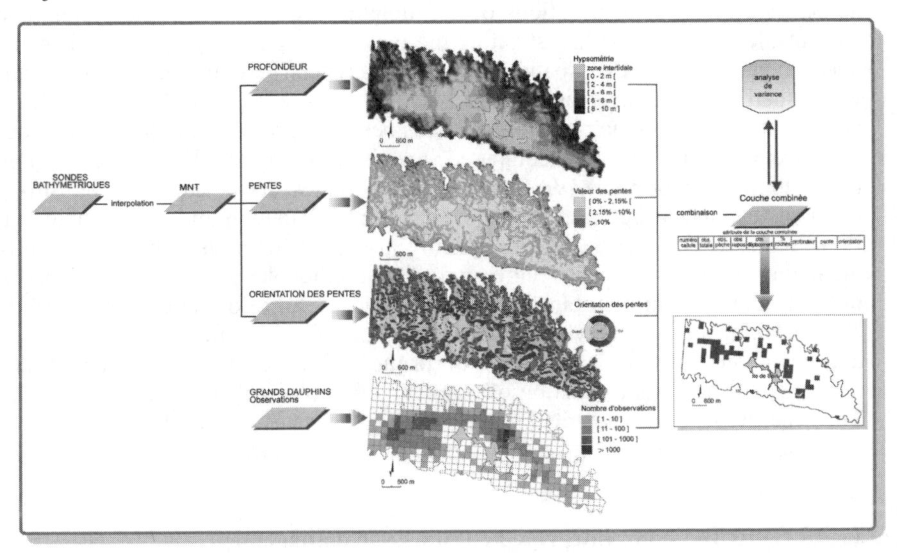

Figure 11.3 Investigating factors influencing spatio-temporal distribution of bottlenose dolphin (*Tursiops truncatus*)

In contrast to these successes on-shore, the implementation of GIS is more difficult in the marine environment due to the peculiarities of maritime space-time scales, which tend to be generally wider, more complex and less well-known than those of terrestrial ecosystems. A regional approach to marine environments is often required, and data collected at several spatial and temporal scales need to be integrated, for example data relating to the dynamic component of biological and hydrological parameters that may occur daily (like tides), seasonally or annually (like lunar and biological cycles), or that may only be noticeable over several years and decades, such as long term changes (Holligan, 1994). This complexity implies a need to link the GIS to models for the production of dynamic data, i.e. hydrodynamic modelling, and for the analysis of the functioning of the whole system and its subsystems (Capobianco, 1999).

As has already been mentioned above, within the Iroise Sea one area of data analysis aims to investigate the environmental factors that may influence the spatio-temporal distribution of the bottlenose dolphins (see Figure 11.3). In France, the increasing interest in natural heritage has encouraged development of resource management efforts that maintain a high degree of biodiversity. In this context, a Marine National Park project has been proposed for the Iroise Sea, which would include both the areas of coastal bottlenose dolphins. A spatial and temporal approach to the dolphin habitats is required to define a conservation strategy based on habitat features, more than on spatial distribution of groups. Such an approach has been developed in the United States through the Gap Analysis Program conducted by the U.S. Geological Survey (Davis *et al.*, 1990), and is being adapted for the Iroise Sea. Within this wider framework, a more localised study is focusing on a particular group of dolphins observed around Sein Island, and which has been monitored since 1992 (Gourmelon *et al.*, 2000). The annual and seasonal distributions of this dolphin group reveal some preferential sites inside its vital area. Investigating the factors leading to this perceived spatial determinism requires that spatio-temporal parameters be considered, and is thus well-suited to GIS analysis. Meanwhile, at a more regional spatial scale, a classification of bathymetric parameters has been performed to produce a sea-floor stratification that may be compared to observations of bottlenose dolphin distributions (Le Berre *et al.*, 2002). At both scales, local and regional, strong relationships appear between bottlenose dolphin distribution and topographic sea-floor features. The GIS is subsequently used for dolphin habitat suitability modelling, based on logistic regression. This methodology provides a means of mapping the dolphin distribution, and providing useful information for the future National Marine Park. Nevertheless, distribution of prey with depth of slope and hydrodynamic activity is probably of prime influence. The next step of this research will be the introduction of physical and biological modelling in the GIS. The synergy between numerical models and GIS is necessary for the production of dynamic data useful in such an investigation.

11.3.3 Contribution to integrated coastal zone management

A large part of the Iroise coastal zone belongs to protected areas of various types (natural reserves, hunting reserves, biosphere reserve, proposed national marine park, etc.) for which there is a strong need for scientific long-term monitoring, and also of concrete strategies and policies for management. In such contexts, maps are essential, both for decision-making by field staff, and also for information communication aimed at the stakeholders involved (decision-makers, managers, scientists, etc.). The integration of data from various sources within a consistent framework, as provided by a GIS, improve their accessibility and their availability. GIS also provides efficient functions for rapid production of thematic or synthetic maps and other visualisations, designed for different purpose. Above all, GIS shows great potential for the design and the production of good quality maps tailored to specific uses, such as decision-support during a specific critical event such as an oil-spill, particularly if the mapping process is automated. For instance, in the environmental monitoring programmes implemented for the Iroise sea, the GIS is used for the preparation of sampling strategies, for the detection of spatial changes, for the assessment of future evolution, and for impact simulations.

The assessment and simulation of impacts or changes that affect coastal areas are generally made within the scope of management strategies and policies (Capobianco, 1999). Spatio-temporal modelling, coupled with geographical databases may be very efficient for the production of realistic syntheses of an ecosystem's evolution after different disturbances. In such an integrated and multidisciplinary context, the design of operational tools for decision support depends on the upholding of links between numerical geographical information and other databases, on the integration of spatio-temporal modelling and on the production of user-friendly interfaces that facilitate access to information.

In order to test the ability of *BIGiroise* as an operational tool, the geographical information available for the tidal zone of a limited part of the Iroise Sea was used as a test bed for the development of a model dedicated to oil spill contingency (Le Berre, 1999). It takes into account the relevant data for assessing the sensitivity of coastal areas in order to prepare and to plan the strategies of intervention and to provide useful information for the monitoring of damaged sites (see Figure 11.4). In a practical way, shore sensitivity is assessed through morpho-sedimentary, ecological and socio-economical indices (Michel and Dahlin, 1997) but, for the production of a relevant information for operational needs, vulnerable areas exposed to the pollution risk must be identified. The assessment of risk requires that meteo-oceanic and physico-chemical features of the pollutant be taken into account. This can only be achieved within a spatio-temporal model. In this operational context, the GIS may provide information, especially maps, useful for helping decision-making, but this is only a part of a more comprehensive facility that should be implemented. An efficient system for such purposes requires significant extensions such as oil spill drift models, links with thematic databases (i.e. about pollutant composition) and the production of users interfaces (Howlett *et al.*, 1997).

11.4 PLAN FOR AN INSTITUTIONAL COASTAL ZONE INFORMATION SYSTEM

The research programme described herein demonstrates the contributions and the limits of the GIS implemented for the Iroise Sea. The major issues are now its timelessness and the improvement of its ability to produce a better knowledge of environmental and socio-economical processes in the aim to provide useful information for decision-making. It implies in a next step an integration into a more efficient framework (see Figure 11.4). An integrated environmental coastal zone management system, for assessing coastal zone changes and functioning, should integrate remote sensing, modelling, GIS and GIS-based decision support systems (Van Zuidam *et al.*, 1998).

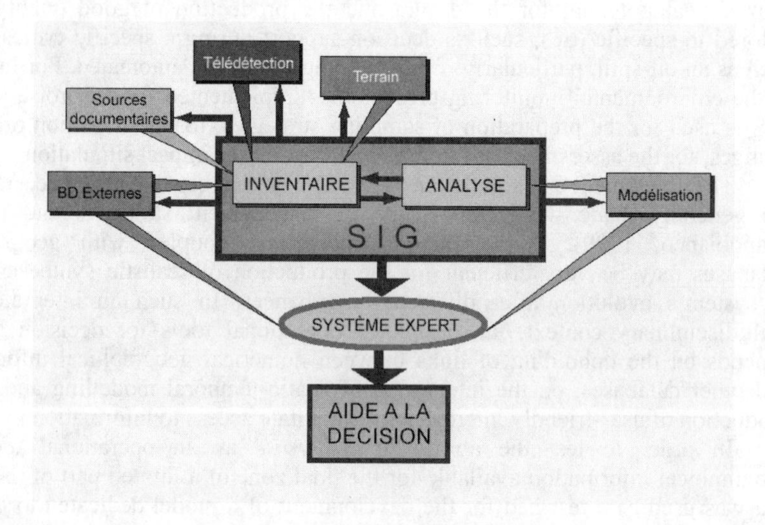

Figure 11.4 Development of an operational GIS

11.4.1 Context and objectives

The GIS developed for the Iroise Sea is an essential component of the infrastructure established to support environmental research programmes conducted by the "Géosystèmes" laboratory. The GIS is used for scientific analysis, to improve the protocols for field data collection, and in some cases to provide decision-making information such as maps and statistics. Due to its environmental characters, the Iroise Sea is the core area of most of these research programmes, sometimes spanning several decades. A proposal has been submitted by the component research units within the laboratory, to the national authorities, to establish a Coastal Domain Observatory with the aim of recording and monitoring long-term changes in a western European coastal area strongly influenced by both climate fluctuations and anthropogenic impacts. The proposal is currently undergoing evaluation for possible funding. In this context, a Coastal Environment Information System (Système d'Information des Environnements Côtière, SIEC) connected with an intensive computing centre is planned, to reinforce the synergy of the Institute, especially by providing facilities for gathering remote sensing data, image processing, GIS and dynamic modelling.

This common service will have five main functions:

- **Collection of data and integration of existing datasets** produced by the IUEM, concerning physical, biological and anthropogenic parameters, and creation of a consistent GIS database, based on the geographical reference information approved by the French National Committee for Geographical Information (Conseil National pour l'Information Géographique, CNIG, see http://www.cnig.fr);
- **Develop and maintain global data exchange conventions** in order to make geographical reference information produced by organisations such as IGN, SHOM, Ifremer or Météo-France available to the IUEM research teams;
- **Information production** by the digitisation of maps, or by data processing of numerical data (interpolation, image analysis, etc.);
- **Metadata production** for supplying a catalogue to assist internal and external data exchange and diffusion between data producers (laboratories, institutions) and users (laboratories, marine professionals, policy and decision-makers);
- **Education and training** of the IUEM staff and students in Geographical Information Sciences.

The feasibility of this joint project has been tested from an inventory of the whole data produced in the IUEM research labs.

11.4.2 The first stage: data dictionary and metadata production

Within the IUEM, the needs for data cataloguing are both external and internal. For internal uses, the maintenance of an up-to-date description of datasets is a guarantee of its long-term viability, and may improve collaboration and exchanges between research teams as it avoid redundancies in research projects. As a research institute, IUEM is an important data producer and user, and has a strategic need for being identified as a data provider by different kinds of external users, especially in a context of the growth of data and information communication across the World Wide Web. The Institute's data broadcasting policy within the SIEC implies a strong compliancy with existing metadata standards, in order to guarantee their robustness and encourage their exchange (Bartlett *et al.*, 2000).

At present, two major metadata standards are used by the international community of geographic information providers: the European standard (CEN/TC 287) and the International standard (ISO/TC211). Correspondence between these two is not perfect, especially in terms of data ownership rights, but the general hierarchical structure of the data descriptions allows the conversion from one standard to the other. As the contribution and established collaborations of the IUEM to international research programs takes place mainly at the European level, the European standard has been adopted for SIEC. The metadata are integrated and managed with Report V2.0, a CEN/TC287-compliant metadata software system developed by the French Ministry of Equipment. At the laboratory level, although the geodatabase was developed initially with ArcInfo software (where ArcCatalog is compliant with the ISO standard), Report is also used for producing and managing the metadata catalogue until such time as ArcCatalog can take into account the European metadata standard.

In fact, the conversion of standardised metadata to specific formats stays unavoidable when supplying information to Web geodatabase repertories or clearinghouses such as Bosco (http://www.bosco.tm.fr, for French coastal data) or EDMED (http://www.sea-search.net/edmed/welcome.html for data produced during MAST research programs) for example. The need to be able to undertake this kind of specific task will be taken in account when developing the SIEC structure.

11.5 CONCLUSION

The experience acquired after ten years of coastal GIS development shows that even if the databases are useful in many scientific and management applications, their prospects depend on a better standardisation of geographical reference data and their metadata, and on a better integration of the GIS into environmental coastal zone management systems. In the future, data collected and used in Brest will be transferred into the SIEC, a coastal environment information system planned at the IUEM level. In this long-term framework, specific funding and a proper organisational context will be allocated to this information system. Currently, the inventory of data available to the concerned laboratories, and the metadata production, are continuing. The next steps of this joint project will concern the design and the implementation of this multidisciplinary coastal GIS (Muller *et al.*, 2000). From specific scientific applications, the GIS now has to evolve towards an institutional system designed and developed by a team based on future users and data providers.

11.6 ACKNOWLEDGEMENT

The authors would like to pay tribute to the late François Cuq, Director of the Géomer (Géosystèmes in the past) CNRS laboratory and of the Coastal observatory of the European Institute for Marine Studies, who died in May 2003, for his strong involvement in the birth and the growth of a GIS team of geographers in Brest, and for his huge contribution to the development of applied geomatics to coastal zone management in France.

11.7 REFERENCES

Allain, S., Guillaumont, B., Le Visage, C., Loubersac, L., and Populus, J., 2000, Données géographiques de référence en domaine littoral. In *Coastgis'99: Geomatics and coastal environment*, edited by Populus, J. and Loubersac, L. (Brest: IFREMER-SHOM), pp. 67-79.

Bartlett, D., Fowler, C., Longhorn, R., Cuq, F., and Loubersac, L., 2000, Coastal GIS at the turn of the century. In *Coastgis'99: Geomatics and coastal environment*, edited by Populus, J. and Loubersac, L. (Brest: IFREMER-SHOM), pp. 307-318.

Burbridge, P. and Humphrey, S., 1999, On the integration of science and management in coastal management research. *Journal of Coastal Conservation,* **5**, pp. 103-104.

Capobianco, M., 1999, On the integrated modelling of coastal changes. *Journal of Coastal Conservation*, **5**, pp. 113-124.

Crain, I.K. and Mc Donald, C.L., 1984, From land inventory to land management. *Cartographica*, **21**, pp. 40-46.

Davis, F.W., Stoms, D.M., Estes, J.E., Scepan, J., and Scott, J.M., 1990, An information systems approach to the preservation of biological diversity. *International Journal of Geographical Information Systems*, **1**, pp. 55-78.

De Sède, M.H. and Thiérault, M., 1996, La représentation systémique du territoire: un concept structurant pour les SIRS institutionnels. *Revue internationale de Géomatique*, **1/1996**, pp. 27-50.

ENR/OELM, 2000, *Diagnostic de territoire de la Côte d'Opale*. Publication Espace Naturel Régional / Observatoire de l'Environnement Littoral et Marin, http://www.enr-littoral.com.

Fabbri, K.P., 1998, A methodology for supporting decision-making in integrated coastal zone management. *Ocean & Coastal Management*, **39**, pp. 51-62.

Gourmelon, F., Bioret F., and Le Berre, I., 2001, Historic land-use changes and implications for management of a small protected island. *Journal of Coastal Conservation*, **7**, pp. 41-48.

Gourmelon, F., Liret, C., and Bonnet, M., 2000, Approche géomatique de l'habitat grand dauphin en mer d'Iroise. In *Coastgis'99: Geomatics and coastal environment*, edited by Populus, J. and Loubersac, L. (Brest: IFREMER-SHOM), pp. 186-197.

Gourmelon, F., Bioret, F., Brigand, L., Cuq, F., Hily, C., Jean, F., Le Berre, I., and Le Demezet, M., 1995, *Atlas de la Réserve de Biosphère de la Mer d'Iroise : inventaire numérique des milieux terrestres, intertidaux et marins* (Hanvec: Parc Naturel Régional d'Armorique).

Guillaumont, B. and Durand, C., 2000, The integration and management of regulatory data in a GIS: an applied analysis of the French coasts. In *Coastgis'99: Geomatics and coastal environment*, edited by Populus, J. and Loubersac, L. (Brest: IFREMER-SHOM), pp. 269-283.

Haines-Young, R., Green, D.R., and Cousins, S.H., 1993, *Landscape ecology and GIS,* (London: Taylor & Francis).

Holligan, P.M., 1994, *Land Ocean Interaction in the Coastal Zone (LOICZ) : Implementation Plan*. IGBP.

Howlett, E., Anderson, E., and Spaulding, M.L., 1997, Environmental and geographical data management tools for oil spill modelling applications. In *Proceedings of the 20th Arctic and Marine Oilspill Program (AMOP)*, Environment Canada, (Vancouver), pp. 893-908.

Journaux, A., 1985, *Cartographie intégrée de l'environnement un outil pour la recherche et pour l'aménagement*. MAB-UNESCO, ed. UGI.

Le Berre, I., 1997, *Réserve de biosphère de la Mer d'Iroise : carte de synthèse Conseil Général du Finistère*. MAB-UNESCO.

Le Berre, I., 1999, *Mise au point de méthodes d'analyse et de représentation des interactions complexes en milieu littoral.* Thèse de Doctorat, Géographie, Brest, UBO.

Le Berre, I., Gourmelon, F., and Liret, C., 2002, Modélisation bathymétrique de la mer d'Iroise, application à l'étude du grand dauphin côtier. *Revue internationale de Géomatique*, 12/2002, pp. 337-354.

Le Berre, I., Meyrat, J., and Pastol, Y., 2000, Application des données hydrographiques à l'étude synthétique de l'environnement côtier: exemple d'un SIG sur le littoral du Finistère (France). In *Coastgis'99: Geomatics and coastal environment*, edited by Populus, J. and Loubersac, L. (Brest: IFREMER-SHOM), pp. 233-244.

Michel, J. and Dahlin, J., 1997, *Guidelines for developing digital ESI atlases and databases*. NOAA.

Muller, F., Donnay, J.P., De Cauwer, K., Schwind, L., Devolder M., and Scory, S., 2000, Design of an oceanographic database. In *Coastgis'99: Geomatics and coastal environment*, edited by Populus, J. and Loubersac, L. (Brest: IFREMER-SHOM), pp. 124-133.

Van Zuidam, R.A., Farifteh, J., Eleveld, M.A. ,and Cheng, T., 1998, Developments in remote sensing, dynamic modelling and GIS applications for integrated coastal zone management. *Journal of Coastal Conservation*, **4**, pp. 191-202.

CHAPTER TWELVE

Cultural Intermixing, the Diffusion of GIS and its Application to Coastal Management in Developing Countries

Darius Bartlett and R. Sudarshana

12.1 INTRODUCTION

Whether viewed as system or as science, the origins and development of GIS are essentially rooted in Western (largely "Anglo-Saxon") geographies, sciences, and technologies. Indeed, Coppock and Rhind are even more specific, and point to the "dominant contribution of North America to the development and implementation of GIS up to the mid- and late-1980s" (Coppock and Rhind, 1991). In addition, the overwhelming majority of international scientific journals and educational texts in the discipline similarly originate in Europe and North America. They are also mostly anglophone, although recent years have seen the gradual emergence of a corpus of GIS literature in non-English, though still predominantly European, languages (e.g., for the French language, see Collet, 1992; Didier and Bouveyron, 1993; Pantazis and Donnay, 1996; Pornon and Hortefeux, 1992; as well as references in Bourcier, this volume, Populus *et al.*, this volume, and Gourmelon and Le Berre, this volume).

While the origins of GIS and most other Information and Communications Technologies (ICT) lie in the West, they are increasingly being applied to other parts of the world, and their diffusion may be seen as a step towards the eventual creation of global information infrastructures. This transfer of GIS and ICTs, as well as the know-how to apply them, is generally heralded as a "good thing" (Nag, 1987; Taylor, 1991; Hastings and Clark, 1991; Yeh, 1991; Salem, 1994; Alhusein, 1994; Metz *et al.*, 2000; Nwilo, this volume), and a growing corpus of literature attests to the benefits that can accrue to receiving countries from the acquisition and application of spatial information technologies.

Less well-publicised is the potential for cultural and other impacts on non-western societies as these tools and technologies become more widespread. In this paper, we explore some of the issues that we believe merit closer study in this

0-41531-972-2/04/$0.00+$1.50

regard, drawing by way of example on the use of GIS for coastal zone management in the Indian sub-continent.

12.2 GIS DIFFUSION TO THE NON-“WESTERN” WORLD

The process of technology transfer is arguably as old as humanity itself. In the past fifty years, however, it has gradually gained increasing importance as an element in the quest for sustainable development, reduction of vulnerability to environmental hazard, and the opening up of new markets. In short, technology transfer may be seen as a cornerstone of globalisation.

Based on our own experiences in a number of technology and know-how transfer initiatives, as well as a critical review of the literature, we have come to the conclusion that much technology transfer, particularly from north to south, and from west to east, appears often to be predicated on a number of important, though often untested assumptions.

1. **Technology transfer is driven primarily by philanthropic notions of serving “the public interest.”** In reality, much transfer of technology is driven by very different forces, such as the marketing efforts of large (mostly Northern) corporations, including GIS vendors, or the interests of donor governments seeking to obtain or retain political power and influence, pursue military strategies, influence global geopolitics, etc. These latter forces are rarely motivated by altruism, and may frequently seek to attach political or other conditions to the transfer: Alhusain (1994), for example, argued that the implementation of GIS in many developing countries is “still in its primitive stages… due to many wrong practices common in these countries… One of the most serious practices in these countries is that many of them - with some exceptions - are non-democratic societies.” In order to help overcome these perceived deficiencies, he advocates that the transfer of GIS and other technologies should be explicitly linked to political change: “International organisations such as ISPRS and the UN… can play very important role in encouraging the process of transferring the IGIS (Integrated GIS) technology to developing countries … *and, particularly, the democratic practicing [sic] within these societies should be highly appreciated and encouraged.*” (Alhusain, 1994, emphasis added).

2. **The notion that “new” technology equals “better” technology** and, linked to this, an implicit equation suggested between acquisition of technology and generally improved quality of life. This can lead to the acquisition of technologies becoming seen as a desirable and worthwhile goal in its own right, rather than as a means to an end. Thus the allure of GIS may be as much based on the wish to acquire a high-visibility “status” product, linking the owner with the developed world, as well as being a route towards accelerated national development. As Apfell-Marglin (1996) explains it, “The First World is ‘developed’ and the Third World is ‘developing’ or ‘underdeveloped’. In these phrases the *telos* of development stands revealed and the superior results are there already, luring everyone ‘forward.’” Particularly in recent years, uneven diffusion and take-up of spatial information systems, as well as the

Internet, and other Information and Communication Technologies, have led to the expression of fears among sections of the developing world that the technology "connects people who mostly belong to developed nations [and] excludes the major population from the developing nations" and even that "the developing countries will suffer from unequal distribution of scientific knowledge" and might thus be deprived of opportunities for development (Shanmugavelan, 2000). This fear of "being left behind" adds to the pressures on decision-makers in developing countries to embrace new technologies.

3. **Technology transfers will, indeed should, only consist of one-way (generally North-South or West-East) flows.** As currently practiced, the diffusion of GIS to the developing world exemplifies and reinforces this belief. However, there is in practice no reason why this should be the only pathway available, and there may be circumstances in which a reverse flow – that is, a transfer of ideas, technologies and methods, from South to North or from East to West - could be not only possible, but perhaps even desirable. The countries of the South may be less developed than those of the North, but they are not totally lacking in their own expertise, wisdom and ways of doing things: as the late president of Tanzania, Julius Nyerere, argued in a 1968 speech, "Knowledge does not only come out of books… We have wisdom in our own past and in those who still carry the traditional knowledge accumulated in that tribal past…we would be stupid indeed if we allowed the development of our economies to destroy the human and cultural values which African societies had built up over centuries" (quoted in Swantz and Tripp, 1996).

4. **Delivery of knowledge and capabilities to the recipient country is sufficient**, and subsequent diffusion of these tools or techniques within the country can be left to take care of itself. In practice, greater focus on internal diffusion is needed, especially from the often more cosmopolitan and progressive urban centres to the traditionally more conservative, and particularly undeveloped small towns and rural settings.

These observations, we believe, carry potentially important implications for the manner in which GIS is diffusing to, and being applied to assist coastal management tasks within, particularly developing world countries. The two areas where we believe these issues to be especially relevant are in the formalisation and representation of conceptual models of the coast within GIS databases; and in the incorporation of GIS within the decision-making process. Each of these will be examined in more detail in the remainder of this paper.

12.3 CONCEPTUAL MODELS AND REPRESENTATION OF SPACE

GIS consists of a nested and interlinked series of conceptual, logical and data models (Peuquet, 1984) that unite the human user, the computing hardware and software, and the data and operations being performed, into a more-or-less integrated and synergistic whole. The resulting system is a metaphor which we hope embodies all the elements and properties of what we perceive as important in the real world, while discarding the unimportant (Bartlett, 2000).

Measurement, description and apportionment of land appears to have always been an important feature of human society, to the extent that spatial thinking is probably a fundamental element of human intelligence (Chrisman, 1997). The *representation* of geographic information in some (carto)graphic form may also be seen in the histories of civilisations the world over (e.g. Noble, 1981; Thrower, 1996; Burrough and McDonnell, 1997; Godlewska, 1997; Harley and Woodward, 1989; Wilford, 1981). The Chinese, under Yü Kung, and the Greeks, under Anaximander, were reputedly making detailed maps and surveys as early as the sixth century B.C. (Thrower 1996). The diffusion of Buddhism from northern India into China was also taking place at about this time, and there were regular trading links and exchanges between China, India and the Roman Empire along the "Silk Road", as well as perhaps by sea. It may be assumed that much of the early European cartographic thinking that was emergent at this time was derived from, or at least influenced by, eastern philosophies and ideas.

The intervening centuries, however, have seen the ascendancy of western map-making. This cartography, while it almost certainly borrows unconsciously from other traditions, is nonetheless rooted firmly in western science, western world-views and western ideas of effective information communication. The net result has been the effective globalisation and hegemony of a cartography based on Euclidean distances, Cartesian space, the conceptual separation of time and space, and cartographic design rules based on western aesthetics. It is this cartography, enhanced through the application of Boolean logic and implemented through the medium of binary arithmetic, that we find embedded in the representation of geographic space within current GIS. These European (and, partly by derivation North American) scientific-cartographic paradigms have thus become the *de facto* worldwide standard for geographic data modelling, to the effective exclusion of all possible alternative models (Bartlett, 2000).

12.3.1 Geography, GIS and metaphysics

While the metaphors and paradigms embedded in GIS are thus western in origin, the applications of GIS are increasingly extending to other cultures and traditions. Little attention appears to have been given, until now, to the implications underlying this essentially top-down convergence of western technologies with non-western thinking and modes of practice.

White (1967) has observed that human ecology is "deeply conditioned by beliefs about our nature and destiny - that is by religion," while Robinson (1982, quoted in Singh, 1999) reminds us that, "geographers and most other researchers, consciously or unconsciously, have tended to assimilate a scientific perspective which in turn is based partly upon certain metaphysical assumptions concerning the nature of relationship between man and environment." In practice we implicitly know that western perspectives on science, which are largely based on Judaeo-Christian ethics and the application of Newtonian physics and mechanics, are merely subsets of an enormous range of possible world views and approaches to human ecology. There are fundamental and very important differences in the "metaphysical assumptions" and views of human-environment relationships, between the major religions and cultures of the world. Ipso facto, alternative world views and mental maps must also exist although, given their diverse cultural and geographic origins, these "maps" may take very different form and appearance from the types of cartography more usually encountered in the West. For example, if the late Bruce Chatwin's hypotheses are to be believed, even the Aboriginal "songlines" may represent a set of "musical maps" of the Australian continent, and a mapping of space relayed through oral tradition rather than via any graphic medium (Chatwin, 1988).

As is well known, coastal zone management remains an area of potential GIS application where major challenges are encountered. Many of these difficulties are due to pervasive inherent weaknesses in the conceptual and data models used by these systems, that mitigate against effective multi-scalar representation of highly dynamic objects with indeterminate boundaries. Although progress is being made towards resolving these issues, it is perhaps worth asking whether any solutions or insights to better representation of coastal space might be found through exploring alternative, non-western perspectives and conceptual models, and seeking to embody these within our computer representations? Such an exploration would require knowledge of computer technology to be interfaced with concepts and methods drawn from the social sciences including, perhaps, cultural and social anthropology. To the best of our knowledge, no such collaboration between these very distinct disciplines has yet been undertaken.

The Indian sub-continent would appear to be a particularly promising place to begin such research. As well as being an ideal "laboratory" for testing and further refining any leads that might suggest themselves, due to the extent, diversity and importance of the coastal zone, in India the predominant Hindu-Buddhist philosophies interface with western, Judaeo-Christian traditions, and the products of these, in virtually every sector and aspect of daily life. India is also a land of strong contrasts between traditional, frequently feudal-based rural societies on the one hand and, on the other, a burgeoning Information Technology sector – including its own space programme and orbiting remote sensing platforms – to match anything in the West.

12.4 THE APPLICATION OF GIS TO COASTAL ZONE MANAGEMENT

One of the primary reasons for wishing to model coastal space in a GIS framework is so that the technology may be used for enhanced decision-making and territorial management. Humanity has always had a close relationship with the coast, with at least 40% of the world's population now living within the coastal zone (Carter, 1988; Bartlett, D.J. and Carter, R.W.G. 1990). The oceans, and especially the continental margins, provide resources, living space, industry, and the locale for a growing leisure and tourism activity, all of which may contribute significantly to the social and economic development of society. The coastal zone may also be an important milieu for confrontation and intermixing of cultures, particularly as a growing global population competes for ever scarcer resources, and as globalisation brings yet more "exotic" coastal regions – often located within the developing world – into easy reach of leisure- or adventure-seeking holiday-makers, especially from the more affluent nations of the world.

Finally, the coastal zone is one of the world's more hazardous regions in which to live and work (Carter, 1988). Specific dangers include the risks of flooding and shoreline erosion, which are themselves often caused or exacerbated by extreme weather events (hurricanes, cyclones) or geotectonic disturbances (e.g. tsunami arising from earthquakes) (Burton *et al.*, 1993; Smith 1996). On top of these lie the indirect impacts and second-order consequences of disaster, arising through disruption of economic and social systems. The nature of these consequences may vary between urban and rural areas, as well as between developed and less-developed countries but, wherever they occur, their effects can be far-reaching and severe (Smith, 1996). Continued growth of world population, allied to recent warnings about possible near-future changes in global climate and accompanying rises in sea-level, increases in storm frequency and impact, etc., suggest the coast is likely to become more, rather than less, hazardous in the future (e.g. Titus, 1987; Carter, 1988; Devoy, 1992).

12.4.1 Western approaches to coastal management

The challenge of coastal management is to reconcile human and non-human uses of the shore. Unfortunately, this goal has more often remained elusive than achieved, prompting Carter to remark that humanity "has an uneasy relationship with the coast. Throughout history, we have tried to ignore it, adjust to it, tame it or control it, more often than not, unsuccessfully" (Carter, 1988). Until comparatively recently, coastal management in the West (and in those countries influenced by western thinking) has tended towards increasing reliance on engineering methods, and the application of various technologies – including GIS – in order to resolve specific problems.

This technocentric approach finds several echoes in Western (Judaeo-Christian) approaches to nature generally, and traditionally sees nature/the world as being provided for human use. It is based on an essentially linear view of time, and on reductionist thinking where the progress of science is directed towards ever deeper understanding of ever smaller parts of the picture, with comparatively little attention being given to the whole (Koestler, 1969). We find that this philosophy of "nature to be commanded" has impacted on the terminology used to define and delineate coastal management objectives with, for example, recurring use of the language of conflict and expressions such as "protecting the coast", "armouring the line of the shore", "defending the beach", and even "reclaiming land" (when, in many cases, what is being referred to is not a re-taking of that which was once lost to the sea, but instead a carving out and colonisation of new territories we never had in the first place!). Carter conveys the flavour of this ethos very well when he quotes Sir John Rennie's Presidential address to the Institute of Civil Engineers, London, in 1845: "Where can Man find nobler or more elevated pursuits than to interpose a barrier against the raging ocean (Carter, 1986)?"

While this particular sentiment was expressed one and a half centuries ago, it is still encountered regrettably frequently to the present day, and inevitably makes its way into the strategies, algorithms and models adopted when applying GIS to the coast. In the West, most coastal management problems are likely to be articulated in the language as alluded to above; the questions for which answers are sought will be framed by the assumptions and beliefs espoused by the society the coastal manager is part of; and the science policies adopted in order to resolve the situation will likewise be determined largely by the cultural and institutional contexts. Thus, traditional applications of GIS to coastal management have likewise drawn heavily on civil engineering and CAD-based approaches, the goal of which is most frequently to "stabilise" the line of the shore. Social, cultural and, indeed, environmental aspects of the problem have often tended to take a subsidiary role.

12.4.2 Eastern approaches to coastal management

In the East, and especially in India, the situation is much more complex. Many Asian countries were under European (in the case of India mainly British) rule for a few centuries, and the pre-existing feudal system of resource use was confronted with bureaucratic administration based on European principles. Thus we see a convergence of "East" and "West". Under colonial influences, management decisions were made, and resources categorised, largely on the basis of revenue unit polygons. Such units have spatial expression, and are relatively easy to represent in a traditional GIS. However, operating according to a completely different geography, the feudal system still exists or has re-emerged (in some places in the form of Mafioso-style organised crime) in many parts of post-colonial India, and decisions concerning the fate of a resource are now no longer expressed as per-revenue units that are easily interfaceable with GIS. Debt repayment systems based on lines of bartering, species segregation, accounting of fish catch instead of geographic accounting, *etc*., are all local values that stand distinctly different in comparison to the pattern of accountability in the west. The complexity of applying GIS to such management contexts lies in the fact that the local governments attempt to keep records in the Western pattern, while resource management is grossly influenced by indigenous, informal and feudal beliefs inaccessible to the procedures of modern information systems.

One of the basic assumptions underpinning both Western-style coastal management, and also implementation of a GIS, is the laminar structure of environmental management process. This philosophy of "layers" projects the nature of the problem, as well as the knowledge, decisions and actions needed to solve it, in a stratified manner. Within GIS, these "layers" are most frequently recognised in the thematic stratification of information. They also, however, find expression in the institutional contexts that frame uses of the technology. The layers in an organisation are generally distinguished by expertise, experience, hierarchical influence and a divisional or departmental approach to achieving work objectives. They can only be established and interlinked by the social process of *co-operation*. But whereas co-operation is a frequently encountered social value of Western society, *subjugation* and conflict at a human level are the dominant social values of the East, arising from the roots of the civilisations concerned.

This carries important implications for the use and transmission of information, since the co-operative laminar approach depends more on mechanical and modular interpretation of information through an iterative process, while the subjugative approach lends itself to single-window interpretative mode based more on human bias, instincts and an individualistic ethos. Hence, it may be argued that machine-based knowledge and decision-making will always find less acceptability in the East in comparison to the West. This applies in particular to the process of automated information extraction, e.g. by GIS, which finds broad acceptability in the Western mind whereas manual charting and information extraction methods are much more common in the East. Again, in the latter, responsibility for the validation of information and interpretation is hierarchical, and finds its acceptance once more in the unquestioned or less questioned 'subjugation' social ethos.

As was seen above, human affairs in Asian society tend to be based strictly on a feudal approach to responsibility and authority, and thus based on an ethos of domination by one group or individual over another. When it comes to human-nature interactions, however, a somewhat different ethos prevails. Here, when compared to Judaeo-Christian philosophies, Eastern traditions appear to give far greater emphasis to a fundamental interconnectedness of things; they place great stress on the importance of cycles (and also see time as a circular rather than a linear concept); and teach a need to accept and adjust to change rather than resist it. In short, they would appear to be much more inherently compatible with the goals of sustainability and integrated coastal zone management.

Assuming that the GIS has to cater to the process of coastal decision-making, such issues again become important, and it is once more necessary to debate on the background social realities of decision-making. We know that decisions are either made on a Long Term Perspective (LTP) or with a Short Term Perspective (STP). LTP aims at qualitative values in deliverables and is prompted by slowly emerging needs, with fewer and diminishing contingency methods built in to the system of delivery. However, STP has the urge and need to cater rapidly and in a quantitative manner, since speed and precision of delivery may be critical to sustainability and survival. These contrasting perspectives are also discernible between East and West: Western societies tend to be more stable and secure, and hence can afford both the resources and the time required for longer-term planning. In less-developed countries, however, the concerns and the priorities are usually very much more immediate and focused on issues of survival and daily sustenance. These will inevitably have implications for the design of appropriate information systems. Hence, system modelling and both the design and implementation of a GIS needs to be very meticulous and detailed under the LTP mode while it only has to be at a practically manageable level to yield the effective STP that is vital to many Eastern societies. It is also a reality to work with STP in many politically less stable social systems. Such being the case, there is a need to revisit basic GIS principles and practices for the East, to cater to faster but less precise (i.e. gross) decision-making.

12.5 CONCLUSION

Ultimately, integrated coastal zone management should be *informed* coastal zone management. This evidently requires access to appropriate, timely and reliable data and information which, in turn, suggests an important role for GIS and other suitable information technologies. However, it is worth bearing in mind the cautionary message of Henry Nix who, in his Keynote address to the 1990 AURISA conference, pointed out that:

> *"Data does not equal information; information does not equal knowledge; and, most importantly of all, knowledge does not equal wisdom. We have oceans of data, rivers of information, small puddles of knowledge, and the odd drop of wisdom"* (Nix, 1990)

If we truly desire to know the coast, and manage its resources wisely, we urgently need to look beyond the confines of our current technologies and our scientific (and social) paradigms.

We do not intend to suggest here that Indian, or for that matter any other, cultures and philosophies hold *all* the answers to solving current problems of coastal GIS. Nor do we wish to suggest that GIS and other "Western" technologies have no role to play in the developing world. Instead, it is the intention of this paper to argue for a greater receptivity to other ideas and approaches; a greater willingness to step outside the boundaries of western-dominated methods of science; and finally, a greater appreciation that introduction of GIS as part of wider technology transfer programmes needs to be carefully attuned to the cultural and other social contexts of the receiving communities. We believe that if the diffusion of GIS around the world is conducted in a spirit of genuine collaboration and desire to share experiences among and between cultures, then nature, as well as humanity, is likely to be the richer for the endeavour.

12.6 ACKNOWLEDGEMENTS

This paper evolved out of discussions between the authors that took place during and subsequent to a workshop on "Subtle Issues in Coastal Management" (Sudarshana *et al.*, 2000), which was held in Dehra Dun, India, in February 2000. Darius Bartlett gratefully acknowledges generous funding from the Embassy of India, Dublin, which was crucial in enabling his participation in this workshop.

12.7 REFERENCES

Abdel-Kader, 1994, Requirements for Implementing Regional GIS Systems to serve Developing Countries, in *Proceedings of an International Workshop on Requirements for Integrated Geographic Information Systems.* 2–3 February (New Orleans, Louisiana, USA), pp. 75–87.

Alhusain, O., 1994, Implementation of Integrated Geographic Information Systems in Developing Countries - A Special Need, in *Proceedings of an International Workshop on Requirements for Integrated Geographic Information Systems.* 2–3 February (New Orleans, Louisiana, USA), pp. 89–94.

Apfell-Marglin, F., 1996, Introduction: Rationality and the World. In *Decolonizing Knowledge. From Development to Dialogue,* edited by Apfell-Marglin, F. and Marglin, S.A., (Oxford, England: Clarendon Press).

Bartlett, D.J., 2000, Cultures, Coasts and Computers: Potential and Pitfalls in Applying GIS for Coastal Zone Management, In *Hidden Risks: Subtle Issues in Coastal Management,* edited by Sudarshana, R., Mitra, D., Mishra, A.K., Roy,

P.S. and Rao, D.P., Dehra Dun, India: Indian Institute for Remote Sensing (National Remote Sensing Agency) pp. 33–44.

Bartlett, D.J. and Carter, R.W.G., 1990, Seascape ecology: The landscape ecology of the coastal zone. *Ekologia (CSFR),* **10,** pp. 43–53.

Burrough, P. and McDonnell, R., 1997, *Principles of Geographical Information Systems,* (Oxford: Oxford University Press).

Carter, R.W.G., 1988, *Coastal Environments,* (London: Academic Press).

Chatwin, B., 1988, *The Songlines,* (Harmandsworth, Essex: Penguin).

Chrisman, N.R., 1987, Design of information systems based on social and cultural goals. *Photogrammetric Engineering and Remote Sensing*, **53**, pp. 1367–1370.

Collet, C., 1992, *Systèmes d'information géographique en mode image*, (Lausanne: Presses Polytechniques et Universitaires Romandes).

Didier, M. and Bouveyron, C., 1993, *Guide économique et méthodologique des SIG*, (Paris: Hermes).

Godlewska, A., 1997, The idea of the Map. In *Ten Geographic Ideas that Changed the World*, edited by Hanson, S., (New Jersey: Rutgers University Press), pp. 15–39.

Harley, B. and Woodward, D. (eds.), 1989, *History of Cartography*, (Chicago: University of Chicago Press).

Koestler, A., 1969, Beyond atomism and holism – the concept of the holon. In *Beyond Reductionism: New Perspectives on the Life Sciences (The Alpach Symposium)* edited by Koestler, A. and Smythies, J.R., (London: Hutchinson).

Metz, B., Davidson, O., and Van Wie, L.L. (eds.), 2000, *IPCC Special Report on Methodological and Technological Issues in Technology Transfer*, (Cambridge, UK: Cambridge University Press), in prep.

Nix, H., 1990, *A national geographic information system - an achievable objective?* Keynote address, AURISA.

Pantazis, D.N. and Donnay, J.-P., 1996, *La Conception de SIG: Méthode et formalisme,* (Paris: Hermes).

Peuquet, D., 1984, A conceptual framework and comparison of spatial data models. *Cartographica,* **21**, pp. 66–113.

Pornon, H. and Hortefeux, C., 1992, *Systèmes d'Information Géographique pour petites communes: Guide méthodologique*, (Paris: Les Ed. du STU).

Salem, B.B., 1994, Scientific Applications of GIS in Egypt: A commentary on the current status. In *Proceedings of an International Workshop on Requirements for Integrated Geographic Information Systems.* 2-3 February (New Orleans, Louisiana, USA), pp. 29–34.

Shanmugavelan, M., 2000, IT in developing nations. *Intermedia* **28**(1), (International Institute of Communication, UK).

Singh, R.P.B., 1999, Nature and cosmic integrity: A search in Hindu geographical thought. In *Nature and Identity in Cross-Cultural Perspective,* edited by Buttimer, A. and Wallin, L. (Dordrecht, the Netherlands: Kluwer) pp. 69–86.

Smith, K., 1996, *Environmental Hazards. Assessing Risk and Reducing Disaster,* (London: Routledge).

Sudarshana, R., Mitra, D., Mishra, A.K., Roy, P.S., and Rao, D.P., 2000, *Subtle Issues in Coastal Management,* (Dehra Dun, India: Indian Institute of Remote Sensing).

Swantz, M.-L. and Tripp, A.-M., 1996, Development for 'Big Fish' or for 'Small Fish'?: A study of Contrasts in Tanzania's Fishing Sector. In *Decolonizing*

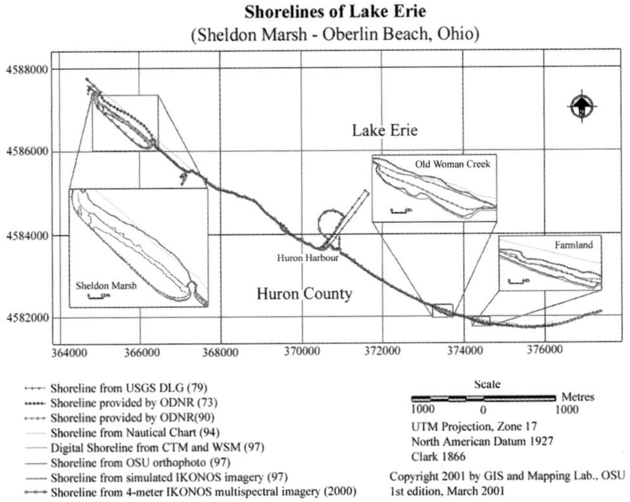

Shorelines of Lake Erie
(Sheldon Marsh - Oberlin Beach, Ohio)

4588000
4586000
4584000
4582000
364000 366000 368000 370000 372000 374000 376000

Lake Erie
Old Woman Creek
Farmland
Huron Harbour
Huron County
Sheldon Marsh

Shoreline from USGS DLG (79)
Shoreline provided by ODNR (73)
Shoreline provided by ODNR(90)
Shoreline from Nautical Chart (94)
Digital Shoreline from CTM and WSM (97)
Shoreline from OSU orthophoto (97)
Shoreline from simulated IKONOS imagery (97)
Shoreline from 4-meter IKONOS multispectral imagery (2000)

Scale
1000 0 1000 Metres
UTM Projection, Zone 17
North American Datum 1927
Clark 1866
Copyright 2001 by GIS and Mapping Lab., OSU
1st edition, March 2001

Plate 3.1 Shorelines compared and analysed in Chapter 3, Li et al.

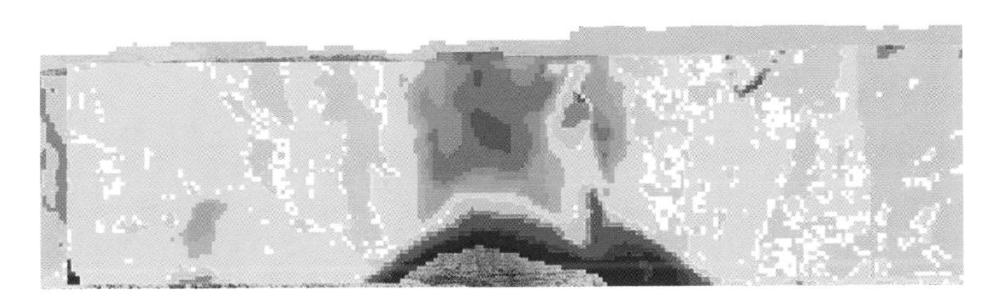

Plate 4.1 Changes between August 2000 and February 2001.

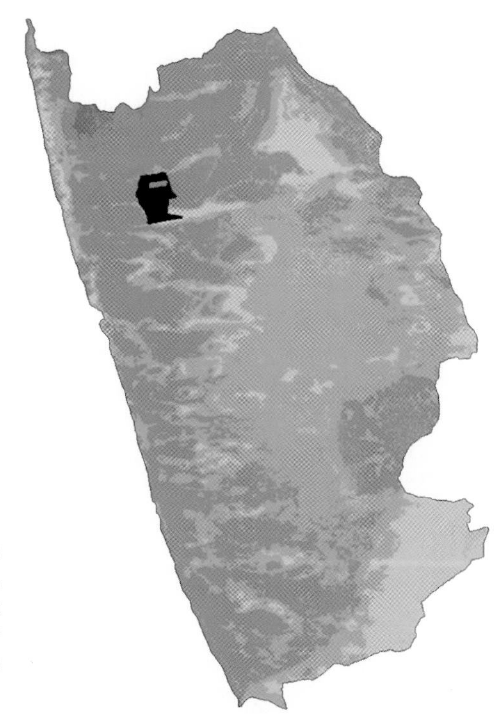

Plate 4.2 The Digital Elevation Model (DEM) for Kenfig NNR. Elevations are represented by different shades of green. Lower to higher ground is graded from darker to lighter shades.

Plate 4.3 Slope data for Kenfig NNR. Slope "steepness" is represented by different shades of red, with darker shades representing steeper slopes. Clusters of red in circular shapes are individual trees.

Plate 4.4 Aspect data for Kenfig NNR. Orientation of the slopes is represented by different shades of brown. Lighter shades represent a northerly orientation, and darker shades indicate a southerly direction.

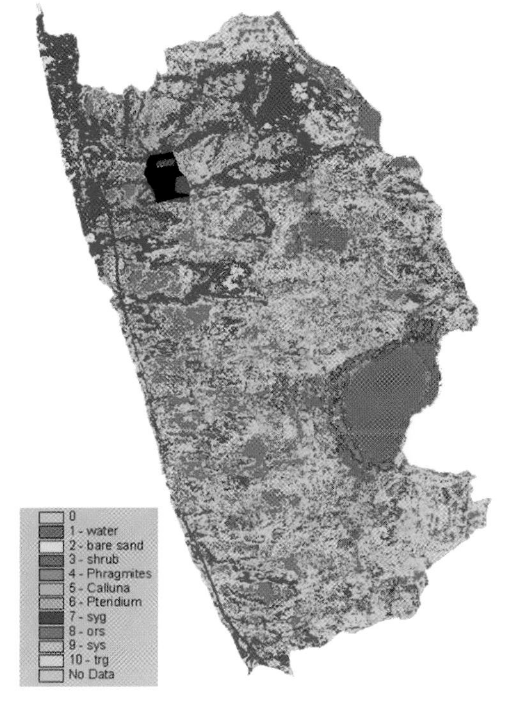

Plate 5.5 The classified habitat map of Kenfig NNR using CASI data.

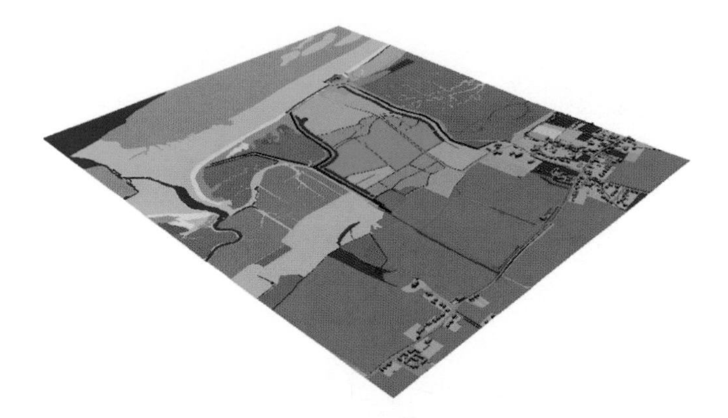

PLATE 8.1 ArcView 3D Analyst view of the site prior to the scheme. © Crown Copyright Ordnance Survey. An EDINA Digimap/JISC supplied service.

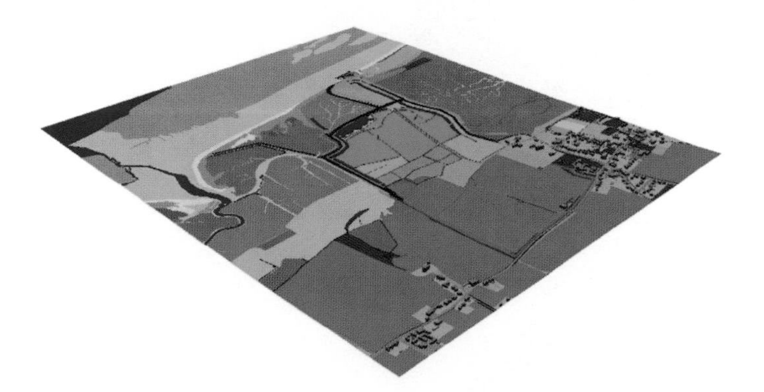

PLATE 8.2 ArcView 3D Analyst view of the site following construction of the scheme. © Crown Copyright Ordnance Survey. An EDINA Digimap/JISC supplied service.

PLATE 8.3 Word Construction Set view looking south over Brancaster West Marshes prior to scheme construction. © Crown Copyright Ordnance Survey. An EDINA Digimap/JISC supplied service.

PLATE 8.4 World Construction Set view looking south over Brancaster West Marshes following construction. © Crown Copyright Ordnance Survey. An EDINA Digimap/JISC supplied service.

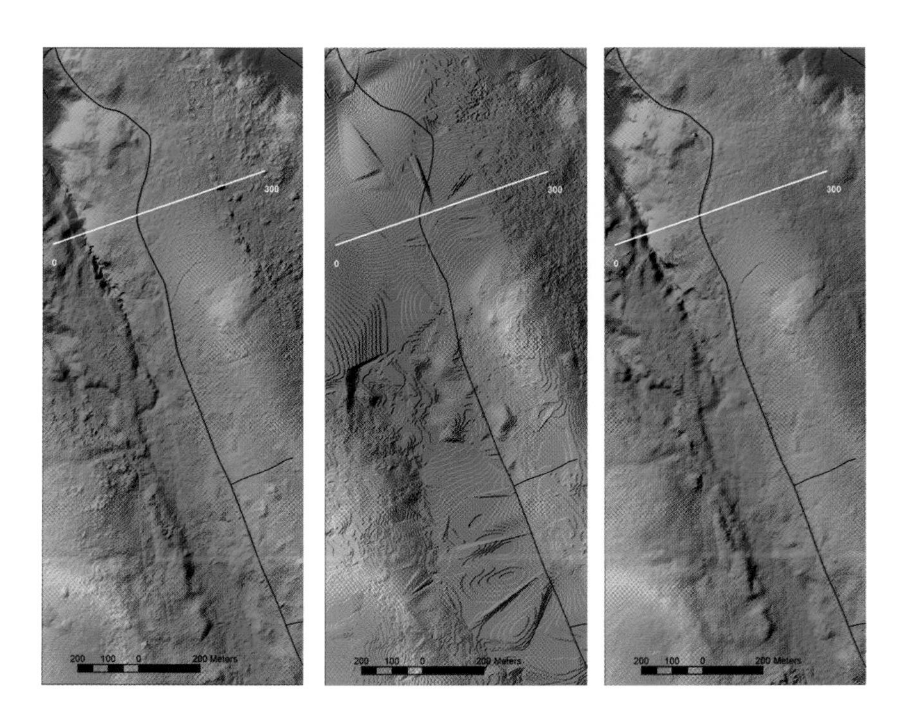

PLATE 15.1 Mosaic of 1 m CASI at low tide (left image), 2 m LIDAR DSM at low tide (centre image), and 3 m CASI at high tide (right image) for Port Lorne along the Bay of Fundy. Overall image is approximately 4 km across.

PLATE 15.2 Close-up of flood levels of one-in-twenty year (yellow) and one-in-one-hundred year (red) storm events with property boundaries (purple) that intersect the 100-year flood level displayed over an Iknonos image. Economic impact of such an event can be assessed by using GIS overlay functions to determine the value of infrastructure and properties affected. Includes material copyright Space Imaging, LCC.

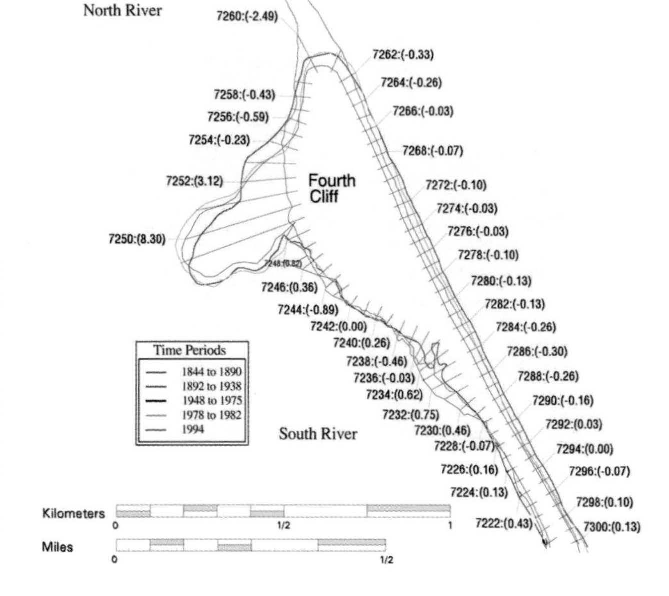

PLATE 16.1 Enlarged section of a shoreline change map for Scituate, Massachusetts, displaying historical shorelines, transects, and linear regression rate. (For clarity at page size, the orthophotograph underlay is not shown.)

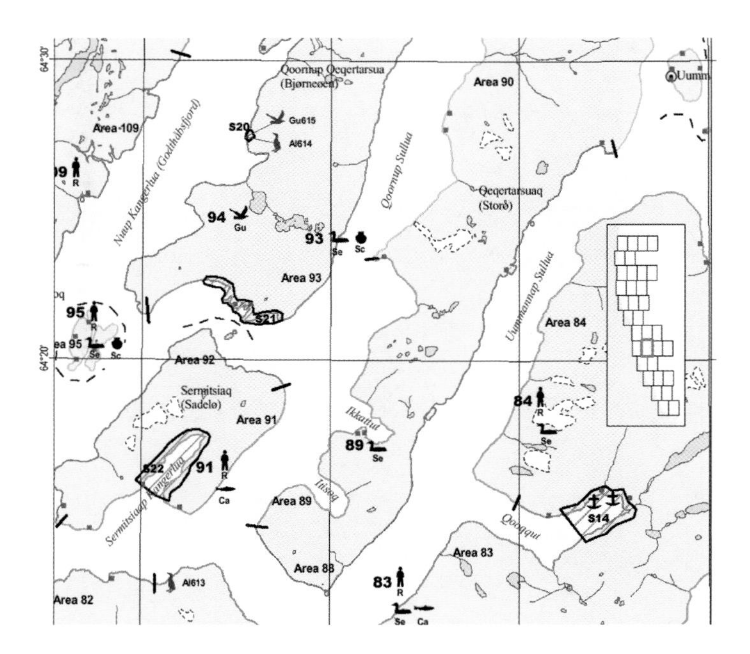

Figure 20.1A An example from the Atlas showing part of a shoreline sensitivity map. The shoreline areas (segments) have four different colours, indicating the level of sensitivity to oil spill. The black symbols indicate the important biological elements and resource use at the shoreline area, the blue symbols shows the site specific bird colonies.

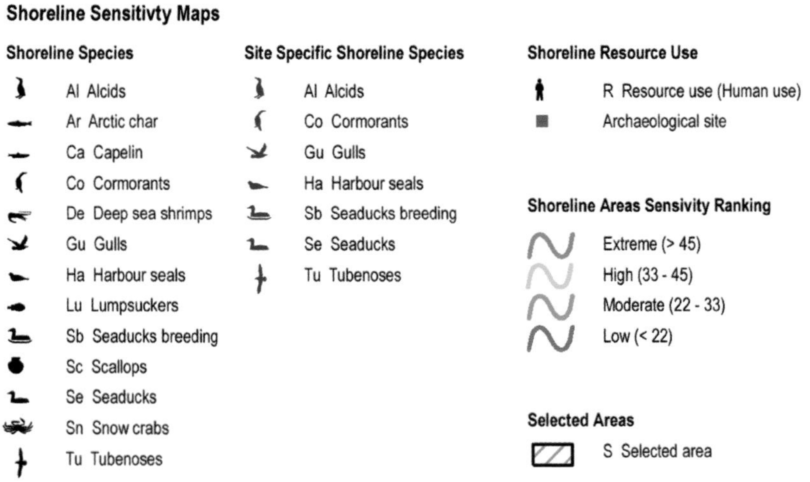

Shoreline Sensitivty Maps

Shoreline Species

- Al Alcids
- Ar Arctic char
- Ca Capelin
- Co Cormorants
- De Deep sea shrimps
- Gu Gulls
- Ha Harbour seals
- Lu Lumpsuckers
- Sb Seaducks breeding
- Sc Scallops
- Se Seaducks
- Sn Snow crabs
- Tu Tubenoses

Site Specific Shoreline Species

- Al Alcids
- Co Cormorants
- Gu Gulls
- Ha Harbour seals
- Sb Seaducks breeding
- Se Seaducks
- Tu Tubenoses

Shoreline Resource Use

- R Resource use (Human use)
- Archaeological site

Shoreline Areas Sensivity Ranking

- Extreme (> 45)
- High (33 - 45)
- Moderate (22 - 33)
- Low (< 22)

Selected Areas

- S Selected area

Figure 20.1B The legend for the shoreline sensitivity map. Symbols are displayed with the species name and the abbreviation for the name.

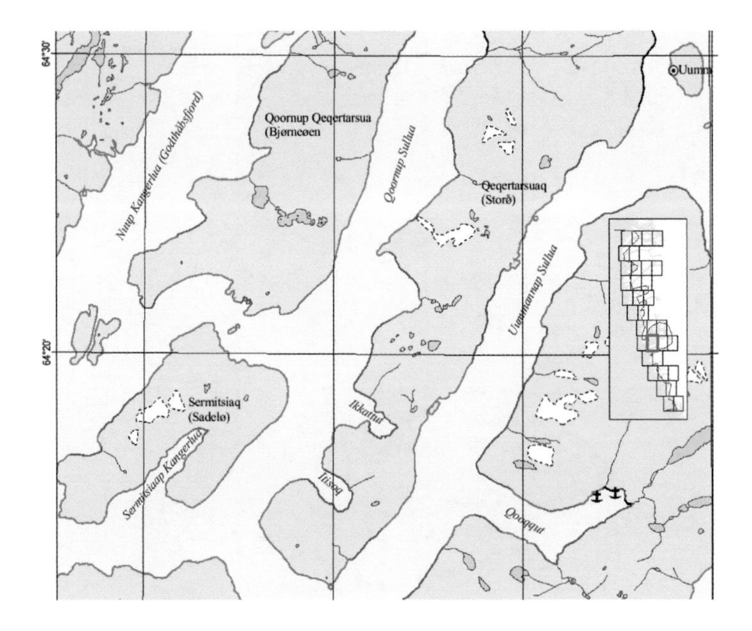

Figure 20.2A An example from the Atlas showing the Physical Environment and Logistics map. The map shows the shoretype classification and where harbours/anchorages are located.

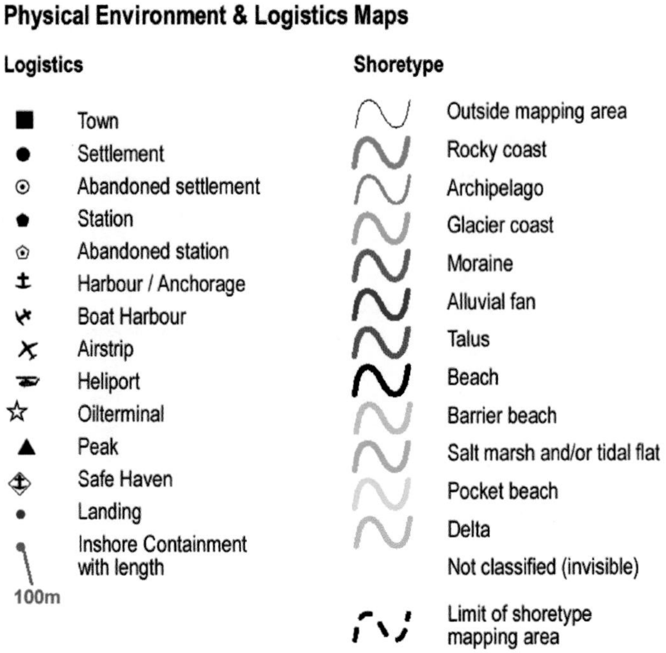

Physical Environment & Logistics Maps

Logistics

- ■ Town
- ● Settlement
- ⊙ Abandoned settlement
- Station
- Abandoned station
- Harbour / Anchorage
- Boat Harbour
- Airstrip
- Heliport
- ☆ Oilterminal
- ▲ Peak
- Safe Haven
- Landing
- Inshore Containment with length (100m)

Shoretype

- Outside mapping area
- Rocky coast
- Archipelago
- Glacier coast
- Moraine
- Alluvial fan
- Talus
- Beach
- Barrier beach
- Salt marsh and/or tidal flat
- Pocket beach
- Delta
- Not classified (invisible)
- Limit of shoretype mapping area

Figure 20.2B The legend for the Physical Environment and Logistics map. The twelve geomorphologic shoretypes all have individual colours.

Knowledge. From Development to Dialogue edited by Apfell-Marglin, F. and Marglin, S.A., (Oxford, England: Clarendon Press).
Thrower, N.J.W., 1996, *Maps and Civilisation. Cartography in Culture and Society*, (Chicago: University of Chicago Press).
White, L. Jr., 1967, The historical roots of our ecological crisis. *Sciences,* **155**.
Wilford, J.N., 1981, *The Mapmakers*, (New York: Random House).
Wright, D. and Bartlett, D.J. (eds.), 2000, *Marine and Coastal Geographical Information Systems,* (London: Taylor & Francis).
Yeh, A.G.O., 1991, The Development and Application of Geographic Information Systems in Urban and Regional Planning in the Developing Countries. *International Journal of Geographical Information Systems*, **1**, pp. 5–27.

CHAPTER THIRTEEN

The Use of GIS to Enhance Communications of Cultural and Natural Resources and Contamination

John A. Lindsay, Thomas J. Simon, Aquilina D. Lestenkof, and Phillip A. Zavadil

13.1 INTRODUCTION

Five Pribilof Islands, volcanic in origin and remotely located in the Bering Sea, are home to Aleuts and the breeding grounds to 70% of the world's northern fur seal population and numerous seabird species. The once uninhabited islands were first occupied by the Russians in 1786 and later became controlled by the U.S. Government. At the time of the Russian incursions into the Aleutian Islands, some have estimated the Aleut population at 15,000-18,000. First Russia and then the U.S. relied on the forced labour of Aleuts relocated to the Pribilof Islands from the Aleutian Islands chain to harvest sea otters, fur seals, and arctic fox. From their first arrival and through much of the 20th century, *Unangan* (people) or Pribilof Aleut people relied on the islands' and Bering Sea's natural resources, including fur seals, sea lions, whales, arctic fox, walrus, sea bird species, and a variety of plants for customary traditional purposes and subsistence. Even today natural resources are vital to the survival interests of *Unangan* on the Pribilofs. *Unangan* of St. Paul and St. George Islands are the world's single largest ethnic Aleut community whose world population approximates 3,200 individuals. Fur seal, endangered Steller sea lion, introduced reindeer, halibut, crab and other marine invertebrate species, and plants, such as crowberry, continue to play a significant role in customary traditional practices, economic development, and the maintenance of cultural and ecological harmony.

Settlement terms under the Alaska Native Claims Settlement Act (ANCSA) of 1971 required the National Oceanic and Atmospheric Administration (NOAA), as the most recent of the former federal land managing agencies for these islands, to transfer more than 95% of the land area to the local Aleuts. A 1976 Memorandum of Understanding and a 1984 Transfer of Property Agreement (TOPA) incorporated the details of the property transfer between NOAA and various local entities. The settlement and subsequent legislations including the Pribilof Islands Environmental Restoration Act of 1995 and the Pribilof Islands Transition Act of 2000 required NOAA to restore the islands' environmental

0-41531-972-2/04/$0.00+$1.50

integrity compromised by U.S. Government activities supporting its commercial fur sealing enterprise. Environmental concerns included numerous releases of petroleum fuel products associated with the disposal of used oil, overfilling of storage tanks, corrosion of storage tanks, pipelines, and barrels, as well as landfills for household wastes, construction and demolition debris, scrap metal and junked vehicles, boats, barges, and aircraft. In addition, military activities during World War II contributed to soil and groundwater contamination. More than ninety sites on the two inhabited islands, St. George and St. Paul, required evaluation and potential restoration in order to complete the land transfer. The islands' remoteness, approximately 2,200 statute miles from NOAA's base of operations in Seattle, Washington requires a minimum of eight hours in travel time. Weather extremes involving fog, snow, and wind frequently cause flight cancellation. Improving communications through such multimedia technologies as Internet, GIS, and video were identified early as absolutely critical to the restoration project's success.

In the context of public land transfer to private sector entities and the restoration of lands and structures located within a National Historic Landmark and District, the National Historic Preservation Act (NHPA) further mandates NOAA to mitigate for these actions. Mitigation efforts at a minimum will provide a historical legacy associated with the federal ownership period for the benefit of future generations. The legacy is expected to include a book, video documentaries, and a detailed Geographic Information Systems (GIS) project. This chapter focuses on the GIS project. The NOAA Pribilof Islands Environmental Restoration Project Office (PPO) entered into collaboration with the St. Paul Tribal Government's Ecosystem Conservation Office to build a GIS project that included not only extant conditions, but also traditional cultural and natural resource features. This chapter presents an ongoing project that began in 1999, the conclusion of which is not expected for several more years.

13.2 GIS APPLICATION

The environmental restoration project began nearly two decades ago. Numerous written reports included site locations and analytical results from hundreds of soil samples. Unfortunately, site boundaries and sample point locations have not been easy to recover because of poor referencing and a relatively monotonous environment that becomes overgrown with waist high vegetation or eroded by violent winds. A decision was made at the outset to map all features using differential global positioning systems (DGPS) and a GIS. This approach would allow the relocation of sites and sampling points in years distant if necessary, as well as aid in other efforts, such as defining the vertical and horizontal extent of contamination, and the calculation of volumes of soils requiring excavation. The PPO also recognized the importance of mapping historical, natural, and cultural resource features, such as gravesites, to aid the communities in their future management of the islands, and subsequently to comply with the NHPA and the Coastal Zone Management Act.

Alaska's high latitudes and the islands' remoteness challenge GIS application. Normally available geo-referenced base maps, such as topographic quadrangles and nautical navigation charts, are scarce, small scale, or nonexistent.

Early project acquisition and prosecution of DGPS data acquired with Trimble Pro XRS and corrected with post processing layered over existing electronic versions of nautical charts revealed positional errors requiring exhaustive research to rectify.

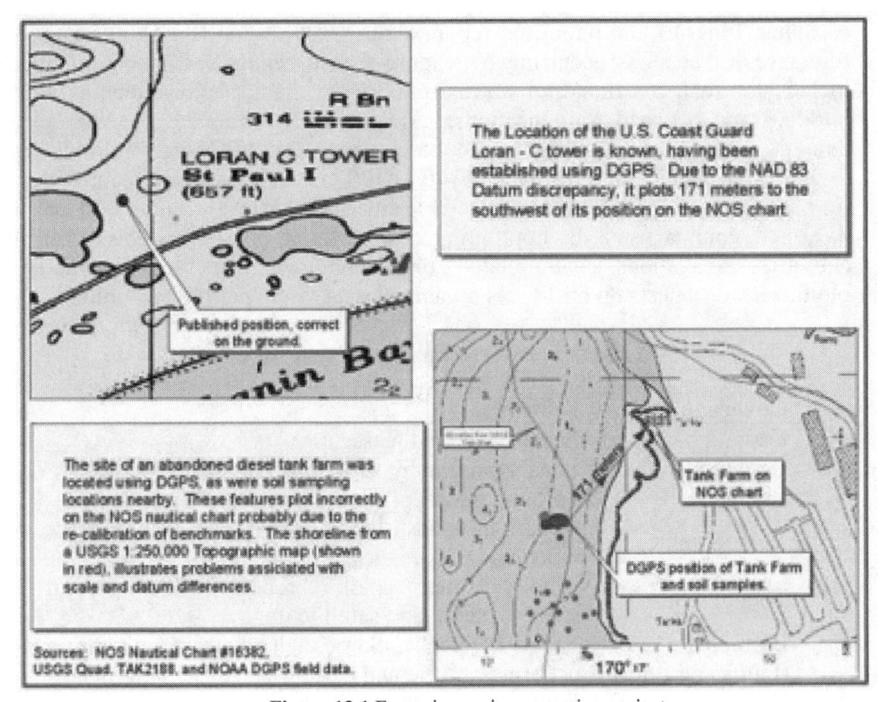

Figure 13.1 Errors in previous mapping projects.

The project resurveyed the two principal islands, St. Paul and St. George, to first order following National Geodetic Survey protocols. The results revealed a transcriptional error in an earlier version of the cartographic process that translated into an approximate 180-meter error (Figure 13.1). New benchmarks were installed on the islands, and NOAA/NOS issued new nautical navigation charts in March 2001. The chart scale, however, was still too small (1:50,000) for practical use. The NOAA project entered into a cost share collaboration with the US Geological Survey to provide Digital Elevation Models (DEMs) at 1:24,000 scale, topographic quadrangles at 1:25,000 scale with 10-meter contour intervals, and Digital Ortho Quarter Quads (DOQQ's) at 1:12,000 scale using pre-existing (1993) colour aerial photography. In an effort to acquire recent high-resolution imagery, NOAA contracted with a satellite vendor to provide 1-meter panchromatic and 4-meter multi-spectral imagery. The PPO is considering the application of interferometric synthetic aperture radar (IFSAR) imagery of the islands for use in establishing hydrographic controls to further improve island charting and to monitor long-term Aeolian erosion processes that have a potential to impact current restoration efforts. These products, when combined with historical maps and project data, will provide a comprehensive view of the islands and their changes over time.

The Pribilof Islands GIS mapping project uses ESRI's ArcView version 3.2. Federal Geographic Data Committee compliant metadata allows rapid data searches. The project focuses on several coverages and features (Table 13.1). As noted above, base maps include NOAA navigational charts as well as airphoto- and satellite-derived DEMs. Historical maps acquired from various sources including libraries and published reports dating back to the late 1800's provide a reference for changes occurring over more than a century. Changes of interest include fur seal, sea lion, and walrus rookery and haul out locations, as well as building and off-road trail locations. The maps are scanned at 1200 dpi and overlaid on registered and rectified base maps. The PPO uses Trimble Total Station RTK (real time kinematic) 5700 DGPS survey equipment to provide real time horizontal spatial accuracy of ±1 cm and vertical accuracy of ±2 cm. Acquired data is used for contouring and creation of 3-D models of landfill activities, excavations, contaminated soil stockpile volumes, and potentially for plotting seabird nests on cliff faces combined with laser spotting technology.

Table 13.1 Pribilof Islands GIS coverages and features

Coverage	Features
Geographic	Islands and land masses
Geologic	Soils, stratigraphy
Chemical	Soil and water
Demographic	Buildings, roads, debris sites; historic and current video and still images
Topographic	Cinder cones, landfill relief, DGPS ground surveys, aerial and satellite imagery
Biological (fauna and flora)	Fauna, flora, rookeries, haul outs, migration patterns, historic and current video and still images
Oceanographic	Bathymetry, currents, ice pack
Archaeological	Habitations, burial sites, historic and current video and still images

The authors' previous experience with marine GIS projects (e.g. Lindsay *et al.,* 1999; Butman and Lindsay, 2000) revealed the robustness of ArcView GIS. Using hotlinks the GIS project is enabled to include multimedia features, such as still, raster (laser line scan), and video imagery and sound recordings. The PPO uses Canon XL-1 digital and Panasonic 720p High Definition (HD) electronic media cameras and digital still cameras to document field activities. HD is down-converted to standard definition. Standard Def and digital videotape postproduction relies on MacIntosh G4 editing suites with Final Cut Pro software and DVD Studio Pro authoring software. Real GIS data are imported into a Hewlett Packard X4000 with XSI animation software to create textured, lighted, and animated production quality 3D models for viewer friendly but scientifically accurate renditions of real data. Historical video on Betacam and VHS tapes are converted to digital format using the same hardware and software. The GIS project itself and video mini-documentaries are placed onto DVD (digital video disk) using super drives. (Please note that any use of trade, product, or firm names is for descriptive purposes only and does not imply endorsement by the U.S. Government.)

13.3 ENVIRONMENTAL RESTORATION

The U.S. Government initiated environmental restoration activities on the Pribilof Islands in 1983 following the final decision to transfer island lands from the public to the private sector. Former operations managers, residents, and historical reports aided in the identification of contaminant and debris sites. From a mapping perspective the primary emphasis of the environmental restoration activities on the Pribilof Islands has been on the locations of land features, boundaries, and the acquisition of soil and water chemical contaminants data. Demographic land feature focus has been on ownership, including federal land tracts and parcels targeted for transfer under the ANSCA and TOPA, topography, cultural uses associated with waste disposal practices, debris and contaminated waste site boundaries, building locations, utility locations (water, electrical, fuel transfer, cable, telephone and television lines), and grave site locations. Pribilof Islands' soil and groundwater chemistry has typically been evaluated at sites where spills or releases of petroleum products were suspected or known to have occurred, such as landfills and building sites.

A significant commercial infrastructure was built on St. George and St. Paul Islands to harvest and cure northern fur seal skins between 1872 and 1984. Numerous buildings were constructed for brine and salt curing, barrel making, skin packing, and seal by-product production, such as oil and animal meal. Other structures included electrical generation plants, gasoline stations, a radio station, fuel storage and transfer, commercial store, storage and utility buildings, a school, health clinic, quarries, and roads. As the value of the harvesting grew, so did the sophistication of the infrastructure and the use of petroleum based fuels for electrical generation and motorized vehicles. Storage tanks of various sizes ranging from about 500 gallons to 200,000 gallons and fuel lines serviced the islands' energy needs. Until about 1970, these tanks were filled from 55-gallon barrels brought ashore by bidarrahs, a native rowing craft with Russian influenced origins that were covered by sea-lion skins, and eventually by fuel barges with floating transfer lines.

Environmental restoration efforts prior to 1999 lacked either resolve and/or management experienced in restoration; consequently their efforts resulted in restoration of readily visible debris that included six thousand tons of surface debris, thousands of barrels either empty or containing used oil, and some underground storage tanks. Contaminated soils, groundwater, and landfills were left virtually untouched. In addition, initial efforts relied on conventional surveying and mapping techniques that archived features on paper maps without providing adequate geo-references. Subsequent field activities failed to recover site boundaries and soil sampling points. Short field seasons punctuated by years of inactivity resulted in as many as four characterizations at some sites in this treeless and knee-high grass environment in part due to the lack of geo-referencing and unique visual cues such as building foundations, stressed vegetation, soil staining, or surface debris. In an effort to capture these data, the PPO scanned existing paper maps, periodic air photos dating back to 1948, geo-referencing these when possible, and including them in the GIS project along with textual and tabular information. Since 1999, the PPO has relied on DGPS survey systems to log data points and vectors, to catalogue site features, and to relocate sites and sampling points. For the first time Pribilof Islands data was entered into a GIS project that permitted near real-time project map creation and distribution. Maps were spontaneously created in the field office for planning and discussion purposes. Subsequent maps demonstrated the extent of soil and groundwater contamination, and served to increase the efficiencies of removal action decisions.

For example, diesel fuel contaminated sites are first verified by laboratory soil chemical analyses. Using GIS created maps depicting analytical results, and site historical and current features as guides, a removal or excavation is initiated within the bounds defined about these analytical results. These analyses seldom define within several meters the horizontal and vertical extent of contamination. Therefore, following initial excavation of contaminated soils, the project relies on thin layer chromatography (TLC) soil screening to verify whether remaining soils have diesel fuel contamination. The project has determined that field applied TLC based on standards offers a high probability (>95%) of accuracy within a sliding scale (e.g. 0-100 mg/kg, 100-250 mg/kg). TLC samples and their horizontal and vertical positions can be collected, processed, and the results entered into a GIS within approximately two hours. The sample results are overlaid onto registered and rectified historical air photo(s) depicting the historical structures, such as buildings, fuel tanks and fuel lines linked as sources of the contamination but long since dismantled. The composite photomap allows the cleanup team to substantiate the TLC results in context with the extent of the excavation and likely sources. If the map depicts a sample result with likely low levels of contamination, and no historical source was within close proximity to the spot, the cleanup team has received greater certainty about stopping further excavation. When excavation is considered complete, additional soil samples are taken for fixed laboratory analysis. The photomaps created during these field exercises can be displayed on laptop computer screens, printed on photographic papers, projected onto walls and plasma screens, and transmitted via Internet to far off locations.

High resolution (sub-meter scale) topographic maps of quarries and mineral stockpiles are created with the DGPS and GIS. On the islands, the project is required to purchase sand, scoria, and rock for use as backfill in excavations of contaminated soils, making haul-roads over the tundra, and capping landfills. Owners of the mineral rights require the purchaser to conduct pre- and post-surveys of the quarries to first approve and then verify quantities removed to establish a cost or royalty. The projects survey grade GPS in combination with the GIS are used to create the survey maps.

As mentioned above, the driving force behind the environmental restoration is the land transfer from the American government to local entities. These lands in the form of lots, parcels, and tracts were initially surveyed more than two decades ago using traditional surveying techniques by a separate federal agency. The available plats are in paper form and they are not easily applied in discussions involving numerous individuals and multiple issues that may involve structures, and contamination of soil and groundwater. In addition some shoreline changes have occurred since the original surveys. The project scanned and registered best fits of these plats and entered them into the GIS project for use as overlays with air and satellite photos. These enhanced maps from the original plats, hotlinked to still or video imagery, greatly improved communications between on- and off-island officials.

13.4 NATURAL RESOURCE NAMING CONVENTIONS

On and around the Pribilof Islands, marine mammals and seabirds dominate the environment. Only three terrestrial mammals are indigenous to the Pribilofs; they include the arctic fox, the black-footed lemming on St. George, and the Pribilof shrew on St. Paul. While English naming is the predominant convention followed today, some *Unangan* elders still refer to various species and/or their life stages in either *Unangam tunuu* (our people's language) or Aleut language, or Russian. For example, the arctic fox is called *aygagux* in *Unangam tunuu*. Fur seals are identified by sex and life stage in one or more languages. GIS mapping features can incorporate the various naming conventions applied to these species as appropriate.

13. 5 LOCATION NAMES

Present Pribilof Islands location names are also a mixture of *Unangam tunuu*, Russian, and English. For example on St. Paul Island, Lukanin Bay and beach, also spelled Lukannon, was named after the *promyshlenik* [fur hunter] who along with another, Kaiekov, killed 5000 sea otters during the first year of occupation in that area. Zolotoi Beach is literally interpreted as "Golden Sands" (Martin, 1960). Other present-day place names demonstrate American influence. On St. Paul, Hutchinson Hill acquired its name after Hayward Hutchinson, the first American to secure the commercial rights to the islands' fur seals.

The northern fur seal is the most intensively studied mammal on the islands. Records and maps of rookery locations, and haul out areas date back to the Russian period. Henry W. Elliot and others (*cf.* Elliot, 1881; Townsend, 1896; Osgood *et al.,* 1915) provided some of the earliest American rookery and haul out area maps. Rookery names as with other points of interest varied over time. For example, today's Northeast Point rookery on St. Paul was known during the Russian period as Novastoshnah Rookery while in *Unangam tunuu* it is *Chaxax*.

During NOAA's restoration activities the Aleut community expressed desire for greater recognition for their contribution to the islands' history. Although the literature is not devoid of *Unangam tunuu* place names, such names are uncommon on maps. Leaders on both islands had already begun to ascribe *Unangam tunuu* (Aleut place names) but official maps were wanting. NOAA enlisted the assistance of the U.S. Geological Survey (USGS) to procure digital aerial mosaics of the islands for the construction of base maps and the first large scale (1:25,000) topographic relief maps of the islands. These maps were released as a publication in March 2003. NOAA and USGS encouraged the local residents to document *Unangam tunuu* on these topographic maps and the USGS worked to secure their official status by having the place names reviewed and accepted as Geographic Names by the Alaska Historical Commission.

13.6 POPULATION DISTRIBUTIONS

While the Russian discoverers, beginning with the navigator Captain Gerassim Pribylov, may have recorded various marine mammal population statistics, locating such records will require more concerted efforts than we have so far undertaken. But early American records clearly indicate that many of the aforementioned species at least once flourished on the islands.

Pack ice is said to influence the comings and goings of a variety of species, but most notably, walrus, arctic fox, and polar bear. Mastodon bones and molars have been recovered from lava tubes on the islands, suggesting either arrival *via* land bridge or pack ice. Purportedly in 1835, long lasting pack ice around St. Paul prevented haul out by pregnant fur seals, which resulted in the loss of both females and potential pups (Martin 1960). The pack ice in the early 1960's is reported by locals to have brought a polar bear to St. Paul, as well as several other times in the past. During 2000, pack ice was responsible for the occurrence of 6 to 7 arctic foxes on Walrus Island, a significant seabird rookery. These foxes are largely responsible for depopulating sea bird colonies, especially murres, on this small rocky island. Presently, NOAA has pack ice distributions in GIS format dating back to 1972. Ice pack incursions about the Pribilofs are being included in the GIS Project. More detailed information can be obtained through a weblink to NOAA's National Environment Satellite Data and Information Service's National Ice Center (www.natice.noaa.gov).

As noted elsewhere, fur seals received the attention of both the Russians and the Americans. Official duties of government agencies from both countries included mapping rookeries and haul out areas and maintaining various population statistics. With the passage of time, the influence of humans and the alteration of habitat has resulted in rookeries being modified or lost. For example, Webster's Point and the Salt Lagoon Rookeries are no more. GIS layering of historical areas is useful for the study of such changes, and the project also takes into account location names as ascribed over time within and between cultures.

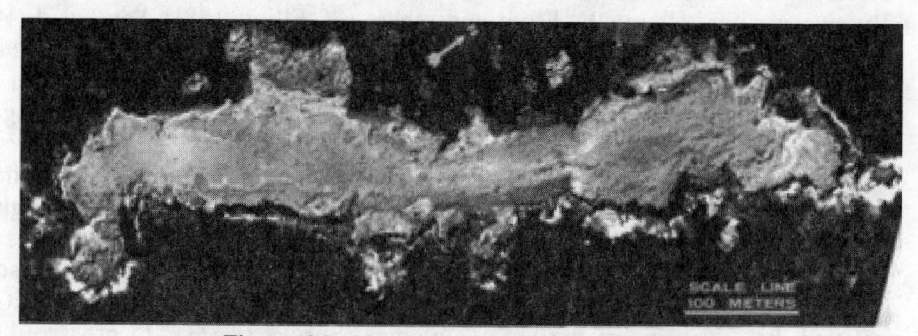

Figure 13.2 1948 aerial survey of Walrus Island.

The NOAA National Marine Mammal Lab has accumulated numerous aerial photographs of seal rookeries on St. Paul. The PPO scanned, registered, and rectified images from 1948, 1967, and the 1980's and included them in the GIS project.

Elliott noted (1875) only "a few sea-lions" bred on Walrus Island, although thousands could be seen among the fur seals at Northeast Point on St. Paul. Osgood *et al.* (1915) reported only two breeding rookeries: one of these was at Sea Lion Neck on St. Paul and the other near Garden Cove on St. George. But they also noted major haul out areas on all five Pribilof Islands. An aerial survey in 1948 found 1,258 adult sea lions on Walrus Island, but the resolution was insufficient to determine the number of pups if any. The aerial photos have been registered and rectified for inclusion into the GIS project. During a rare visit to Walrus Island in 2000, our contingent observed 109 sea lions of which 11 were pups. During the same period fewer than 10 sea lions were observed on Otter Island. Today, despite the significantly reduced numbers, Walrus Island is the major sea lion rookery in the Pribilofs. Eleven species of seabirds nest on the Pribilof Islands. The red-legged kittiwake, as one example, occurs only in the Bering Sea and it has major rookeries on the Pribilofs. Only one shore bird, the rock sandpiper, breeds on the Pribilofs. The U.S. Fish & Wildlife Service monitors seabird nesting on the islands, but no electronic maps are available to date.

13.7 CULTURE AND SUBSISTENCE

Veltre and Veltre (1981) undertook the first focused comprehensive evaluation of the subsistence natural resources on the Pribilofs, although much of their work was acquired from historical documents dating back to the 1800's. They point out that the sealing industry "had significant ramifications on subsistence in the Pribilofs." Table 13.3 provides a listing of some representative natural resources on or about the Pribilof Islands and the customary traditional uses of each.

Two species have been virtually eliminated from the Pribilof Islands, the Pacific walrus and the sea otter. Neither animal apparently played much of a subsistence role on the Pribilof Islands in part because of their demise by the promyshleniks who sought the valuable ivory and pelts, respectively. Their historical haul out areas are included in the GIS project.

A third species, the Steller sea lion, is close to extinction on the Pribilof Islands as discussed above. Scans of historical maps depicting sea lion rookeries and haul out areas are layered onto the GIS project and verified observations of sea lions on shore will be logged into the project. Today, the sub-adult male sea lion constitutes a subsistence resource and is hunted primarily from the shore. Some hunting does occur on the water at St. George Island, however. Sea lion are hunted for their meat, blubber, internal organs, and fore flipper, although in earlier years the animal provided several customary traditional uses (Table 13.2). Historically, the dried flesh and hides of sea lions along with fur seals has long been a staple on the Pribilofs. The sea lion provided essential meat and clothing, as well as covering for the bidarrah.

The northern fur seal continues to play a major subsistence role on the Pribilof Islands with several hundred being killed on St. Paul and St. George each year. Until recently, the U.S. Government allowed a subsistence take of 2,500 young bachelor males to be harvested each year. Today, the Tribal Governments on each island and the NOAA National Marine Fisheries Service (NMFS) are co-managing the subsistence harvest of fur seal. Locations authorized for subsistence sealing by year and favourite hunting areas are being included in the GIS project. Several recent seal midden locations are being mapped as well.

Links from the GIS maps are used to display more detailed information on the food subsistence value offered by the northern fur seal. A link from the map to a photograph (Scheffer 1948) that laid out the various body parts provides even more detailed information (Figure 13.3).

Table 13. 2 Species list using either Aleut or English name found on or about the Pribilof Islands and their customary traditional use.

Subsistence Resource	Customary Traditional Use
qawax	Hides: baidarkas and bidarrahs (skin boats) Intestines: waterproof garments Throats: boot tops Fore-flipper palms: boot soles Stomachs: oil containers Whiskers: traded to Chinese Sinews from back: "thread" for sewing Gall bladder: healing sores Carcass, liver, lungs, lard: food (Elliott, 1881)
itxayan	Carcass: food
aygagux	Skin: sold for it fur value
kagayax	Eggs: food Carcass: food
chuchiigix	Carcass: food
sikiita	Skins and feathers: parkas Eggs: food Carcass: food
saquudax	Plant stem: food
qayux	Berries: food
Salmonberry	Berries: food
aguganax	Roe: food
waygix	*Body: food*

Today, as in the past, many seabirds have customary traditional uses. Millions of seabirds flock to the islands each year for breeding. Henry W. Elliott was the first American on the Pribilof Islands that accounted for their natural history and cultural subsistence through prolific writings extending over 20 years. He noted that large flocks of *chuchiigix* (*Aethia pusilla*) arrived in the spring and were harvested by *Unangan* with hand scoop nets. *Chuchiigix*, known as the least auklet, make nests in crevices among boulders. *Chuchiigix* are still hunted today by using a long bamboo pole to strike the small birds out of the sky (bamboo often washes up on the shores of the islands). Veltre and Veltre (1981) point out that this is one of the few species that can be hunted by children. Elliott (1881) reported that least auklets were replaced in late July by large flocks of red-legged turnstones that were also eaten by *Unangan*. Unlike the *chuchiigix*, the turnstones bred elsewhere, but they stopped on the Pribilofs to feed on the "flesh flies and their eggs, which swarm over the killing grounds." The killing grounds are not nearly as extensive as they once were, but whether the abundance of turnstones on the islands was affected is not known.

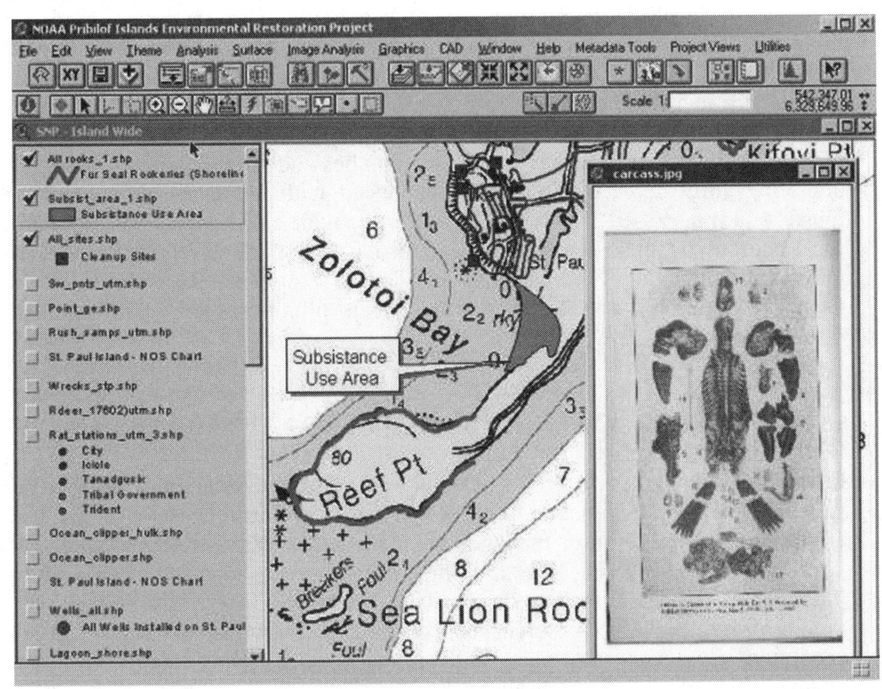

Figure 13.3 Example of hyperlinked media accessible from within the GIS: body parts and subsistence values of a northern fur seal.

Egg gathering was a major annual activity on the Pribilofs, although it is much less so now. Primary egg gathering areas included the steep rocky bluffs and the tabletop flat Walrus Island. According to Elliott, the primary egg-gathering season on Walrus Island was June-July (Elliott 1875). Historical egg gathering locations are included in the GIS project.

The Pribilof Islands are treeless, except for a low-to-the-ground-willow species. Several plant species afforded both cultural and subsistence value. For example, Elliot (1875) commented that only two berry plants, *Empetrum nigrum* and *Rubus chamemorus*, are readily harvested by the Aleuts during the end of August and first week in September. These species are still much sought after and often families spend summer afternoons harvesting the berries. *Saquudax* (wild celery) seemed to have had more importance in the past than it does now. A silent 1930's black and white film archived in the Oregon Historical Society shows Aleut children harvesting *saquudax* and demonstrating how it is eaten. A digitized version can be viewed via a hotlink when the distribution plots for the plant are active in the GIS project.

Heretofore, little documentation logged by the U.S. Government, Aleut community, or Russian Orthodox Church is known to exist as to the burial of deceased islanders. Some gravesites are marked, while many are known only orally and most are not recalled at all. As a community service and as part of the interest in historical documentation, NOAA received permission from the local churches to develop a GIS project of the cemeteries. Using a survey grade GPS, more than 700 gravesites on St. Paul and St. George Islands were entered into the GIS project. Digital images of each location along with individual statistics (individual names and photo, birth and death dates, and short biographical sketches as available) are linked to the geographic locations. Printed hard copy maps were provided for use by the local churches and municipalities. Using DVD authoring software, the GIS project enhanced with animation and Aleut choir music was transcribed onto a DVD. NOAA provided the respective DVDs to the local church councils. In turn, the St. Paul Island council is relying on the Tribal Government for long-term maintenance of the GIS project. Recently, the Russian Orthodox Church Diocese of Alaska began promoting this GIS application throughout the Diocese.

13.8 LONG TERM NATURAL RESOURCE MANAGEMENT

Currently, Tribal Government of St. Paul's Ecosystem Conservation Office (ECO) is using GIS in combination with traditional ecological knowledge for use in environmental and resource management. Due to the increased human interactions with the Pribilofs' Bering Sea environment, the need to monitor and record these interactions and their impacts on St. Paul and neighbouring islands has become vital. Monitoring of commercial and customary traditional activities, the island ecosystem (i.e. numerous plant and animal species, lands, and environmental changes), in addition to ecological disaster prevention planning are essential in safeguarding customary traditional practices and for protecting and conserving the health of the islands. Critical response planning to prepare the community to effect immediate responses to future ecological perturbations is also vital as competition increases between traditional and commercial interests for the dwindling Bering Sea fishery resources.

The impacts associated with regional oil spills, off-road all terrain vehicle trailbreaking, and commercial fishing along with knowledge about decreasing populations of northern fur seals, Steller sea lions, and various seabird species serve to highlight a need for a balanced harmony between the subsistence users and other interests. Harmonious relations can be achieved through communication, planning, negotiation, and mutual work towards an understanding and respect for differences among groups. To this end, ECO has worked on ongoing development of *Tanalix Amgignax* (Island Sentinel/Guard) Program in coordination with the Pribilof Islands Stewardship Program, PPO, NMFS, and USFWS. This program includes the following elements:

- Promote "caretaking" and stewardship of the Pribilof Islands' many life systems.
- Provide a central forum that actively deals with environmental issues.
- Monitor activities that have potential adverse implications on this area of the Bering Sea environment.
- Promote a proactive position on behalf of all life systems indigenous to the Pribilof Islands Bering Sea area for immediate monitoring.

- Exercise the Aleuts' inherent right to be responsible for the islands' environment.
- Share caretakership concerns and activities with others, expecting visitors to be on their "best behavior" while visiting St. Paul Island and the surrounding sea.
- Promote environmental and cultural awareness.

Some specific *Tanalix Amgignax* activities that also involve GIS include: preventing rats from coming to the island by actively setting and maintaining rat traps in and around the harbour area and sanitary landfill; monitoring the islands' shoreline and interior for natural process impacts (weather and wildlife) and disturbances, such as vegetative changes by overgrazing reindeer, shoreline erosion, and derelict fishing gear (i.e. nets, buoys, line, etc.) that wash up on shore; removal of fishing gear from around fur seals' necks; and sharing the importance of taking care of the Bering Sea environment with the community and visitors alike. Rat trapping locations are entered into the GIS project. If and when any rats are captured this information will be similarly logged into the project. GIS mapping of derelict fishing gear not only provides information to managers for directing shoreline cleaning up efforts, but also documents the periodicity of cleanup efforts along shorelines, the relative quantities and types of debris recovered, and identifies areas where oceanographic conditions work to focus debris.

Tanalix Amgignax Program is a prime example of the use of GIS in bringing together traditional ecological knowledge and "western science." NOAA's contribution to the ECO GIS project will foster the dialogue among the many interests that the *Tanalix Amgignax Program* seeks to bring together.

13.9 CONCLUSIONS

GIS application in the Pribilof Island Environmental Restoration Project afforded insights and opportunities not contemplated at the outset of the project. The use of DGPS technology to simply map contaminated and debris site locations, and sampling points led to numerous significant contributions. One of the more routine contributions involved using JPEG and TIFF files cut from the GIS project to make maps for documents on topics such as site investigations, corrective action plans, corrective action reports, and requests to oversight regulators for no further action. Among other significant contributions came new nautical charts with improved near-shore confidence by local fishers in GPS navigation systems; map overlays of historical northern fur seal and sea lion rookeries for comparison with current and ever changing population and environmental conditions; estimates of shoreline erosion rates; improved control management of shoreline marine debris; near real time mapping of contaminant excavations allowing for on the ground decisions; sub-meter scale topographic maps to assist engineers in landfill closeout design and calculating royalty payments from sand and scoria (used as excavation backfill and capping solid wastes) mining; improved communications among officials who lacked on-island experience yet hold responsibility for land transfers following cleanup; and sorting through conflicting boundary descriptions arising from a variety of official documents over an approximate 25-year period.

A collaborative spirit between the government and Aleut entities grew out of training opportunities provided by NOAA in DGPS and GIS, access to NOAA DGPS equipment for ECO Projects, and contributions of GIS software to the City of St. Paul and ECO. Subsequent collaborative geo-spatial projects further improved overall communications. DGPS and GIS enabled local Aleuts to express their own cultural heritage by applying Aleut location names to the first ever large-scale topographic quadrangles of the islands. In terms of community relations, the most significant contribution afforded by DGPS and GIS came from the mapping of the islands' gravesites. Unwittingly, this single DGPS/GIS exercise offered up as a community service in context with a federal mandate to preserve historical information provided the strongest bond and trust between the islands' Aleut population and U.S. Government agents.

Acquisition of precision geo-spatial data in real time is one of many challenges in the higher latitudes. Geodetic control points, such as those offered by Continuously Operating Reference Stations (CORS), are far and few between in America. Alternatively, one can, as done in this project, survey in a base station to provide short-range control. For long-term applications base stations must be maintained against vandalism, climatic events, and curious animals such as foxes that chew through cables. Repeater stations serve to overcome multipath interference such as caused by hilly terrain and buildings. Once these obstacles are overcome, real-time DGPS in combination with GIS provide the investigator with tremendous power to work multiple field situations near real time as well as providing for long-term documentation.

13.10 REFERENCES

Butman, B. and Lindsay, J., 2000, *A Marine GIS Library for Massachusetts Bay: Focusing on Disposal Sites, Contaminated Sediments, and Sea Floor Mapping.* U.S. Geological Survey Open-File Report 99-439.

Elliott, H.W., 1875, *A Report Upon the Condition of Affairs in the Territory of Alaska* (Washington: Government Printing Office).

Elliott, H.W., 1881, *The Seal Islands of Alaska,* Bulletin Tenth Census of the United States. Dept. of the Interior (Washington Government Printing Office).

Lindsay, J.A., Simon, T, Graettinger, G., and Bailey C., 1999, A Global Offshore Hazardous Materials Sites GIS (GOHMS-GIS). In *Proceedings of Coastal Zone '99 Conference*, San Diego, California, pp. 392-394.

Martin, F., 1960, *Sea Bears: The Story of the Fur Seal.* (Philadelphia: Chilton Company).

Osgood, W.H., Preble, E.A., and Parker, G.H., 1915, The Fur Seal and Other Life of the Pribilof Islands, Alaska, in 1914. *Bulletin of the Bureau of Commercial Fisheries*, Vol. XXXIV, 1914; Document No. 820 (Washington: Government Printing Office).

Scheffer, V.B., 1948, Use of Fur-Seal Carcasses by Natives of the Pribilof Islands, Alaska. *The Pacific Northwest Quarterly*, **39** (2), pp. 131-132.

Townsend, C.H., 1896, Condition of Seal Life on the Rookeries of the Pribilof Islands. In *Reports of Agents, Officers, and Persons, Acting Under the Authority of the Secretary of the Treasury, Condition of Seal Life on the Rookeries of the Pribilof Islands and to Pelagic Sealing in Bering Sea and the North Pacific*

Ocean in the Years 1893-1895. 54th Congress, 1st Session, Senate, Document 137, Part 2 (Washington: Government Printing Office).

Veltre, D.W. and Veltre, M.J., 1981, *A Preliminary Baseline Study of Subsistence Resource Utilization in The Pribilof Islands*. Alaska Department of Fish and Game, Division of Subsistence (Contract 81-119) Technical Report Number 57.

CHAPTER FOURTEEN

GIS Applications In Coastal Management: A View from the Developing World

Peter C. Nwilo

14.1 INTRODUCTION

Geographic Information Systems have revolutionised the way spatial data/information is acquired, stored, managed, and displayed. In the past the acquisition of spatial data was done mainly with analogue theodolite and levels, and the recording and storage of these data were done in analogue manner. Data management was a very cumbersome exercise and the display was in the form of paper maps. Recent developments in information and digital technologies, communication and satellite technology, and improved computing speed and computer hard disk space, have impacted tremendously the way spatial data is acquired, stored, managed, and displayed. The development has created a revolution in spatial information technology in a way never experienced before. Today, virtually every human activity is spatial. Some schools of thought stipulate that in countries such as the United States and the United Kingdom as much as 75-80% of all the activities is spatial. There is correlation between geospatial information technology development and economic development. Developed countries of the world are associated with high spatial information technology development. Conversely, poor economic development is associated with low spatial information technology development as witnessed in developing countries of the world.

The development of Geographic Information Systems and associated technologies has created and is still creating employment opportunities all over the world. It has also led to changes of curricula of training institutions, and new training opportunities are being created. Transfer, sharing access to and development of spatial data standards have become major issues. We now discuss spatial data infrastructure as a major requirement for development, as has long been done with regards to other infrastructures such as roads, rail and waterways, electricity, and water supplies.

Geographic information has been employed in several fields such as management of disasters, vehicle tracking systems, forestry, utilities management, oil exploration, environmental management, health management, census, and

0-41531-972-2/04/$0.00+$1.50

mineral resources development, governance (EIS Africa, 2002) and – the focus of this paper – coastal management.

14.1.1 Coastal Areas

The coast is an area that witnesses a substantial amount of physical and economic activities. It is a zone of great significance in the sense that intense agriculture, business and recreational activities take place here. Settlements have been established in these areas and international trade and communication originated in the coastal fringes (Nwilo, 1995; Nwilo *et al.*, 1995). The coastal zone houses about two thirds of the world's population and this percentage is expected to increase to 75% by the year 2020 (UNEP, 1992; Hanson & Lindh, 1993). Most of the important cities of the world, such as Alexandria, Lagos, London, Rotterdam, Shanghai, Tokyo and Venice, are located within the coastal environment, while two thirds of the world's cities having a population of over 2.5 million are within the coastal zone (Borrego, 1994). The coastal environment is very rich in both living and non-living resources. For example, a lot of crude oil and gas exploration and exploitation take place with the coast.

Wetlands are known to be a very productive ecosystem. Most of these wetlands are located in the coastal areas. They are the sanctuaries to many global endangered species and also can assist in dampening the effect of flooding from normal sea flooding, storm surges and sea level rise. Fishes that are caught in the upwelling zone of the Canary Island coast are hatched within the mangrove ecosystem of the Niger Delta and later migrate to the upwelling area. Similar relationships are found in most of the world's major fishing waters.

The coast is therefore a very important zone. Having discussed the importance of the coastal areas, it is now necessary to define the 'coastal zone.' Bird (1967; 1985 & 1993) defines the coast as a zone of varying width including the shore and extending the crest of cliff, the head of a tidal estuary, or the solid ground that lies behind the coastal lagoon, dunes and swamps. Komar (1976) on the other hand defines the coastal zone to include the littoral zone and extending further inland to include the sea cliffs, any marine terraces, dune fields and so on; the seaward limit is limitless. For most studies the coastal zone refers to that zone that is affected by what happens on the sea and on land. The coast is therefore very important for the existence of human life.

14.1.2 Integrated Coastal Area Management

Integrated Coastal Areas Management is defined as a dynamic process in which a coordinated strategy is developed and implemented for the allocation of environmental, socio-cultural and institutional resources to achieve the conservation and sustainable use of the coastal area (Coastal Area Management and Planning Network, 1989; Nwilo, 1995). Planning for sustainable resource management is based on weighing priorities, translating these priorities into policies and finally defining goals, identifying responsibilities for each step and establishing a time frame for action and review (Nwilo, 1995). There is no one 'right' way to manage coastal areas. However the design of a coastal management programme or policy for a nation or a region should take into consideration the legal instrument for the implementation of the program as well as the instrumental frame work for policy implementation (Emovon, 1991).

In order to practice effective coastal management, planners need to understand the way the natural environment and human activities are interconnected to form a system. Key aspects of the system include the following information themes:

i. Biological: This includes type and extent of ecosystem, primary productivity, specie diversity and abundance, nursery grounds and life cycles;
ii. Physical: This includes topography, geology, temperature, salinity, nutrients, tides, sea level and current, meteorology, sediment types and distribution, flooding and erosion/ accretion;
iii. Socio-economic: This includes human population distribution and growth, economic activities and land use;
iv. Legal and Institutional: Land tenure system, resource use rights, relevant laws and regulations, responsible agencies and availability of financial and human resources (Borrego, 1994).

Each of these activities is greatly influenced by activities within and beyond the coastal zone. It is for this reason that the resolution of conflicts in the use of coastal resources requires a broad perspective on the environmental process and interaction among human activities.

14.2 THE GUINEA CURRENT LARGE MARINE ECOSYSTEM PROJECT

14.2.1 The Concept of Large Marine Ecosystem Project

Large Marine Ecosystems are regions of the ocean space encompassing coastal areas from river basins and estuaries to the seaward boundaries of the continental shelves, and the outer margins of the world's current system. These marine regions are considered in terms of their ecological unity based on their distributive bathymetry, hydrographic, productivity, and trophic linkages (Sherman *et al.*, 1993; UNIDO, 2002). The world's 64 LMEs are the most productive ecosystems, and together produce 95% of marine fishery biomass yields. It is also in the LMEs that most of the global ocean pollution, fisheries over-exploitation and coastal habitat alteration take place.

The LME ocean management approach focuses on the sustainable development of the ocean resources and requires a paradigm shift from a small spatial scale to a larger one and from a short-term to a long-term perspective. It requires countries bordering the LME to set priorities on how to tackle their common transboundary issues. Since the early 1990s, the developing countries have approached the Global Environment Facility (GEF), the United Nations Industrial Development Organisation (UNIDO) and other United Nation implementing agencies for technical and scientific assistance in restoring and protecting their coastal and marine ecosystem. GEF agreed to provide guidance and funding in addressing these issues within the framework of sustainable development. It recommended the use of the Large Marine Ecosystem (LME) concept and their contributing freshwater basins as the geographic focus for addressing the issues (UNIDO, 2002).

14.2.2 The Guinea Current Large Marine Ecosystem Project Implementation

The Guinea Current Large Marine Ecosystem Project is funded by the Global Environment Facility (GEF) and managed by the United Nations Industrial Development Organisation (UNIDO). The phase I of this project was started in 1995 and ended in 1999 with the countries of Cameroon, Nigeria, Benin Republic, Togo, Ghana and Côte d'Ivoire participating. Apart from GEF and UNIDO, some form of scientific and technical support came from the United Nations Education and Scientific Cultural Organisation (UNESCO), other UN agencies, and the US National Oceanic and Atmospheric Administration (NOAA). The participating countries provided in country support in the form of logistics such as provision of offices and transportation vehicles. The execution of the project was in form of modules. The modules that were covered in the project included Integrated Coastal Areas Management (ICAM), Geographic Information System (GIS), Mangroves, Industrial Pollution, Policy Issues, Fisheries, Coastal Erosion, Plankton and other pollution. The structure made provision for a National Project Director in each of the participating countries and a National Expert for each of the modules. All the national experts were reporting to an International Expert who co-ordinated each of the modules. The project had a Project Coordinator who was based in Abidjan, Côte d'Ivoire. The office of the Project Coordinator also served as the Secretariat for the project. In each of the participating countries, there was a Focal Point Agency and Focal Point Institution.

In Nigeria, the now-defunct Federal Environmental Protection Agency was the focal point agency, while the Nigerian Institute for Oceanography and Marine Research was the focal point institution. There were other participating institutions such as the University of Lagos. In the other participating countries, the focal point agency was the Ministry of Environment. Efforts were made in the project to collaborate with relevant institutions.

The second phase of the project is bringing in an additional 10 countries. These include Guinea Bissau, Guinea, Sierra Leone, Liberia, Congo Brazzaville, Democratic Republic of Congo and Angola. These countries, in addition to the original six, are in partnership activities for sustainable development of the Guinea Current Large Marine Ecosystem (GCLME). The addition of these to the original six countries meant that the full extent of the Gulf of Guinea Current Large Marine Ecosystem is covered. This stage of the project will commence in January 2004 (UNIDO, 2002).

All participating countries are making contributions to a Transboundary Diagnosis Programme (TAD) and a Strategic Action Programme (SAP) for the project (UNIDO, 2002).

14.2.3 The ICAM Module

The essence of the Guinea Current Large Marine Ecosystem project is the sustainable management of the living resources of the Gulf of Guinea region. The region was experiencing a serious environmental stress due to urbanisation, pollution from industries and domestic wastes, population explosion, oil spillage, deforestation, erosion, and over-fishing. The fish stocks were depleting at a fast rate and the socio-economic lives of the inhabitants are adversely affected. The lagoons that dot the coast from Côte d'Ivoire to Nigeria had become heavily polluted. Most of the problems highlighted above are transboundary in nature. There was therefore a need to reverse the situation and address these problems from a regional perspective rather than from a national perspective.

Under the module, an experienced ICAM professional prepared guidelines for implementation of ICAM in the countries. The National Experts prepared a country profile for each of the six participating countries in the first phase. The country profile is a baseline of the situation in each of the participating country's coastal area. The information used in preparing the country profile was obtained from existing literature and from relevant institutions. This was followed by a three-day workshop where stakeholders discussed the profile and made recommendations where necessary. A final copy of the country profile was made after effecting the necessary corrections from stakeholders.

An outcome of the workshop was the setting of the National Integrated Coastal Area Committee for each of the participating countries. Another success of the workshop and activities of the Guinea Current LME phase 1 is the creation of awareness on the importance of Integrated Coastal Area Management: prior to the Guinea Current LME project, there was no National ICAM Committee in any of the six participating countries.

As part of the efforts to assess the health of the Guinea Current LME, a fish trawl survey was carried out from Côte d'Ivoire to Cameroon. Results from the survey confirmed the fear that the fish stock, together with the plankton, was being depleted.

Figure 14.1 shows the Gulf of Guinea Region while Figure 14.2 shows the countries that took part in the first phase of the project

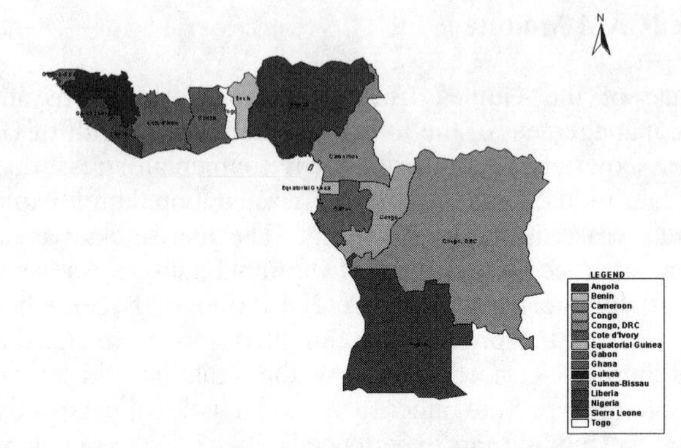

Figure 14.1 The Guinea Current Region

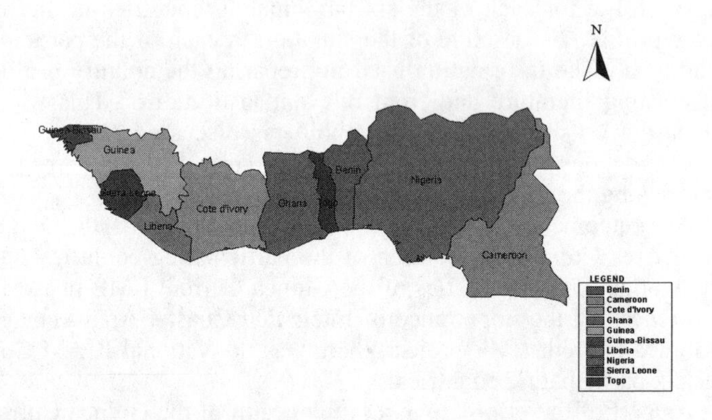

Figure 14.2 Countries that participated in the GCLME Project, Phase I

14.2.4 Application of GIS to The Project

As was the case in the Integrated Coastal Area Management Module, there was a National Expert for the GIS Module in each of the participating countries in phase I of the project. The author served in this position for Nigeria. The project started in 1995 but the National experts were engaged in 1997 for a two-year period. The contract was on part time basis.

The major goal of the GIS application in the project was to use GIS as an information and decision support system. Decision-makers are not always in a position to read all the volumes of materials that are passed to them. The graphic displays that are provided by GIS will invariably reduce the time that would have been spent in reading several volumes of materials, and assist in taking decisions that affect a project quickly. For example, it is easy to visualise the way a mangrove ecosystem of a region is changing over time through graphical displays.

14.2.5 Capacity Building for the GIS Module

The project provided for National Experts to attend Training workshops and conferences. A Project Expert Committee Workshop was organised by the Project Coordinator's office in Abidjan in April 1997 for all the National GIS Experts. This workshop touched on GIS application in the project and in particular the issue of metadata. Since the project is regional in nature the issue of harmonisation of different mapping datums in the region was discussed extensively. The National Experts provided information on institutions using GIS in the different participating countries. It was also a forum for experts to get to know each other. Another GIS Expert meeting took place in Abidjan between 29-30 September 1998 to deliberate on issues affecting the project.

14.2.6 SIMM Coast Workshop

SIMM is an acronym for sustainable integrated management and so SIMM Coast stands for sustainable integrated management of Coast. This workshop was organised by Prof. McGlade, formerly of the Centre for Marine Research, University of Plymouth, United Kingdom. It was essentially on the application of artificial intelligence in integrated coastal management. All the GIS National Experts were sponsored by UNIDO to participate in the workshop. This workshop took place at the Food and Agricultural Organisation office in Accra, Ghana between July 30th and 31st, 1998.

14.2.7 The Accra Declaration

The first meeting of the Ministerial Committee of the Gulf of Guinea Large Marine Ecosystem (GOG-LME) Project took place in Accra, Ghana, on 9th and 10th of July, 1998. Five Ministers with responsibility for the Environment respectively in Benin, Cameroon, Côte d'Ivoire, Ghana and Togo and the Director General/Chief Executive of the Federal Environmental Protection Agency of Nigeria attended the meeting.

After the deliberations of the ministers, which were based on extensive and substantive preparations, the Committee of Ministers adopted the Accra Declaration on Environmentally Sustainable Development of the Large Marine Ecosystem of the Gulf of Guinea. Part of the decision reached was that efforts shall be made to initiate, encourage and work synergistically with current and/or programmed national and international programmes on integrated coastal zone management in the region. The national concerns of flooding, and pollution caused by hydrocarbons, toxic chemical products, fisheries productivity and over-exploitation and, above all, coastal erosion, call for the special attention of donors. Also, data and information networking between the GOG-LME countries should be improved. National and Regional databases on the coastal and marine environment should be established using the Geographical Information System (GIS) to support decision-making, to be available to all users.

14.2.8 Results from the Phase I of the GCLME Project

As part of the success story of the GCLME project, guidelines for integrated coastal areas management and planning have been developed for Benin, Cameroon, Côte d'Ivoire, Ghana, Nigeria and Togo. Country coastal profiles were similarly developed and major issues and options for national and regional actions were identified.

The Gulf of Guinea Large Marine Ecosystem (GOG/LME) project has organised and facilitated a series of activities that have enabled each country to take the appropriate steps toward the adoption of Integrated Coastal Areas Management Plans, including strengthening legal and policy frameworks. In addition, these plans and policies are available for incorporation into National Environmental Action Plans. ICAM has proven to be the most effective way to involve the grassroots stakeholders in a meaningful way and could be the cornerstone for the GOG/LME potential Phase II Project.

Indications that member countries will embrace and make this effort sustainable are beginning to appear. First, National Steering Committees have been established by legal instruments. Second, the individuals who are participating in the programme range from top Ministry officials to citizens in NGOs and CBOs, indicating the existence of multilevel support.

An important parallel development with the effort to put National ICAM Plans in place has been the process of establishment of information management and decision-making systems such as Geographic Information Systems (GIS) at both the national and regional levels. ArcView 3.1 GIS software was chosen as the GIS platform for the project. It is hoped that in the second phase of the project, ArcGIS 8.3 will be the platform.

Some of the maps, which were developed from existing information in Nigeria during the execution of the project, are shown in Figures 14.3 and 14.4.

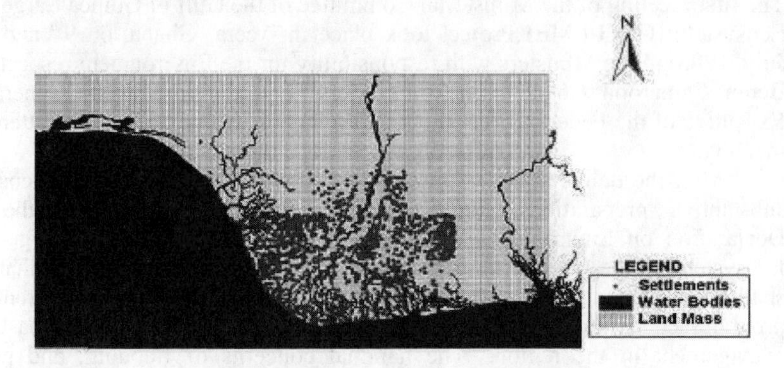

Figure 14.3 Settlements in the Niger Delta

Figure 14.4 Mangrove Ecosystem in the Niger Delta

14.3 OTHER COASTAL AREAS PROJECTS IN AFRICA

14.3.1 Niger Delta Environmental Survey

The Niger Delta is an environmentally sensitive and fragile region owing to its peculiar natural physical setting and its distinctive ecological features and functions which over the years have drawn the attention of conservationists especially on the need for sustainable development and the protection of its biodiversity (NDES, 1997). Considerable changes are occurring in the ecological environment and in the socio-economic setting of the Niger Delta as a result of both natural and another anthropogenic transformations that include: upstream dam construction, coastal zone modification, urbanisation, deforestation, agriculture, fishing, industrial development, population pressure, and crude oil exploration and exploitation. The Niger Delta is therefore under increasing pressure from rapidly deteriorating ecological and economic conditions, social dislocation and tension in communities, problems that are not being addressed by current policy and behaviour patterns.

It is for this reason that the Niger Delta Environmental Survey (NDES) was set up. The NDES, in concert with communities and other stakeholders, undertook a comprehensive survey of the Niger Delta. It established the causes of ecological and socio-economic changes over time and induced corrective actions by encouraging relevant stakeholders to address specific environmental and related economic problems aimed at improving the quality of life or the people and achieving sustainable development in the region. Part of the objectives of the survey was to generate data and information on the Niger Delta including the establishment of a Geographic Information System (GIS). The data and information can be used in formulating strategies and plans for effective natural resource management towards the sustainable use of resources, in order to protect the environment and the livelihood of peoples of the region (NDES, 1997).

Analogue maps of the Niger Delta at scale of 1/25,000 were digitised and updated with remotely sensed images. The map information is now in a GIS environment, and is used in decision-making for the planning and management of the Niger Delta.

14.3.2 Secretariat for Eastern African Coastal Area Management (SEACAM) Project

SEACAM means secretariat for Eastern African Coastal Areas Management. It was set up in August, 1997 in Maputo, Mozambique, with the purpose of accelerating the implementation of integrated coastal zone management in the region as put forth in the Arusha Resolution of 1993 and Seychelles statement of 1996 on ICZM. The Secretariat is hosted by the Ministry of Environmental Affairs for Mozambique.

The objective was to assist Eastern African Countries to implement ICZM following up the Arusha resolution and the Seychelles statement on capacity building and information dissemination. It was supported by the Government of Finland, Danish International Development Assistance (DANCED), the World Bank and UNESCO. A number of other donors provide important support to select activities that match their programme areas.

In 1996, the Western Indian Ocean Marine Science Association organised an expert and practitioners workshop in Taga, Tanzania on integrated coastal area management, which provided an input into the second ministerial policy conference on integrated coastal and zone management in Eastern Africa and Island States held in Seychelles, October, 1996. It is not certain whether GIS was used in this project.

14.3.3 Lake Victoria Environmental Management Project

For decades, East African scientists, policy makers, stakeholders and others around the world noted the escalating environmental degradation of Lake Victoria, which is an international water body. It was observed that increased population is a major contributing factor. The total population of the riparian communities currently stands at 30 million.

As a result of the degradation the riparian countries to Lake Victoria, which are Kenya, Tanzania and Uganda, initiated discussions immediately after the United Nations Conference on Environment and Development (UNCED) to broaden regional cooperation in environmental and social issues affecting Lake Victoria Basin. This led to the signing of a triplicate agreement in 1994. The project attracted financial support from the International Development Association (IDA) and Global Environmental Trust Fund (GEF) to the tune of US $70.00 million. The participating countries agreed to contribute $7.2M as counterpart funds. Apart from other achievements of the project, mapping of the present land use/land cover and soil erosion hazard in the Lake basin is almost completed. Data on pollution loading are being collected from selected micro-catchments and provide useful information on soil nutrients and water losses. Human capacity is being developed. An important component of the project is community participation in the preparation and implementation of the project. It is not clear whether GIS was used as an information management tool but it is obvious that there was mapping which could, if still in analogue format, be converted to digital format and managed in a GIS environment.

14.3.4 Information Technology for Coastal Zone Management of the Nile Delta

This is an initiative for development and implementation of an integrated coastal analysis and monitoring system. A workshop was organised between 7 – 10 April 2003 aimed at addressing issues such as coastal erosion, pollution, monitoring and fisheries management. It will serve as a model for information system implementation in the developing world context, where the need for operational coastal management is often great but for which information infrastructure and knowledge base are frequently lacking.

14.3.5 Eastern African Coastal Resource Maps Project

The Eastern African Coastal and Marine Environment Resources Database and the Atlas was founded within the framework of the 1985 UNEP brokered Eastern African Action Plan for the protection, management and development of the Marine Environment of the Eastern African Region (EAF). The countries of Comoros, La Réunion, Kenya, Madagascar, Mauritius, Mozambique, Seychelles, Somalia and Tanzania participated in the project.

The project was initiated in 1993 by the Ocean and Coastal Area Programme Activity Centre (OCA/PAC) in co-operation with the Global Activity Centre (GRID/PAC-Nairobi). The project was scheduled to run for five years. The first phase of the project only focused on Kenya while Phase II and III (1995-1996) concentrated on Comoros, Mozambique, Seychelles and Tanzania. Phase III and IV (1996-1997) concentrated on Madagascar, Mauritius, La Réunion and Somalia. Phase V is intended to round up all activities at the National Level.

The production of the digital database led to wiser use of resources and helped reduce wastage of non-renewable resources. The main objective of the project is to collect existing information on coastal resources of the region and to summarise the same in country map sheets managed in a GIS environment (Okemwa, 1995).

Scientists from the participating countries took part in the collaboration of the information and the development of the database. Information in the database covers parameters including but not limited to climate, hydrology, oceanography, coastal types, geomorphology, geology, tourist infrastructure, important ecosystems such as mangroves systems, wetlands, estuaries, rangelands and mineral resources.

The maps will be useful in studying the vulnerability of the coastline and indeed as a versatile tool in making strategic intervention in cases of oil spills.

14.4 CONCLUSIONS

Various coastal management projects in Africa have been discussed above. It is observed that most of these projects were conceptualised and executed by the developed countries, mainly European countries, while the African scientists only played supportive roles. Two projects in the opinion of the author stand out as clear exceptions to this situation. These are the Guinea Current Large Marine Ecosystem project and the Niger Delta Environmental Survey.

As earlier stated, the Guinea Current Large Marine Ecosystem project is funded by the Global Environment Facility and managed by the United Nations Industrial Development Organisation. Initially, the project was conceptualised by the governments of Nigeria, Ghana, Togo, Cameroon, Benin and Côté d'Ivoire. It was later realised that there was a need to include all the countries bordering the Guinea Current and this increased the number of participating countries to sixteen.

The other project, the Niger Delta Environmental Survey, was conceptualised by the oil prospecting companies in Nigeria. Its mission statement is: "In concert with communities and other stakeholders to undertake a comprehensive environmental survey of the Niger Delta, establish the causes of ecological and socio-economic changes over time and induce corrective action by encouraging relevant stakeholders to environmental and related socio-economic problems identified in the survey to improve the quality of life of the people and achieve sustainable development in the region" (Niger Delta Environmental Survey, 1996). Incidentally, these two projects are in Nigeria.

There is therefore a need to involve the African scientists and researchers from the conceptualisation stage of projects on the Continent. This is believed to be more beneficial to these countries and their scientists, particularly in the area of capacity building.

14.5 REFERENCES

Bird, E.C.F., 1967. *Coasts: An Introduction to Systematic Geomorphology*, (Cambridge, Mass.: The MIT Press).

Bird, E.C.F., 1985. *Coastline Changes. A Global Review*, (Chichester, England: John Wiley and Sons).

Bird, E.C.F., 1993. *Submerging Coasts: The Effect of a Rising Sea Level on Coastal Environments*, (Chichester, England: John Wiley and Sons).

Borrego, C., 1994, Sustainable Development of Coastal Environment: Why is it Important? In *Littoral 94*, edited by de Cavellio, S. and Gomes, V., pp 11-23.

Coastal Area Management and Planning Network, 1989, *The Status of Integrated Coastal Zone Management: A Global Assessment,* Workshop on Coastal Zone Management, Rosenteil School of Marine Sciences , University of Miami, USA.

EIS Africa, 2002, *Geoinformation Supports Decision-Making. An EIS Position Paper*

Emovon, E.U., 1991. National Science and Technology Policy and the Nigerian Environment. In *The Making of the Nigerian Environmental Policy*, edited by Aina, E.O.A. and Adedipe, N.O., pp 71-78.

Hanson, H., and Lindh, G., 1993, Coastal Erosion - An Escalating Environmental Threat. *Ambio,* **22**, (1), pp. 189-195.
Komar, P.D., 1976, *Beach Processes and Sedimentation*, (New Jersey: Prentice Hall).
NDES, 1996: Terms of Reference
NDES, 1997: Final Report Phase 1, Volume 1: Environmental and Socio-Economic Characteristics.
Nwilo, P.C., 1995, *Sea Level Variations and the Impacts along the Coastal Areas of Nigeria,* Ph.D Thesis, University of Salford, Salford, UK, 229p (Unpublished).
Nwilo, P.C., Onuoha A.E., and Pugh Thomas, M., 1995, *Monitoring Sea level/Relative Sea Level Rise in a Developing Country - The Nigerian Experience*, (Intergovernmental Oceanographic Commission).
Sherman, K., Alexander, L.M., and Gold, B.D., 1993, *Large Marine Ecosystem: Stress, Mitigation and Sustainability*, (Washington, D.C.: AAAS Press).
UNESCO, 1995, *Workshop Report No. 105 Supplement: International Conference on Coastal Change 95*, Bordeaux, France.
UNEP, 1992, *The World Environment 1972 - 1992*, edited by Tolba, M.K. and El-Kholy O.A., (London: Chapman & Hall).

CHAPTER FIFTEEN

High-Resolution Elevation and Image Data Within the Bay of Fundy Coastal Zone, Nova Scotia, Canada

Tim Webster, Montfield Christian, Charles Sangster, and Dennis Kingston

15.1 INTRODUCTION

The Applied Geomatics Research Group (AGRG) is a component of the Centre of Geographic Sciences (COGS) located in the Annapolis Valley, Nova Scotia. Its mandate is the application of geomatics technology for environmental research within Maritime Canada. In the fall of 1999 and summer of 2000 a large data acquisition campaign was initiated to collect high-resolution elevation and other remotely sensed image datasets along the Bay of Fundy coastal zone (Figure 15.1 and colour insert following page 164). The purpose of the research was to evaluate their effectiveness in obtaining critical information about the coastal zone, particularly data to be used to assess flood-risk potential associated with storm surge events. Global mean sea level has been increasing between 0.1 and 0.2 meters per century. With increasing greenhouse gases, sea level rise is expected to accelerate and the Intergovernmental Panel on Climate Change predicts that global average sea level may increase by 0.09 to 0.88 meters by 2100, placing the lives and property of an estimated 46 million people at risk (Houghton et al., 2001). The Bay of Fundy is no exception: relative sea-level is rising in this region by an estimated rate of 2.5 cm per century and many coastal areas are becoming more susceptible to flooding from storm events (Stea, Forbes, and Mott, 1992). In addition to sea level rise, storm surge and ocean waves are also factors at the coastline and are carried to higher levels on rising mean sea level. Storm surge in general is defined as the algebraic difference between the observed water level and the predicted astronomical level as one would find in tide tables. With possible increased storminess associated with climate change, the next 100 years will probably see more frequent flooding of coastal zones, and an increase in erosion of coastal features. With the recent increase in the spatial resolution of geomatics data available, both multispectral imagery (Ikonos, Quickbird, CASI) and high accuracy elevation data, landuse planners and policy makers now have access to the information required to manage the coastal zone.

0-41531-972-2/04/$0.00+$1.50

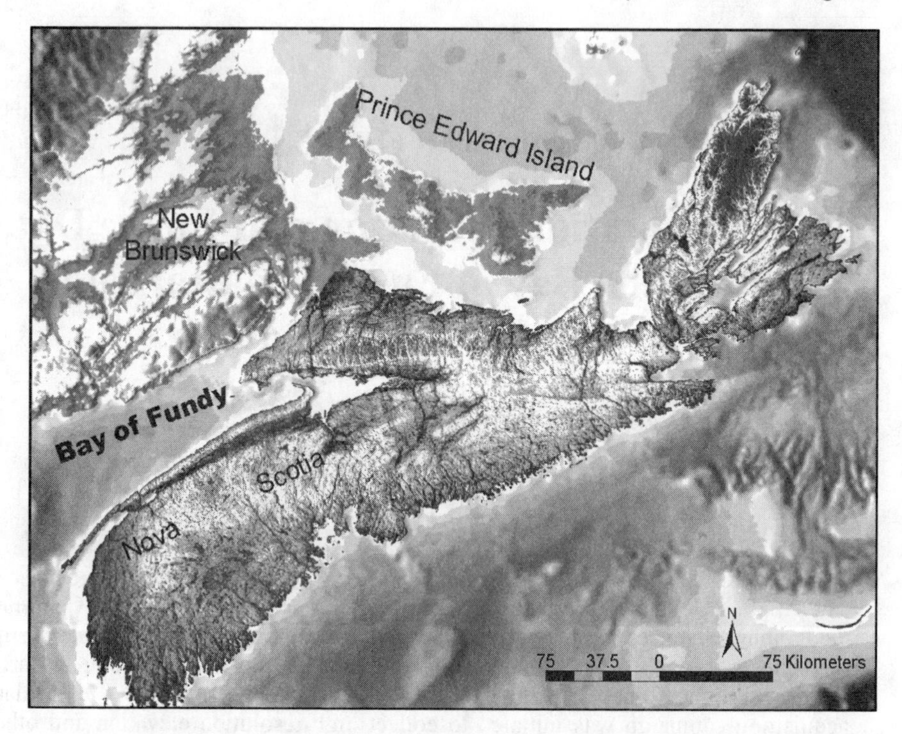

Figure 15.1 Location map, Bay of Fundy and Annapolis Valley (between Bay of Fundy and Nova Scotia label), Nova Scotia, Canada. This image is made up of a Radarsat S-7 mosaic for Nova Scotia, merged with a colour shaded relief map for the rest of Maritime Canada. Radarsat data © 1996, Canadian Space Agency.

LIDAR (Light Detection and Ranging) technology has been employed for a number of years in atmospheric studies (e.g. Post et al., 1996; Mayor & Eloranta, 2001) and as an airborne technique for shallow bathymetric charting (e.g. Guenther et al., 2000), although cost was initially an impediment to widespread acceptance for the latter purpose. The technology can also be used to image the land and water surface (Hwang et al., 2000), as was done in the present study. Terrestrial LIDAR applications have been demonstrated in forestry (Maclean & Krabill, 1986), sea-ice studies (Wadhams et al., 1992), and glacier mass balance investigations (Krabill et al., 1995, 2000; Abdalati & Krabill, 1999). A general overview of airborne laser scanning technology and principles is provided by Wehr and Lohr (1999). Applications to coastal process studies in the USA have been reported by Sallenger et al. (1999), Krabill et al. (1999), and Stockdon et al. (2002), among others. Preliminary trials in Atlantic Canada were reported by O'Reilly (2000) and subsequent experience was described by Webster et al. (2001, 2002, 2003) and McCullough et al. (2002). Most of the coast of the conterminous USA has now been mapped using this technology (Brock et al., 2002). A comprehensive review of the theory and applications of Digital Elevation Models (DEM) covers both terrestrial and marine LIDAR as well as other technologies used for DEM construction such as IFSAR – interferometric airborne synthetic aperture radar – is given by Maune et al. (2001).

Two different data providers were contracted to acquire LIDAR for the region. This chapter will discuss the details of the LIDAR systems, mission planning, data validation, and processing. The LIDAR DEM was then used in generating flood-risk maps associated with storm surge events along the coastal zone. This information has been passed to the local planning commissions to aid in management of the coastal zone. Another outcome of the project is that one of the data providers has significantly improved their acquisition system and approach to quality assurance when collecting LIDAR data.

15.2 DATA ACQUISITION AND PLANNING

Data that were acquired during the fall of 1999 were airborne polarimetric synthetic aperture radar (PSAR) data from the Convair 580 aircraft operated by Environment Canada, and satellite imagery from Radarsat-1, and Landsat-7. The polarimetric SAR was a simulation of Radarsat-2, Canada's second earth observation satellite planned for launch in 2005. Data acquired in the summer of 2000 campaign included high-resolution airborne data from the Compact Airborne Spectrographic Imager (CASI), and LIDAR systems, and more satellite imagery from Ikonos, Radarsat-1, and Landsat-7 (Table 15.1).

Table 15.1 Types of data collected during 1999, 2000 summers.

Data Type/Sensor	Resolution (m)	Attribute
PSAR	6	Polarimetric SAR signal, C-Band 5.6 cm, HH, VV, HV polarizations
Radarsat	12 (variable)	C-Band 5.6 cm, HH polarization, variable incidence angle
Landsat	15 pan, 30 mss	Visible, near and mid-infrared imagery
Ikonos	1 pan, 4 mss	Visible and near infrared imagery
CASI	2 (variable)	Visible and near infrared imagery
LIDAR	2 (variable)	Elevation (ground and non-ground)

Polarizations: HH – horizontal transmit, horizontal receive, VV – vertical transmit, vertical receive, HV – horizontal transmit, vertical receive.

The study area consists of the Annapolis Valley region of Nova Scotia, located on the southeast shore of the Bay of Fundy (Figure 15.1). LIDAR and CASI coverage consisted of the entire length of the valley and coastal zone. The satellite coverage was concentrated in the Annapolis and Minas Basins (Figure 15.2, see colour insert).

15.2.1 Acquisition Planning Issues

The Bay of Fundy is famous for its great tidal range, up to 13 m in this area. For such a large study area the timing of the tides vary by approximately 1 hour between Digby, within the Annapolis Basin, and the Minas Basin. Because of the variability of tide times within the study area, three locations were used to predict tide times and sites: Digby, Margaretsville, and the Minas Basin/Cape Blomidon. Acquisition of remotely sensed data at low tide has several applications including: 1. validation of tidal models; 2. determination of inter-tidal slope from derived elevations from the "waterline method" i.e. knowing the water depth at the time of image acquisition allows the land/water line to be used as a topographic isoline; 3. morphological and biological classification of the inter-tidal zone. Tide predictions were acquired for each port via the Internet (http://tbone.biol.sc.edu/tide/sitesel.html).

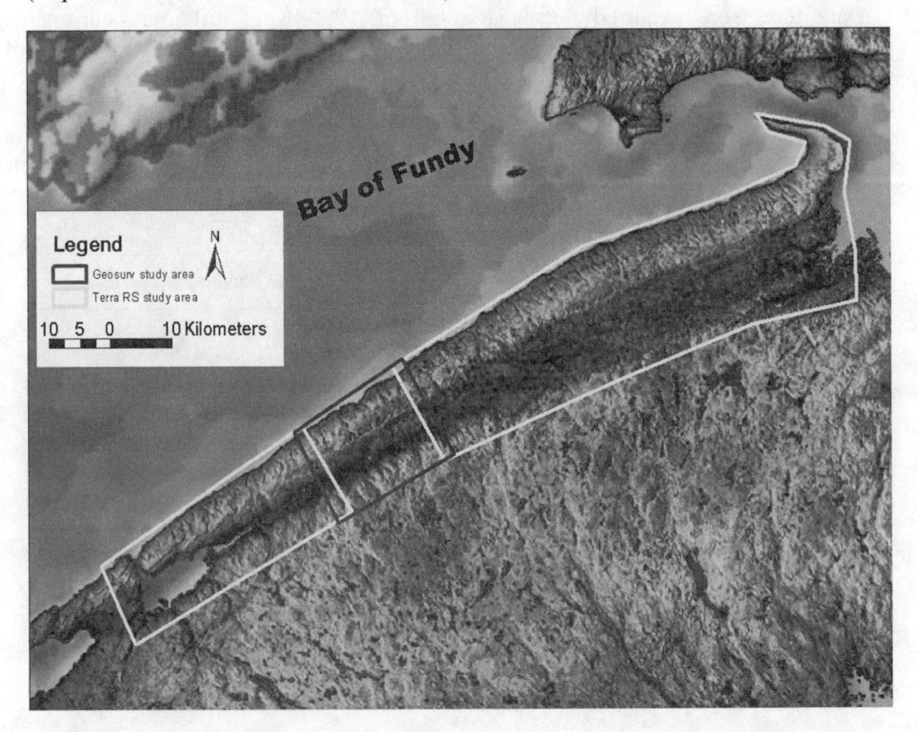

Figure 15.2 Location map of Annapolis and Minas Basin showing LIDAR, and airborne CASI coverage. Minas Basin is located in the upper right of the study area, and Annapolis Basin is located in the lower left of the study area. Radarsat data © 1996, Canadian Space Agency.

Radarsat-1 standard mode 2 scenes and a Landsat 7 scene were acquired near low tide conditions. The Ikonos satellite, which is owned and operated by Space Imaging, is capable of acquiring 1 m panchromatic and 4 m multispectral (3 visible and 1 near infrared band) imagery. Ikonos imagery were acquired near low tide by determining the date of low tide near 11:30 am local time (sun synchronous orbit pass time) and requesting image acquisition from 2 days prior to the low tide's date to 2 days after the low tide's date. This allowed for variable weather conditions (clouds or overcast) and the fact that tide times advance approximately 45 minutes each day. Ikonos orders were sent through a local distributor. In addition to the satellite coverage, the coastal areas were imaged twice with the CASI sensor, once at high tide (3 m resolution) and once at low tide (1 m resolution) (Figure 15.3). The LIDAR survey area was divided into three regions, two flown by Vendor A, and one by Vendor B (Figure 15.2). The coverage for each company overlapped to allow a comparative analysis of the data from each company. All coastal areas were flown near low tide in order to acquire detailed inter-tidal topography. The in-land vegetation state in mid-July is at maximum leaf cover. Both LIDAR providers assured the AGRG that canopy penetration was still possible and that a significant number of laser hits would make it through the canopy to the ground. The data accuracy and data specifications for the LIDAR can be found in Appendix 15.1. The accuracy specifications are discussed in more detail in section 15.4. Details on each LIDAR system are presented in sections 15.3.1 and 15.3.2.

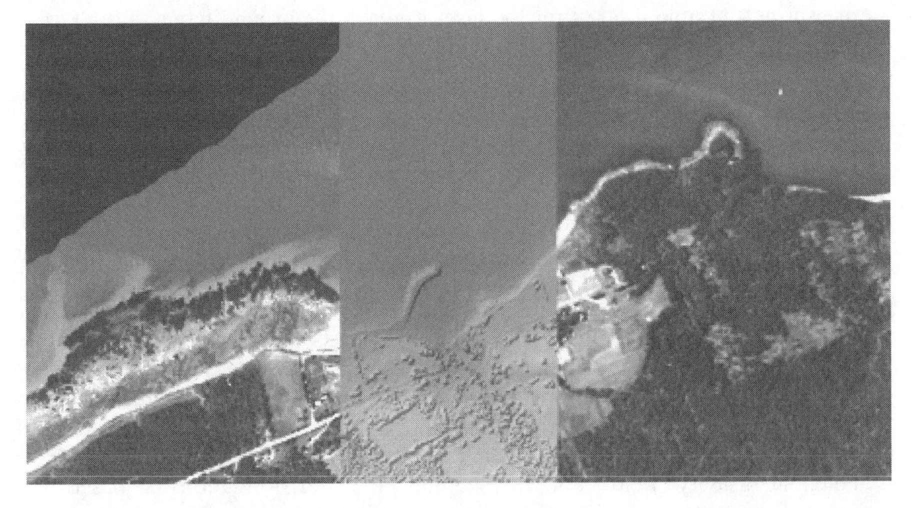

Figure 15.3 Mosaic of 1 m CASI at low tide (left image), 2 m LIDAR DSM at low tide (centre image), and 3 m CASI at high tide (right image) for Port Lorne along the Bay of Fundy. Overall image is approximately 4 km across.

15.2.2 Data Acquisition Issues

LIDAR is an active system providing its own pulse of near-infrared laser radiation and recording the reflected signal; thus it is not dependent on cloud free weather conditions as is the case for traditional aerial photography. However, rain or fog would cause the LIDAR survey to be delayed because the radiation cannot penetrate dense cloud or fog and therefore could not hit the ground. As a safety measure to ensure the laser is at a significant distance from the target, thus the power levels of the laser are not harmful to human eyes, the system will automatically shut down if a laser return is detected to be at too close a range. For example, if the laser were to reflect off of an underlying cloud, high fog, or rain the system would automatically shut down. The LIDAR systems used in this project typically flew at a relatively low altitude, ranging from 300 m to 800 m, thus were often below any clouds. The ground activities for the LIDAR survey consisted of two tasks: 1. provide the aircraft with precise carrier phase GPS base station observations, and 2. collect precise elevation profiles to assist in data validation. Vendor B LIDAR area was estimated at 3 collection days and the Vendor A LIDAR area was estimated between 5 and 7 days.

15.3 LIDAR TECHNOLOGY

The terrestrial LIDAR system consists of an aircraft equipped with a GPS, attitude sensor and active near-infrared laser source and sensor. As the plane advances along the flight path the laser is fired, the pulse is directed toward the ground by an oscillating mirror and the reflected signal from the ground is recorded. By measuring the time it takes for the laser to reach the ground target and return, the range to the target can be accurately determined. The Time Interval Meter (TIM) records the time the pulse is transmitted and when the pulse is returned as well as the angle of the scanning mirror. This information, in combination with differential precise-code GPS and attitude measurements (e.g. pitch, yaw, and roll correction), is used to determine the height of the terrain relative to the ellipsoid. The LIDAR sensors available for this survey could record first or last returns only, while new generation sensors record multiple returns and the intensity of the signal. The Vendor B Optech ALTM 1020 system could record either the first or last laser return. The last return was specified since there is more chance of obtaining ground returns in this mode and a ground DEM was most desirable for our study. The resultant data from a LIDAR survey consists of a series of point location (laser hits) with associated heights above the ellipsoid. Since the ellipsoid based on the World Geodetic System of 1984 (WGS84) is a smooth mathematical surface which approximates the earth, a transformation is used to convert the heights relative to the geoid which is an equipotential surface based on the earth's potential gravity field. For this region, the transformation is based on a Geoid/Ellipsoid separation model known as HT1_01 that allows the data to be referenced to the Canadian Geodetic Vertical Datum of 1928 (CGVD28) defined by the Geodetic Survey of Canada of Natural Resources Canada. With heights now related to the geoid, termed orthometric height, they relate to approximate mean sea level. Since the geoid model is continually being refined, we requested the LIDAR data to contain both ellipsoid heights and orthometric heights (CGVD28). Therefore when new geoid models are released in the future, one can easily recompute the orthometric heights from the ellipsoid heights using the latest separation model.

15.3.1 Vendor B LIDAR Survey

Vendor B used two base stations for aircraft GPS control, the COGS permanent base station and a mobile station set up in the center of the study area to be surveyed. The distance of the survey aircraft from the base station should not exceed 40 km generally, because the GPS error increases with distance from the base station. The location of this central point was calculated based on observations between there and the COGS base station located at the eastern edge of the block (Figure 15.2). In addition to the LIDAR operator, Vendor B had two people on the ground collecting precise elevation values to be used to validate their LIDAR results. In addition to Vendor B, COGS also collected Real Time Kinematic (RTK) GPS elevation values within a limited radius of COGS in Lawrencetown due to the required line of sight radio link between the roving GPS unit and the COGS broadcasting base station. Vendor B collected both fast static point observations and kinematic locations from their vehicle, which required post processing.

Vendor B used a twin engine Piper Navajo airplane that they based out of the Digby airport. Trimble technology similar to that of COGS was used for their GPS collection, thus making data sharing very easy. The attitude information of the aircraft was collected using an Applanix POS Inertial Measurement Unit (IMU). The laser unit from Optech was housed in the camera mount within the aircraft. An Optech ALTM1020 sensor was used in the survey operating at a 5000 Hz laser repetition rate and a 15 Hz scan rate for the mirror. The area was flown at an altitude of 800 m above ground level (AGL) with a scan angle of 18 degrees, producing a swath width of approximately 520 m with raw laser point spacing every 3 m. At this altitude the laser beam had a footprint diameter of 25 cm. This sensor can capture first or last laser returns. For this survey it was set to collect the last return information, thus increasing the probability to hitting the ground in vegetated terrain. In addition to the LIDAR, a digital video camera also collected nadir looking video to be used later to assist in interpreting the laser returns and separating them into ground and non-ground hits.

Vendor B started their survey on July 6 and ended on July 13. They were delayed by some bad weather days, many due to rain and air turbulence. The study area consisted of approximately 64 flight lines, oriented parallel to the coast with two lines running transverse to the coast to be used to cross check the data. After processing the LIDAR data back in Ottawa they detected a problem with some elevation data that could not be resolved. As a result they re-flew several lines on September 1.

15.3.2 Vendor A LIDAR Survey

The Vendor A crew consisted of two people, one on the ground and the LIDAR operator. They used a Bell Ranger helicopter on the belly of which they mounted a pod containing the laser and video camera. Vendor A used Ashtech technology for their GPS collection thus requiring the COGS Trimble GPS data to be translated into an intermediate RINEX file format. The attitude information of the aircraft was collected using a Litton Inertial Reference System (IRS). An IRS is similar to an IMU in function, also known as an Inertial Navigation System (INS). Vendor

A's area of coverage was significantly larger than that of Vendor B and they used the Waterville airport as their base for the eastern block and the Digby airport for the western block. GPS control for Vendor A's local base stations, located at the two airports, was brought in from the COGS base. Vendor A's quality control strategy consisted of firing laser pulses at the base station antenna and comparing the LIDAR-determined height with that derived from GPS observations. They used the permanent COGS base station as a back up for the western extent of the eastern block. For the western block they established another mobile base station location over a provincial High Precision Network (HPN) point. This point is part of the geodetic GPS control network and is located near the Digby runway. Vendor A used a commercial first return laser unit operating at 1047 nm wavelength with a laser pulse repetition of 10,000 Hz and a 10-15 Hz scan rate for the mirror. The area was to be flown at an altitude of 600 m with a scan angle of 50 degrees, producing a swath of approximately 600 m with raw laser point spacing every 3 m. At this altitude the laser beam had a footprint diameter of 1.8 m.

Vendor A started their survey on July 11 and ended it on August 31 for the Nova Scotia study. During the data collection in the eastern block they experienced a power loss problem with the laser. The source of the problem could not be determined and it resulted in less penetration through the vegetation canopy. To remedy the situation they changed the flying altitude from 600 m to 300 m. This also affected the line spacing, since a lower altitude results in a smaller swath of coverage. With this system the relationship is roughly one to one i.e. at 600 m altitude a 600 m swath can be imaged. This caused much less area to be covered than the original estimated time, which in turn led to another problem of pilot fatigue causing more delays in addition to bad weather delays. The loss of power issue was eventually partially resolved by increasing the gain setting on the laser, thus allowing the aircraft to fly at the original altitude of 600 m.

15.4 GIS LIDAR PROCESSING AND VALIDATION

Raw LIDAR data may contain many erroneous hits. Collectively, these are known as 'noise.' Noise is usually filtered out by the vendor prior to delivery of the data using an algorithm. The data set will also contain hits from any number of surfaces (trees, buildings, roads, water, etc.). The vendor will also filter and classify the LIDAR point cloud into ground and non-ground hits (e.g. vegetation, buildings) prior to distribution. Thus two separate files may exist for a single 'tile' of LIDAR data. One file will contain only the ground hits, and one will contain points for all surfaces but the ground. It is important when working with LIDAR data to be aware of the composition of data sets being used. It is also prudent to use some sort of visualization software or technique to check for erroneous points that have escaped the noise filtering process or the ground/non-ground separation process. The LIDAR data was delivered on CD separated into 4 km by 4 km tiles. For both vendors tens of erroneous points per tile were identified by using a threshold of expected height values between –20 and 300 m. These erroneous points were probably a result of laser hits reflecting off of suspended aerosols. Each tile contained approximately 3 million LIDAR points and was in ASCII format. The survey specifications required an average point spacing, or "hit," every 3 metres. The ASCII files were imported into ARC/Info GIS and converted into point coverages. Triangular Irregular Network (TIN) files were built from point coverages of the ground hits. The large study area was processed using multiple tiles in overlapping sections to construct the TINs. The TIN files using ground only points were then gridded to produce a DEM at 2 m resolution using both a 5th order polynomial fitting and a linear interpolation method. In the case of anthropogenic waterfront structures such as wharfs and breakwaters, the linear interpolation method is more suitable, while in rural areas, the quintic approach is more suitable to smooth the data. A Digital Surface Model (DSM) was generated from all of the LIDAR data, ground and non-ground points combined, using a linear interpolation method. A DSM differs from a DEM because it takes into account the height of the vegetation and buildings, while a DEM represents the "bald earth." The overlapping grids were then used to construct a seamless mosaic of the LIDAR surfaces, both a DEM and DSM.

The validation of the LIDAR involved both a comparison of benchmark GPS points to proximal LIDAR points and to the DEM. Two areas of the quality assurance testing involved the investigation of the spatial distribution of the LIDAR points and vegetation penetration, and the vertical accuracy of the LIDAR data utilizing GPS points. The horizontal accuracy of the data was assessed visually by comparing LIDAR map products to the 1:10,000 Nova Scotia Topographic Database.

15.4.1 Vendor B LIDAR Validation

Validation of the vertical accuracy of the Vendor B LIDAR was accomplished by measuring the differences between high precision GPS points and corresponding LIDAR points and by comparing GPS with the interpolated cells of the raster surface constructed from the LIDAR points. This was accomplished through differencing and linear regression of both the GPS elevation values and the LIDAR elevation data.

The first step in the validation process was to obtain quality benchmark data from GPS surveys. To compare the GPS points with the LIDAR surface, GPS accuracy needed to be equal to or better than that of the LIDAR points. This meant that only carrier-phase GPS could be used. This is the most accurate and expensive type of GPS. In the more common differential code GPS it is the code component of the GPS signal that is used to determine the position of the receiver. This code signal has a cycle width that is equivalent to a length of roughly three hundred meters. It is modulated onto a carrier wave that is approximately 20 cm in length. It is this carrier wave that is used to determine the position of the receiver during carrier phase GPS. The shorter carrier wavelengths provide a much finer degree of positioning than the 300 m cycle width of the code signal. This allows a positioning accuracy of within a decimeter using real-time and fast static techniques, or within less than a few centimeters with standard static techniques. Differential code GPS could only provide 1.5 – 0.5 m accuracies that would be insufficient for LIDAR validation. The GPS data used for this validation were collected using the real-time, static and fast static methodologies.

The validation points were compared to the LIDAR surface by a point-in-raster overlay. The validation points were intersected with the LIDAR interpolated ground surface (DEM). The elevation values from the selected LIDAR surface cells were then joined to the validation point's attribute table. At this point the difference between the height values of the validation points and the corresponding LIDAR surface was calculated. In both cases the elevations used were orthometric heights derived from the CGVD28 geoid model. The results of a linear regression between the ground LIDAR surface and the validation points shows a very high correlation coefficient of 99.999%, with a standard error of 0.1869, indicating a very strong agreement between the elevation values from the two data sets.

Table 15.2 contains statistics on the elevation difference values between the LIDAR and the validation points. The vertical specifications in the contract were defined in the following terms: "[t]he vertical accuracy will be within an average of 15 cm of the measured GPS points, and 95% of the data points will not exceed 30 cm in vertical accuracy." As can be seen in the summary table of statistics, all specifications were met. One must be aware of issues that arise from the classification process that separate ground from non-ground points. For example, highway overpasses and bridges may be classified as non-ground points. Therefore any GPS points on and approaching such structures should not be used to compare with the DEM surface. The research group has acquired a Leica System 500 RTK GPS unit in 2003 and has since augmented the GPS checkpoints for this area. An analysis of these new data confirms that the vertical specifications were met.

Table 15.2 Summary of vertical validation of Vendor B LIDAR ground DEM.

Statistics	**Elevation Difference Values (m)**
Mean (specification 1)	0.13
Standard deviation	0.10
Minimum	0.00
Maximum	0.40
Median	0.10
Mode	0.10
Count	934
No. points <= 0.30 meters	915
Percentage points equal to or less than 0.30 meters (specification 2)	97.96%
Are specifications met?	Yes

15.4.2 Vendor A LIDAR Validation

A similar procedure was used to validate the Vendor A data, although it covered a much larger area. After an examination of the LIDAR point files, a distinct pattern was visible where there were no points, and thus no LIDAR returns, for the buildings or roads. This is because the roads and building rooftops are asphalt and low reflectivity to the near-IR laser. This lack of LIDAR return relates back to the power loss experienced in the LIDAR unit during the survey. The lack of returns on the road made it difficult to check the precise relationship to the Nova Scotia Topographic Database but did allow us to determine that the road network generally corresponds to the lack of LIDAR returns along the road.

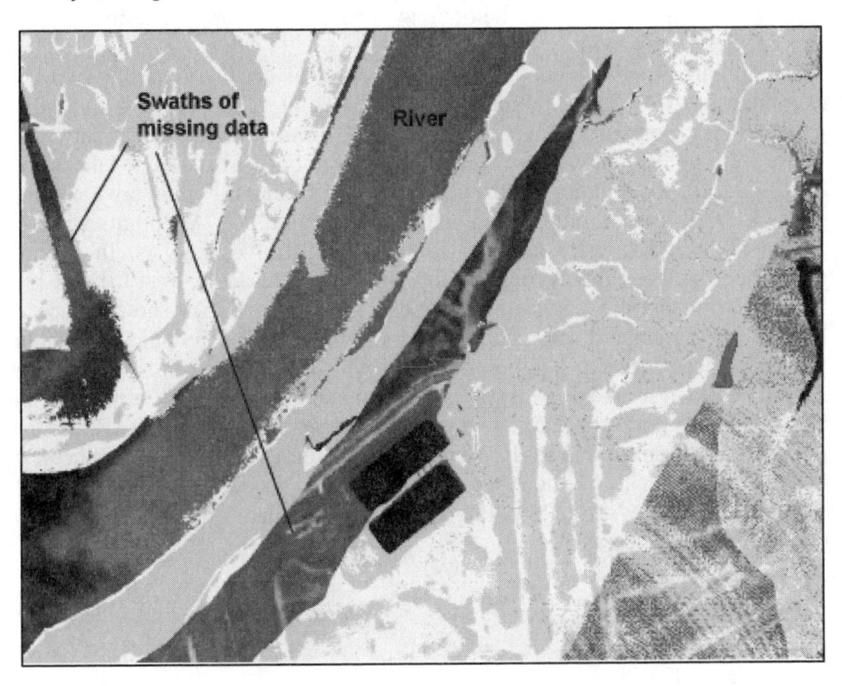

Figure 15.4 Vendor A LIDAR points (darker grey = ground, lighter grey = non-ground) over an Ikonos image for the Grand Pre dyke lands. Notice the missing swaths of data through the centre of the image and to the west. As a result of this missing data the dyke is not represented. Also the classification of ground and non-ground is not correct for the dyke area. Includes material copyright Space Imaging, LCC.

The other problem encountered with the point distribution along the coastal zone involved gaps in the LIDAR coverage. This was probably caused by the adjustment in flying altitude as a result of the power loss issue mentioned earlier. An example of this situation is shown in the area of Grand Pré near the Cornwallis River in the Minas Basin (Figure 15.4). The LIDAR hits for ground and non-ground are overlayed on the orthorectified Ikonos image acquired at low tide. Two areas of missing data swaths are evident, one in the central part of the image that includes the dyke land surrounding the river and the other in the western extent of the image. As a result of this area not being imaged during the LIDAR survey there is a lack of points; thus the ground and non-ground surfaces interpolated from the points do not accurately represent the presence of the dyke area, a critical barrier to coastal flooding. This results in a surface that is not accurate for flood risk modeling. In this case, the LIDAR data were supplemented by adding points manually in order to construct a valid surface model for this area. The Ikonos image was used as a guide for placing artificial points at the top and bottom of both sides of the dyke feature in the areas of missing LIDAR data. Elevation values for the artificial points were based on values obtained from the LIDAR for the top and bottom of the dyke. Also, LIDAR points along the dyke misclassified as non-ground were selected as ground. This is an example of the type of visual quality inspection that should be done in such a project prior to using any derived data for decision-making.

Similarly to the Vendor B data the vertical accuracy of the LIDAR was examined in two ways: 1. comparing LIDAR point values close to the validation points, and 2. comparing the gridded LIDAR ground surface to validation points. There were not as many GPS validation points initially available for the Vendor A data set as for the Vendor B data set. However, three HPN points were compared to the LIDAR ground returns and statistics measured. LIDAR points within 10 m of the HPN points were selected for the comparative analysis. In general the differences between the average orthometric height of the nearest LIDAR hits and the HPN points were between 1 m and 3 m (Figure 15.5). The variance of the LIDAR hits associated with each HPN point were small and could not account for the differences between heights measured with LIDAR and the HPN network. It should be noted most HPN locations are near roads and the lack of LIDAR hits on and near roads does not facilitate a perfect correspondence of LIDAR hits to HPN points. Concurrent to this study, Vendor A also surveyed an area in Prince Edward Island for a Climate Change Action Fund project to derive flood risk maps. An offset of 0.9 m between GPS points and the LIDAR data was detected in that study (Webster et al., in press).

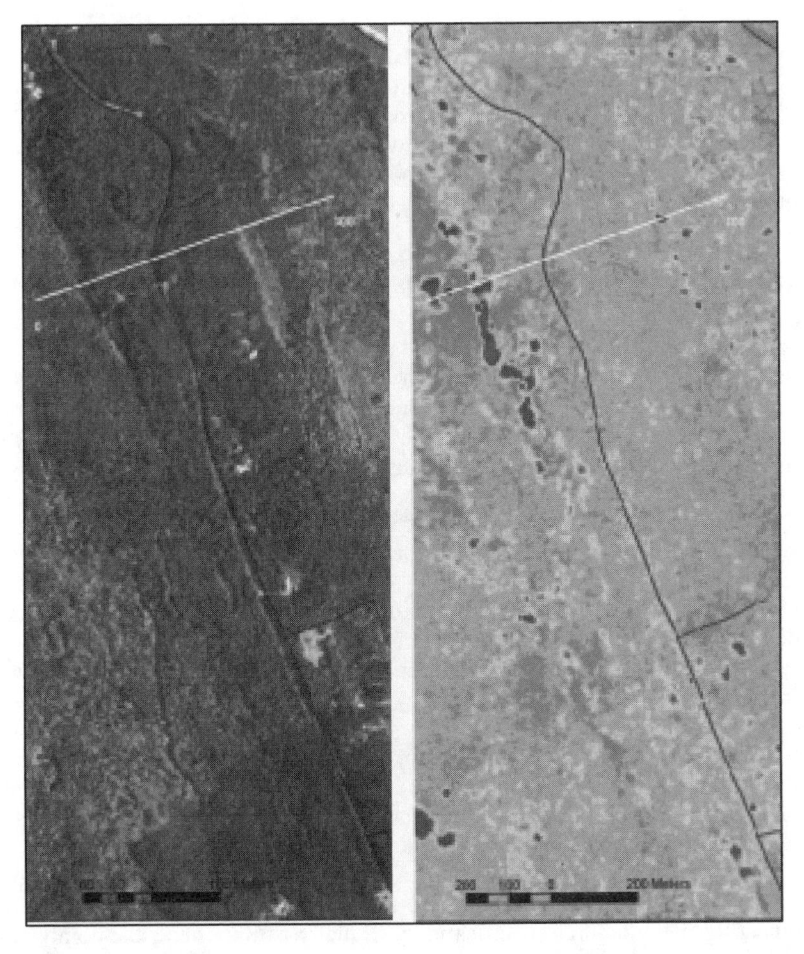

Figure 15.5 Comparison between Vendor B (2000) and Vendor A (2003). Left image is a digital true colour composite (converted to grey for this volume) CASI image; next is the difference map between the Vendor B and Vendor A 2003 DEMs. The main differences occur in the densely vegetated stream valley, and the small mounds (bumps) in the forest region.

As a result of this vertical bias in the LIDAR data, a follow-up GPS campaign was completed in the summer of 2001 by AGRG in the Annapolis Valley and shared with Vendor A in order to understand the nature of the problem. In this case GPS points were collected uniformly across the entire LIDAR swath in open flat areas. This was done to determine if the vertical bias was related to the scan angle of the system. It was finally deduced that the error resulted from the calibration of the system. An important aspect of a LIDAR calibration is the determination of any errors in the raw laser ranges. To properly calculate this value, multiple passes of varying altitudes must be made over known points. Since this aspect of the LIDAR calibration was not properly undertaken in the 2000 survey, an error of approximately 0.9 metres was introduced into the survey

(Webster et al., in press). Since calibration passes were made at an altitude of approximately 300 metres difference from the actual survey, the elevations matched at the calibration sites but did not match over the survey area. Comprehensive tests following the survey showed the need for a laser range bias and a laser range scale factor correction. Several lines of the LIDAR data in the survey area have since been re-processed by Vendor A to confirm these parameters.

15.4.3 Vendor A Re-fly of the LIDAR

As a result of the vertical offset and the limited vegetation penetration issues, Vendor A agreed to re-fly the area in the spring of 2003 at no additional costs. Since the 2000 survey there have been several improvements to their system and deployment strategy as a result of this experience (Shreenan, R. personal communications, 2003). A crucial step to any successful LIDAR survey is to conduct daily calibration passes over objects with known coordinates. Objects with long, straight elevated edges such as bridges and buildings are particularly favoured since they provide a check for both horizontal and vertical alignment. As mentioned earlier, calibration passes for this particular LIDAR survey entailed daily passes over the GPS base station. Although calibration parameters were adequately defined from this information the introduction of elevated man-made features allows a refinement of the calibration computation. This process has a direct impact on the overall accuracy of the survey and Vendor A has since adopted this practice.

The accuracy of a LIDAR system is only as good as the weakest component of the system. Since the 2000 survey, Vendor A has integrated a new Aplannix IMU into their system, which significantly improves the measurement of the roll, pitch and heading components of the aircraft over the previous sensor. The introduction of the new IMU improves the resolution of the aircraft attitude by a factor of five times for the roll and pitch and ten times for the heading (in comparing manufacturer specifications). Equally as important, the components of the IMU have been combined with the navigation sensors (GPS) for the purpose of regulating the IMU errors to those of the navigation sensors. This is commonly referred to as GPS Aided Inertial Navigation Systems and has greatly increased the overall accuracy of the system.

A final step to ensuring that an accurate product has been captured is to compare the processed LIDAR data to points that have been captured at a higher accuracy than the LIDAR. This practice has been adopted by Vendor A and uses the U.S. Federal Geographic Data Committee (FGDC) Geospatial Accuracy Standards (Parts 1-3) (www.fgdc.gov/standards/status/textstats.html) as a guideline for determining overall accuracy of all LIDAR surveys.

Another important step to any successful LIDAR survey is the accurate extraction of features from the LIDAR point cloud. The ability to perform this task accurately and reliably has a direct bearing on the overall accuracy of the survey. As a result of this survey, Vendor A embarked on developing and purchasing new software to reliably handle these tasks. The proprietary software used to process the collected dataset presented in this paper was extremely effective for single line corridor mapping but lacked the algorithms for processing multiple lines of

LIDAR. Also, the software lacked the visualization tools required to adequately detect outliers. The software now employed handles all of these issues as well as provides a host of other processing options including: 1. the removal of large outliers, thus improving the ground extraction algorithm's ability to accurately define ground hits; 2. the LIDAR data is processed according to neighbouring hits rather than the sequence of capture; 3. new hardware which enables tens of millions of points to be processed simultaneously rather than just millions of points; 4. several new visualization techniques have been implemented to help review the processed data and detect anomalies in the ground surface; and 5. video and digital mosaics are now used extensively to help review classified data.

Other improvements in the system include the introduction of a collimator attached to the end of the laser head. This device decreases the divergence of the laser beam to better than 0.3 mrad. This enables weaker reflective signals such as black top pavement and roofs to be captured because of the increased concentrated signal coming from the laser. Additionally, more penetration through the treetops is accomplished with the less divergent beam and therefore more ground hits captured. The foot print diameter in 2000 was 1.8 m, compared to 0.18 m in 2003. This was especially critical to reach the ground with a first return system.

In additional to improvements in the equipment, new field procedures have been implemented to include better field quality control to ensure no gaps exist between LIDAR flight lines. Gaps in the laser data can easily be identified on a daily basis, which can then be re-flown. Also, the laser data storage device has been reconFigured and repackaged which helps to eliminate data loss caused by the vibration of the airframe.

In the spring of 2003 prior to leaf-on conditions, Vendor A re-flew the eastern block at no additional costs and plans are in place to re-fly the western block (Figure 15.2) with a first and last multiple return LIDAR system in the fall of 2003. The preliminary data collected in the spring of 2003 has met the vertical specifications of 30 cm and is a significant improvement in terms of point density, returns from low reflectors (asphalt), and ground penetration in vegetated areas.

15.4.4 Comparison of LIDAR surfaces

When the original LIDAR areas were selected for each data provider, an area of overlap between the regions was maintained in order to compare the quality of the resultant DEMs between the two providers. For each data provider a ground DEM was constructed from the LIDAR points. This included Vendor B (2000), and Vendor A (2000 and most recently 2003 data). The ground surface was based on the construction of a TIN and quintic interpolation to a 2 m DEM grid.

The area of overlap is south of Lawrencetown and consists of a mixture of dense forest within a stream valley to cleared areas on the hilltop. The resulting DEMs are contrasted in Colour Plate 15.1 (following page 164). The problem with the original DEM derived from Vendor A is apparent by the triangular facets. This is a result of few LIDAR shots making it to the ground; therefore the DEM surface is not an accurate representation of the terrain. However, the DEM from the 2003 data that Vendor A re-flew is very similar to that of Vendor B. In fact, many of the small "bumps" on the landscape present in the Vendor B DEM have been removed in the Vendor A 2003 data. Although not verified in the field yet, these bumps are interpreted to probably represent dense brush or other vegetation, based on their size, and do not truly represent the ground elevation. This highlights the improvements of the ground/non-ground classification algorithms from 2000 to 2003, thus enabling the production of a more accurate DEM.

15.5 COASTAL APPLICATIONS OF TERRESTRIAL LIDAR

Although much of this paper has been concerned with operational acquisition and validation of the LIDAR, end applications have been developed for the coastal zone concentrating on flood simulation modelling, inter-tidal slope calculations and merging the terrestrial LIDAR with bathymetry. For the flood-risk mapping, a similar methodology as in the studies by Webster et al. (2001, 2003, in press) have been employed.

Floodplain boundaries are important to emergency management personnel, insurance agents, and developers, as well as ordinary citizens. Flat floodplains are also attractions for further developments. Existing floodplain maps mainly derived from coarse elevation data do not depict actual floodplain boundaries. Major uses of LIDAR data include developing hydrological and floodplain models, telecommunications planning and analysis, transportation assessment, urban planning and natural resource and forest management.

For this study, the Minas Basin area was used because of the increased population and infrastructure that could be vulnerable to coastal flooding. The derived maps were passed to the municipal and town planning authorities to be used to assist in land development policies. In the absence of current land use vector data, Ikonos satellite imagery for the study area was classified to derive land cover types. The imagery was orthorectified using 1:10,000 road vectors and 5 m DEM from 1:10,000 contours. The land cover types derived through a maximum likelihood supervised classification were intersected with the flood levels derived from the LIDAR DEM to determine the percentage of affected areas.

The LIDAR ground points were used to construct the DEM surface. While delineating inundation areas on the derived DEM, some problems were discovered that related to the dykes not being properly represented. As mentioned earlier, some points along the dyke had been removed from the ground file. To depict the actual ground surface, it was necessary to construct a DEM with dykes. The dykes were extracted from the non-ground file by using the coded topographic base map vector database in combination with the Ikonos imagery. These points were in turn used to construct a new surface. As described earlier, where the LIDAR swath was absent points for the dyke and ground area were manually inserted. The heights were obtained by enquiring the LIDAR points adjacent to the missing areas.

Storm surges are the major cause of flooding in coastal areas. Storm surge is an abnormal rise in sea level accompanying a hurricane or other intense storms, and whose height is the difference between the observed level of the sea surface and the level that would have occurred in the absence of the storm. A storm surge at a low tide might go unnoticed but the storm surge at the highest tide might cause heavy coastal destruction. Therefore, the following calculations are performed for surges associated with high tides only.

Table 15.3 Height in meters above Chart Datum at Hantsport, Minas Basin. Source: Canadian Hydrographic Service Chart, HW – high water, LW – low water.

Large Tides		**Average Tides**		**Mean Sea Level (MSL)**
Higher HW	*Lower LW*	*Higher HW*	*Lower LW*	
14.8	-0.1	13	1.2	7.1

Tide levels and land elevation must be referenced to common datum in order to compute coastal flooding inundation areas. Because the LIDAR data and derived DEM are referenced to the geoid (CGVD28-MSL), and the tide levels are referenced to chart datum, which is typically the lowest elevation the largest tide will reach, the tide levels must be converted to reference a mean sea level datum (CGVD28). Thus the chart datum elevation of 7.1 m is subtracted from all the tidal water levels in order to relate them to MSL. Table 15.3 provides the original tidal information derived from the local chart, and table 15.4 shows the computed tidal levels in reference to mean sea level.

Table 15.4 Height in meters relative to Mean Sea Level (MSL) Datum, Hantsport, Minas Basin.

Large Tides		**Average Tides**		**Mean Sea Level**
Higher HW	*Lower LW*	*Higher HW*	*Lower LW*	
7.7	-7.2	5.9	-5.9	0

As can be observed in the above table, the highest tide in Minas Basin is 7.7 m above MSL. The average storm surge measured from all recording stations in the Bay of Fundy is 0.6 m for a one-in-twenty year event and 1.2 m for a one-in-one hundred year event. Surge levels were computed based upon these predictions. Table 15.5 provides the water levels predicted by combining tidal levels and storm surge levels referenced to MSL.

Table 15.5 Storm surge values. Probable return times of storms, one-in-20 year event, one-in-one hundred year event. Source : CHS Chart

Return Period	**Highest tide (m)**	**Storm Surge (m)**	**Resultant water level above MSL (m)**
Twenty-year flood	7.7	0.6	8.3
Hundred-year flood	7.7	1.2	8.9

Dykes have been constructed to protect low-lying coastal agricultural areas from tidal waters. The average dyke height ranges around 8.5 m. This means that the low-lying areas beyond the dykes would be flooded only if the tide level was greater than 8.5 m. It could be deduced that low-lying coastal farmlands are well protected by dykes for floods with a storm return period of twenty years. The dykes are insufficient to stop flooding in a one-hundred-year storm event occurring at the largest tide of 8.9 m water level. These water levels do not take into account relative sea-level rise, which in this region is estimated at 25 cm per century (Stea, Forbes, and Mott, 1992). Therefore, for this project, a 9 m flood level was selected as the extreme limit to derive flood risk maps. The town of Kentville, located near the coast of the Minas Basin, has also used this level for their Municipal Planning Strategy maps.

Flood risk maps were generated using off-the-shelf GIS software tools rather than using flood models such as Mike11. The DEM used in the project has orthometric height values (MSL). The procedure of extracting inundation areas for various flood levels consists of selecting all the cells whose elevation (cell value) is less than or equal to that particular flood level. In order to determine the spatial extent of the flooding associated with a storm surge event, 10 flood levels at 0.5 meter increments were selected with the intention of building an animation. These were 4.5, 5, 5.5, 6, 6.5, 7, 7.5, 8, 8.5, and 9 m above MSL. An Arc Macro Language (AML) script was written for generating flood level grids where cells with values below the flood level are selected resulting in a binary map of flooded areas. In order to check for continuity between the flooded areas and the source of the water, in this case the bay or estuary, the grids were converted to polygons to permit selection of contiguous areas. For an area to be flooded, it must satisfy two conditions; the area must have elevation less than or equal to a surge level, and it must be connected to the water source, for example the Cornwallis River (Figure 15.6).

The process does not consider the effects of culverts and bridges due to lack of data of such infrastructure. However, it is known that many of the culverts and control gates of the smaller estuaries have one-way values to allow water to drain to the bay and control water flowing upstream. The flood levels of a storm return period of twenty (0.6 m, 8.3 m above MSL) and one hundred years (1.2 m, 8.9 m above MSL), with no consideration of sea-level rise, were overlaid on the Ikonos imagery to visualize the areas possibly affected. The calculated surge level is 8.3 m above mean sea level for a storm return period of twenty years and 8.9 m for a storm return period of hundred years. The flood levels generated in this study have been used for GIS overlay analysis to produce reports on the land cover and properties that could potentially be affected by a flood (see example, Colour Plate 15.2). The flood levels have been used to generate a series of visualizations including perspective views and flood animations. A report on the methodology, data processing, and results of the overlay analysis was given to the Kings County Planning Commission to be used for future recommendations.

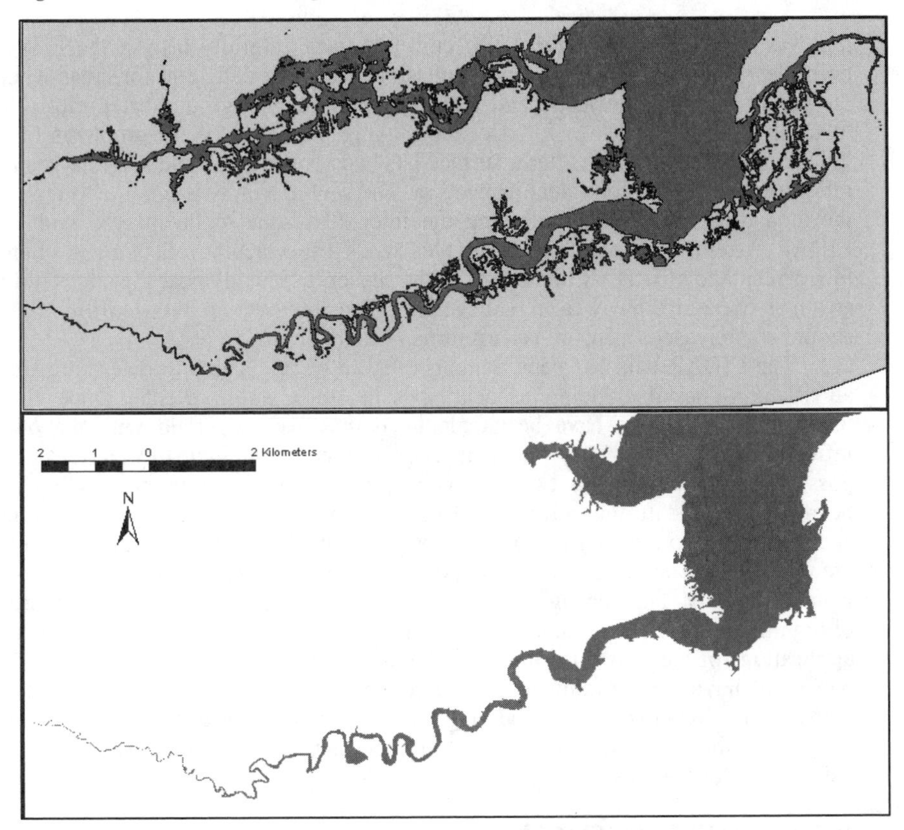

Figure 15.6 Example of how the flood-risk polygons are generated. Top image is the result of coding all cells less than or equal to 4 m. Bottom image shows only the areas that have connection with the bay or estuary. Flood level is 4 m above MSL; average high tide is 5.9 m above MSL.

15.6 CONCLUSIONS

The integration of high-resolution imagery and DEMs derived from LIDAR has several applications within the coastal zone. Here we have discussed the application of these technologies for developing flood-risk maps associated with possible storm surge events. LIDAR is a relatively new technology capable of providing dense elevation points of high precision and accuracy. It is important, however, for users of the data to be sure the data has met the accuracy specifications and independent validation data should be collected for this analysis. The validation of the LIDAR data indicated that the information from Vendor B met the specifications, while that from Vendor A did not for the 2000 survey. However, we worked with Vendor A to determine the cause of the problem and they have re-flown the area at no additional costs in 2003. The preliminary results of the new Vendor A data meet vertical specification of within 30 cm of measured GPS locations. In the case of both data providers, the separation algorithms removed features that the end user may want in the ground datasets, for example highway overpasses and dykes. The end user must carefully inspect the LIDAR point data both for erroneous points, missing swaths of data, and for areas where the points have been misclassified into the ground or non-ground categories prior to using a gridded surface for decision-making. However, these problems often quickly reveal themselves after a surface has been constructed and a colour shaded relief has been made for visual inspection. The Bay of Fundy is unique because of the large tidal range, thus allowing the inter-tidal zone to be imaged with the LIDAR system. Wet smooth mud surfaces act as a spectacular reflector and often a low percentage of LIDAR hits return to the sensor, especially near the edges of the swath as one move away from nadir angles. Thus the overlap between flight lines should be increased to ensure laser returns in such environments.

The LIDAR data has been demonstrated to be applicable to determine areas at risk for coastal flooding associated with storm surge events. Once flood inundation boundaries have been calculated they were overlaid with the other infrastructure and property maps that exist for the municipality to calculate possible economic impacts of such events. High-resolution satellite imagery has been used to update the current land cover maps that were used in the overlay analysis with flood inundation areas. These types of data and information products are critical for land use planners and policy makers in order to manage the coastal zone effectively. This management could include restricting development in areas of high flood risk or mediation to try to minimize the impacts of such events. Other applications of these data for the coastal zone that have not been discussed in this paper but have been examined by the research group include generating slope maps of the inter-tidal zone, validating local tide models for the Minas Basin, and the generation of a seamless terrain model merging the land elevations and bathymetry for the area.

15.7 ACKNOWLEDGEMENTS

The processing of this data and validation examination has been a team effort at the AGRG. Team members include: Steve Dickie, Montfield Christian, Frances MacKinnon, Jamie Spinney, Charles Sangster, Michael Palmer and Trevor Milne. We would like to thank all of them and the other AGRG graduate students who have contributed for their dedication, hard work, and diligence. Also thanks to Dr. Robert Maher and David Colville, research scientists at the AGRG, for review of the manuscript and helpful discussions. We would like to thank Andy Sherin, Darius Bartlett and especially Jennifer Smith for the invitation to include this work in the book and assistance in preparing the manuscript. We are grateful to the staff and field crews from both Geosuv (now Mosaic Mapping) and Vendor A, especially Roger Shreenan of Vendor A, along with all of the other data providers including Hyperspectral Data International, GeoNet, and Radarsat International. We are grateful to the Nova Scotia Geomatics Center and the Municipality of Kings County for providing the base and property information for the study. Funding for the data collection was supported by a research grant from the Canadian Foundation for Innovation of Industry Canada.

APPENDIX 15.1

Terrestrial LIDAR Data Specifications for both vendors.

The LIDAR data will be supplied to the Applied Geomatics Research Group of COGS. The data will be in ASCII format, one row per elevation data point, organized in five columns as follows:

UTM Easting (WGS84)
UTM Northing (WGS84)
Height above Ellipsoid (WGS84)
Orthometric Height above Geoid (HT1_01)
GPS time

Three files will be delivered in this format on CDROM, containing

all data points,
ground points,
non-ground points.

The raw data files and airborne down-looking video will also be provided to the clients in care of COGS in formats they are capable of reading. Appropriate quality control information such as aircraft attitude and speed, quality of GPS positioning, and post-processing error data will be appended to each data point in one or all of the data files.

GPS control points will be occupied within a suitable range of the survey areas to achieve this accuracy and precision. Data will not be collected during increased ionisphereic activity which could degrade the GPS signal. Data will be collected under suitable conditions for video collection, under adequate light with no obscuring cloud beneath the aircraft. Data for coastal inter-tidal areas will be collected within 1 hour of lowest daily daylight tide levels in the survey area of Nova Scotia and within 2 hours of lowest daily tide in the Charlottetown, and North Rustico PEI survey areas.

All laser height, video, GPS, aircraft attitude and speed, and other quality control data will be delivered to the clients in care of COGS.

Full ownership of the data and information including working papers and reports will reside with COGS for data collected in Nova Scotia and the HM The Queen in Right of Canada (as represented by the Minister of Natural Resources) for data collected in PEI.

15.8 REFERENCES

Abdalati, W. and Krabill, W.B., 1999, Calculation of ice velocities in the Jakobshavn Isbrae area using airborne laser altimetry. *Remote Sensing of the Environment*, **67**, pp. 194-204.

Brock, J.C., Wright, C.W., Sallenger, A.H., Krabill, W.B., and Swift, R.N., 2002, Basis and methods of NASA airborne topographic mapper LIDAR surveys for coastal studies. *Journal of Coastal Research*, **18**, pp. 1-13.

Chagnon, R., 2002, Sea-ice climatology. In *Coastal impacts of climate change and sea-level rise on Prince Edward Island.* Edited by Forbes, D.L. and Shaw, R.W. Geological Survey of Canada, Open File 4261, Supporting Document 5, 24 p. (on CD-ROM).

FGDC ([US] Federal Geographic Data Committee), 1998, Geospatial Accuracy Standards, Parts 1 to 3 (www.fgdc.gov/standards/status/textstatus.html, documents FGDC-STD-007.1, FGDC-STD-007.2, FGDC-STD-007.3).

Forbes, D.L. and Manson, G.K., 2002, Coastal geology and shore-zone processes. In *Coastal impacts of climate change and sea-level rise on Prince Edward Island.* Edited by Forbes, D.L. and Shaw, R.W. Geological Survey of Canada, Open File 4261, Supporting Document 9, 84 p. (on CD-ROM).

Forbes, D.L., Shaw, R.W., and Manson, G.K., 2002, Adaptation. In *Coastal impacts of climate change and sea-level rise on Prince Edward Island.* Edited by Forbes, D.L. and Shaw, R.W. Geological Survey of Canada, Open File 4261, Supporting Document 11, 18 p. (on CD-ROM).

Guenther, G.C., Brooks, M.W., and LaRocque, P.E., 2000, New capabilities of the SHOALS airborne LIDAR bathymeter. *Remote Sensing of the Environment*, **73**, pp. 247-255.

Houghton, J.T., Ding, Y., Griggs, D.J., Noguer, M., van der Kinden, P., Dai, X., Maskell, K., and Johnson, C. I., 2001, Eds. IPCC WG1. "Summary for Policy Makers", Climate Change 2001: The Scientific Basis. Contribution of Working Group 1 to the Third Assessment Report of the Intergovernmental Panel on Climate Change. (Cambridge and New York: Cambridge University Press)

Hwang, P.A., Krabill, W.B., Wright, W., Swift, R.N., and Walsh, E.J., 2000, Airborne scanning LIDAR measurement of ocean waves. *Remote Sensing of the Environment*, **73**, pp. 236-246.

King, G., O'Reilly, C., and Varma, H., 2002, High-precision three-dimensional mapping of tidal datums in the southwest Gulf of St. Lawrence. In *Coastal impacts of climate change and sea-level rise on Prince Edward Island.* Edited by Forbes, D.L. and Shaw, R.W. Geological Survey of Canada, Open File 4261, Supporting Document 7, 16 p. (on CD-ROM).

Krabill, W.B. and Martin, C.F., 1987, Aircraft positioning using global positioning system carrier phase data. *Navigation*, **34**, pp. 1-21.

Krabill, W., Abdalati, W., Frederick, E., Manizade, S., Martin, C., Sonntag, J., Swift, R., Thomas, R., Wright, W., and Yungel, J., 2000, Greenland Ice Sheet: high-elevation balance and peripheral thinning. *Science*, **289**, pp. 428-430.

Krabill, W.B., Thomas, R.H., Martin, C.F., Swift, R.N., and Frederick, E.B., 1995, Accuracy of airborne laser altimetry over the Greenland ice sheet. *International Journal of Remote Sensing*, **16**, pp. 1211-1222.

Krabill, W.B., Wright, C., Swift, R., Frederick, E., Manizade, S., Yungel, J., Martin, C., Sonntag, J., Duffy, M., and Brock, J., 1999, Airborne laser mapping of Assateague National Seashore Beach. *Photogrammetric Engineering & Remote Sensing*, **66**, pp. 65-71.

Maclean, G.A. and Krabill, W.B., 1986, Gross-merchantable timber volume estimation using an airborne LIDAR system. *Canadian Journal of Remote Sensing*, **12**, pp. 7-18.

Manson, G.K., Forbes, D.L. and Parkes, G.S., 2002, Wave climatology. In *Coastal impacts of climate change and sea-level rise on Prince Edward Island.* Edited by Forbes, D.L. and Shaw, R.W. Geological Survey of Canada, Open File 4261, Supporting Document 4, 31 p. and 1 attachment (on CD-ROM).

Maune, David, F., 2001, Digital Elevation Model Techniques and Applications: The DEM User Manual. Edited by David F. Maune. *American Society of Photogrammetry and Remote Sensing.*

Mayor, S.D. and Eloranta, E.W., 2001, Two-dimensional vector wind fields from volume imaging LIDAR data. *Journal of Applied Meteorology*, **40**, pp.1331-1346.

McCulloch, M.M., Forbes, D.L., Shaw, R.W., and the CCAF A041 Scientific Team, 2002, *Coastal impacts of climate change and sea-level rise on Prince Edward Island.* Edited by Forbes, D.L. and Shaw, R.W. Geological Survey of Canada, Open File 4261, xxxiv + 62 p. and 11 supporting documents (on CD-ROM).

Milloy, M. and MacDonald, K., 2002, Evaluating the socio-economic impacts of climate change and sea-level rise. In *Coastal impacts of climate change and sea-level rise on Prince Edward Island.* Edited by Forbes, D.L. and Shaw, R.W. Geological Survey of Canada, Open File 4261, Supporting Document 10, 90 p. (on CD-ROM).

O'Reilly, C., 2000, Defining the coastal zone from a hydrographic perspective. *Proceedings, Workshop on risk assessment and disaster mitigation: Enhanced use of risk management in integrated coastal management.* International Ocean Institute, Bermuda. *Backscatter*, April 2000, pp. 20-24

O'Reilly, C.T., Forbes, D.L., and Parkes, G.S., 2003, Mitigation of coastal hazards: adaptation to rising sea levels, storm surges, and shoreline erosion. *Proceedings, 1st Coastal, Estuary and Offshore Engineering Specialty Conference*, Canadian Society of Civil Engineering, Moncton, NB, CSN-410-(1-10), *in press*.

Parkes, G.S. and Ketch, L.A., 2002, Storm-surge climatology. In *Coastal impacts of climate change and sea-level rise on Prince Edward Island.* Edited by Forbes, D.L. and Shaw, R.W. Geological Survey of Canada, Open File 4261, Supporting Document 2, 87 p. (on CD-ROM).

Parkes, G.S., Forbes, D.L., and Ketch, L.A., 2002, Sea-level rise. In *Coastal impacts of climate change and sea-level rise on Prince Edward Island.* Edited by Forbes, D.L. and Shaw, R.W. Geological Survey of Canada, Open File 4261, Supporting Document 1, 33 p. and 5 attachments (on CD-ROM).

Post, M.J., Grund, C.J., Weickmann, A.M., Healy., K.R., and Willis, R.J., 1996, A comparison of the Mt. Pinatubo and El Chichon volcanic events: LIDAR observations at 10.6 and 0.69 mm. *Journal of Geophysical Research*, **101**(D2), pp. 3929-3940.

Sallenger, A.B., Jr., Krabill, W., Brock, J., Swift, R., Jansen, M., Manizade, S., Richmond, B., Hampton, M., and Eslinger, D., 1999, Airborne laser study quantifies El Niño-induced coastal change. *Eos, Transactions, American Geophysical Union*, **80**, pp. 89-92.

Stea, R.R., Forbes, D.L., and Mott, R.J., 1992, Quaternary Geology and Coastal Evolution of Nova Scotia. Field Excursion A-6: Guidebook, Geological Association of Canada, Mineralogical Association of Canada.

Stockdon, H.F., Sallenger, A.H., List, J.H., and Holman, R.A., 2002, Estimation of shoreline position and change using airborne topographic LIDAR data. *Journal of Coastal Research*, **18** (3), pp. 502-513.

Shreenan, R., 2003, Personal communications. Research scientist with Terra Remote Sensing Inc.

Thompson, K., Ritchie, H., Bernier, N.B., Bobanovic, J., Desjardins, S., Pellerin, P., Blanchard, W., Smith, B., and Parkes, G., 2002, Modelling storm surges and flooding risk at Charlottetown. In *Coastal impacts of climate change and sea-level rise on Prince Edward Island.* Edited by Forbes, D.L. and Shaw, R.W. Geological Survey of Canada, Open File 4261, Supporting Document 6, 48 p. (on CD-ROM).

Wadhams, P., Tucker, W.B., III, Krabill, W.B., Swift, R.N., Comiso, J.C., and Davis, N.R., 1992, Relationship between sea ice freeboard and draft in the Arctic Basin and implications for ice thickness monitoring. *Journal of Geophysical Research*, **97** (C12), pp. 20325-20334.

Webster T.L., Dickie, S., O'Reilly, C., Forbes, D.L., Thompson, K., and Parkes, G., 2001, Integration of Diverse Datasets and Knowledge to Produce High Resolution Elevation Flood Risk Maps for Charlottetown, Prince Edward Island, Canada. In *CoastGIS2001 unpublished proceedings*, Halifax, Nova Scotia, Canada.

Webster, T.L., Forbes, D.L., Dickie, S., Colville, R., and Parkes, G.S., 2002, Airborne imaging, digital elevation models and flood maps. In *Coastal impacts of climate change and sea-level rise on Prince Edward Island.* Edited by Forbes, D.L. and Shaw, R.W. Geological Survey of Canada Open File 4261, Supporting Document 8, 36 p. (on CD-ROM).

Webster, T.L., Dickie, S., O'Reilly, C., Forbes, D., Parkes, G., Poole, D., and Quinn, R., 2003, Mapping Storm Surge Flood Risk using a LIDAR-Derived DEM. In *Elevation, a supplement to Geospatial Solutions and GPS World.* May.

Wehr, A. and Lohr, U., 1999, Airborne laser scanning – An introduction and overview. *Journal of Photogrammetry and Remote Sensing*, **54**, pp. 68-82.

CHAPTER SIXTEEN

Mapping and Analysing Historical Shoreline Changes Using GIS

Courtney A. Schupp, E. Robert Thieler and James F. O'Connell

16.1 INTRODUCTION

Understanding coastal change has become increasingly important to the 78 coastal communities in Massachusetts. About half of Massachusetts' 6.2 million residents now live in the coastal zone, and the coastal population is growing rapidly. Erosion threatens beachfront houses and development along much of the state's 2400 km of shoreline, creating a critical need for the government and the public to have accurate, up-to-date information on shoreline change. Prior to this shoreline change update, however, Massachusetts' existing historical shoreline change database had not been updated since 1978. New maps and data displaying long-term and more recent trends of shoreline behaviour increase the capability for sound decision-making and enhance public awareness of coastal change in Massachusetts.

To produce the necessary maps and data, we compiled historical shoreline positions in ArcInfo from a variety of map and aerial photograph sources. We generated 1:10,000 scale maps of the Massachusetts coast that display historical shorelines, rates of change, locations of rate-of-change measurements (transects), and orthophotographs. We also enhanced a Microsoft Access database created by Van Dusen (1996) that contains transects, dates, and rates of change; it can be used to perform spatial queries and to compile regional statistics and trends. The Massachusetts Office of Coastal Zone Management (MCZM) will distribute these maps, along with the data tables and a manual describing how to use the information, to regional government offices and coastal communities. These products will also be made available to the public through the Internet.

0-41531-972-2/04/$0.00+$1.50

16.2 METHODS

16.2.1 Data Sources

Previous projects digitised and assembled historical shoreline data for the Massachusetts coast from 1842 to 1978 into ArcView GIS format (Benoit, 1989; O'Connell, 1997). Those projects obtained shorelines from six different data sources: 1) National Ocean Service (NOS) topographic maps (T-sheets), 2) NOS hydrographic maps (H-sheets), 3) Federal Emergency Management Agency (FEMA) Flood Insurance Study topographic maps, 4) printed orthophotographs, 5) aerial photographs, and 6) digital orthophotographs. The early shorelines (1842 to 1950) were digitised exclusively from NOS T- and H-sheets. Shorelines from the 1970s were compiled by digitising FEMA topographic maps, printed orthophotographs, and aerial photographs. These early data sets were digitised and placed into a GIS-compatible format using the Metric Mapping System (Clow and Leatherman, 1984; Benoit, 1989).

The aerial photography used to generate the 1994 orthophotographs has a nominal scale of 1:48,000 and was taken by the National Ocean Service in September and October of 1994. MCZM provided full-colour digital mosaicked orthophotographs that have a resolution of one meter per pixel. We used these geographically registered orthophotographs to digitise the 1994 shoreline directly within ArcView GIS software.

There are several possibilities for identifying a high-water shoreline. In most locations, we delineated the shoreline using the high-tide wrack line. However, due to the range in geomorphology along the Massachusetts coast, other delineations were sometimes more appropriate, such as the wet/dry interface, algal lines along rocky shores, vegetation changes in salt marshes, and the interface between seawalls and open water. Photographs of some areas, such as bleached-out sandy beaches, were difficult to interpret visually. In these cases, field checks and historical shorelines aided interpretation of the orthophotographs.

To verify the accuracy of orthophotograph rectification, we selected control points along the Massachusetts shore at easily recognizable sites, such as building corners and street intersections. We located these sites in the field and recorded the DGPS coordinates, which we then compared to orthophotograph coordinates. Results show that the orthophotographs comply with National Map Accuracy Standards (U.S. Bureau of the Budget, 1947).

Due to the variety of aerial photograph data sources and analytical techniques used to compile the shoreline positions described here, there are a number of potential sources of error that affect the accuracy of the shoreline positions shown on the shoreline change maps. Since much of the historical database had not been checked previously for geographic accuracy, the historical database was subjected to quality control review. The relatively high geographic accuracy and photographic detail of the 1994 orthophotographs, as evidenced by the accurate display of roads, buildings, and shoreline structures, allowed the identification of errors in the pre-1994 shorelines. Although the pre-existing data set was represented to comply with National Map Accuracy Standards, there are inevitably errors in any large spatial data set that more accurate data bring to light.

Our analysis resulted in the removal of approximately 43 kilometres of poor-quality data that could have contributed to erroneous interpretations of shoreline change.

Analysis of the various sources of error suggests that the individual shoreline positions in our completed data set are generally accurate to within +/- 8.5 meters. Rates of shoreline change derived from these shorelines have a resolution of +/- 0.12 meters/year (Thieler et al., 2001).

16.2.2 Generating the 1994 Shoreline

We used the line tool in ArcView 3.2 to delineate shoreline segments for each orthophotograph at a scale of >= 1:2,000. After saving shapefiles of the 1994 shoreline segments for each orthophotograph, we used the ArcView Geoprocessing Wizard extension to merge all segments into a single shapefile containing the 1994 shoreline. We added the attributes "length" and "year" to the attribute table before using the Geoprocessing Wizard again to merge the 1994 shoreline with the shapefile of historical shorelines.

Some areas had highly migratory shorelines, such as barrier islands where the landward migration over the past 150 years exceeded the width of the island. To ensure that seaward shorelines would only be compared to other seaward shorelines, and not to any landward shorelines, it was necessary to create two edited versions of the shoreline coverages. One coverage included only the seaward edge of migrating areas (for example, only the open-ocean side of a barrier island); the complementary coverage included only landward coasts (for example, only the bay side of a barrier island). This method enabled us to examine barrier island migration in addition to shoreline change.

16.2.3 Preparing the Workspaces

We used a suite of Arc Macro Language (AML) programs and a modified version of programs in the Digital Shoreline Analysis System (DSAS; Danforth and Thieler, 1992; Thieler and Danforth, 1994; Van Dusen, 1996) to pinpoint shoreline positions, cast orthogonal transects along the shoreline, compute a linear regression of the rate of shoreline change at each transect location, collect all of the rates and statistics into an INFO database, and create 1:10,000 scale maps. To achieve this it was necessary to establish a shore-parallel baseline and to create sequentially numbered workspaces containing data for each coastal area.

To establish a point of reference for shoreline change measurement, we used the *buffer* command to draw baseline segments 50 meters landward of and parallel to the general trend of the shorelines. We used a combination of the ArcEdit commands *unsplit, grain*, and *spline* to modify the baseline, placing vertices at 20 meter intervals. Van Dusen (1996) described a similar approach using the *spline, generalize*, and *densify* commands.

We saved each baseline segment, typically about 7 kilometres long, in its own workspace along with a coverage containing the corresponding shorelines. Using polygon coverages, we clipped the larger shoreline coverages, thereby creating smaller coverages to reduce AML processing time.

Once all of the completed shoreline and baseline coverages were saved in the appropriate workspaces, we ran an AML to format the data for the transect-casting and rate calculation software. In each workspace, the AML created a frequency file listing of all of the shoreline years in that area. The AML then used this file with the *reselect* command to select arcs by their year attribute before using the *build* and *ungenerate* commands on the coverage. It also used the frequency file to create a file containing the date for each shoreline. It created formatted text files for the baseline and all shorelines using the generate files.

16.2.4 Digital Shoreline Analysis System

As described by Van Dusen (1996), the basic software used to determine shoreline rates-of-change was a modified version of the DSAS, developed by the U.S. Geological Survey (Danforth and Thieler, 1992). This software is comprised of two C programs. Using the formatted text files, the first program applied a baseline-vertex approach to cast orthogonal transects from each vertex on the baseline to the most seaward shoreline (Figure 16.1 and colour insert following page 169). The program also created a file that contained both the coordinates of points at which the transects crossed each shoreline, and the dates associated with each of those shoreline points. A second C program used this file to compute the rates of change between shorelines at each transect location and to collect all of the rates and statistics into an INFO database. The resulting output file includes four statistical measurements for each transect: the end point rate, average of rates, linear regression rate, and jackknife rate.

These rates of shoreline change (Dolan et al., 1991) were calculated by measuring the differences between shoreline positions through time along a given transect. The end point rate was calculated by dividing the distance of shoreline movement by the time elapsed between the oldest and the most recent shoreline. The average of rates method (Foster and Savage, 1989) involved calculating separate end point rates for all combinations of shorelines and then taking the average of all end point rates. A linear regression rate-of-change statistic was determined by fitting a least squares regression line to all shoreline points for a particular transect; the rate is the slope of the line. The jackknife rate was implemented as an iterative linear regression which calculates a linear regression fit to shoreline data points with all possible combinations of shoreline points, leaving out one point in each iteration. The slopes of the linear regression lines were averaged to yield the jackknife rate.

16.2.5 Compiling the Data

A suite of four AMLs formatted the output files from the two C programs for the maps and database. The first AML created a coverage of the transect lines with associated shoreline change rates. To do this, it built the transect coverage as lines with the *build* command and added formatting symbols and values for the baseline and transect numbers to the attribute table for use in labelling transects on the maps. In order to determine angle and direction of the lines, it then built line and node topology and added coordinates of the transect nodes to the coverage's node attribute table. Next, it imported the data from files created by the rate-calculation program into the ArcInfo tables and established a relate in the workspace for the next step. After joining the area files to the transect attribute table for that area, the AML brought the related information and arcs to a new appended coverage.

At this point, all of the necessary map coordinates and rate-of-change statistics were calculated. A second AML made a database for export to Microsoft Access. This AML employed the *append* command to create a new coverage containing the transects for the area workspaces. It then used the *tics* option with the *append* command to renumber transects in the appended coverage with unique, sequential ID values. After building the line coverage, the AML added a new item to the attribute table to allow reselection of every other transect. It then deleted alternate transects, creating a 40 meter interval for display on the printed maps, and renumbered the transects for labelling in the database and on maps. It also utilized the *additem* and *join* commands to add the town names from a separate coverage to the transect coverage. The final routine sorted the transect coverage by baseline and transect numbers.

We used ArcMap to remove manually the poorly located transects that crossed inlets and jetties. After the manual edits, a third AML added items to the attribute table of the transect coverage and selected the attribute table in ArcTables. It used the ArcEdit command *unsplit* to combine the identification arcs in the transect coverage and then built the transect line coverage. The AML then sorted the transects by transect number in ArcTables. After using the *copyinfo* command on the transect attribute table to create a data file, the AML used *dropitem* to delete attributes such as TNODE that are unnecessary for a user database. It then sorted this data file by transect number and used the *infodbase* command to create a database file. A final AML created the last files needed for the database by indexing the transects according to which map would display them. Using the *infofile* command, it created a DAT file for each map containing the transect number, town name, and town identification number. Finally, it used the *infodbase* command to copy this information into DBASE III+ files.

We imported the INFO database into Microsoft Access and converted it to a standalone application to improve users' ability to perform spatial queries and to compile regional statistics and trends. The database forms, reports, and macros were set up for an interactive display; only the tables needed to be replaced with the new database files and the transect coverage database. We used a macro to update the database with the new statistics and to calculate mean rates and bounding shoreline dates for each transect.

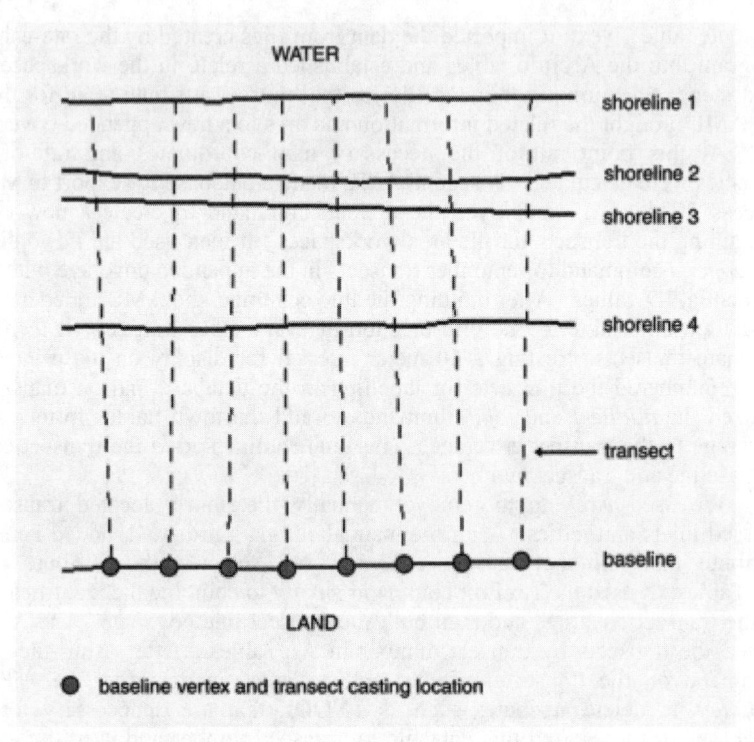

Figure 16.1 The transect casting scheme used to determine locations for the measurement of shoreline rates of change.

16.2.6 Map Production

The final project goal was to generate maps displaying shoreline change. We used an AML to format the maps with a legend and other explanatory information. This AML used the *reselect* command to draw shorelines with a different colour for each era. It also used the command *relate add* to correlate the files created by the rate-calculation program to the existing relate environment, which was saved as an INFO file. The *readselect* command selected every other transect in the writefile of selected transects, and the ArcPlot commands *arcendtext* and *leadertolerance* allowed the transects to be labelled with the associated identification number and linear regression rate of change.

16.3 RESULTS

The series of programs described above produced an Access database that can be searched by map number, community name, or transect number. Transect statistics can be listed in a table or graphed in histograms. The tables and histograms can either be printed or viewed on screen (Figure 16.2 and Figure 16.3).

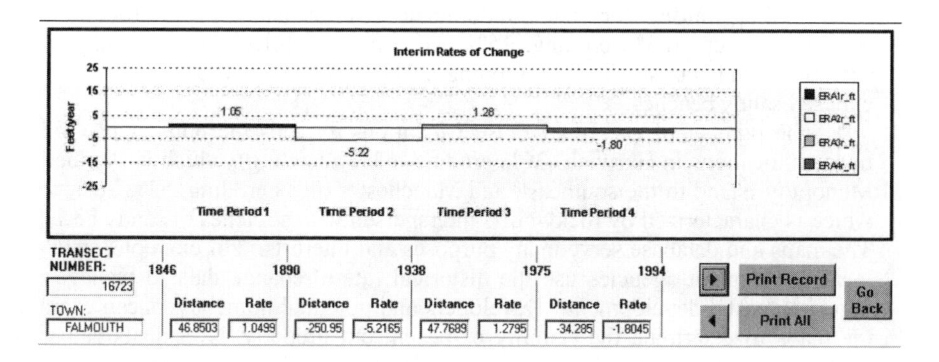

Figure 16.2 Sample histogram depicting interim rates of change.

MASSACHUSETTS HISTORIC SHORELINE CHANGE PROJECT — Transect Statistics for FALMOUTH

Transect Number:	First Time Period (1846–1890): Distance*	Rate**	Second Time Period (1890–1938): Distance*	Rate**	Third Time Period (1938–1975): Distance*	Rate**	Fourth Time Period (1975–1994): Distance*	Rate**	Net (overall) Time Period (1846 to 1994): Net Distance Change*	Summary Statistics: Long Term Annual Shoreline Change Rate~
16719	94.32	2.13	-443.11	-9.22	-94.68	-2.56	-82.84	-4.36	-526.31	-4.17
16720	-168.80	-3.84	-196.72	-4.10	-90.91	-2.46	-79.00	-4.17	-535.43	-3.58
16721	-191.40	-4.36	-192.29	-4.00	-46.85	-1.28	-113.98	-6.00	-544.52	-3.48
16722	-137.17	-3.12	-212.04	-4.43	-20.51	-.56	-104.43	-5.51	-474.15	-3.12
16723	46.85	1.05	-250.95	-5.22	47.77	1.28	-34.28	-1.80	-190.62	-1.57
16724	33.04	.75	-241.08	-5.02	10.01	.26	-63.68	-3.35	-261.71	-2.00
16725	29.53	.66	-232.15	-4.82	6.10	.16	-57.51	-3.02	-254.03	-1.94
16726	24.28	.56	-211.68	-4.40	-6.82	-.20	-59.84	-3.15	-254.07	-1.90
16727	-5.02	-.10	-177.59	-3.71	-24.54	-.66	-46.98	-2.46	-254.13	-1.87
16728	-17.68	-.39	-162.40	-3.38	40.88	1.12	-72.90	-3.84	-212.11	-1.44

Statistics for each interim time period show the distance difference and the derived rate of change. Summary statistics show the overall derived Long Term Annual Shoreline Change Rate.

* units in feet, ** units in feet/year, ~ determined using linear regression.

Go Back | Print

Figure 16.3 Sample transect listing produced by the Access database.

The programs also produced 91 ArcPlot files. The area maps of the Massachusetts coast are at a scale of 1:10,000 and display historical shorelines, transect locations, and long-term rates of change overlaid on the 1994 orthophotographs (shown in Colour Plate 16.1, following page 164).

The resulting database and maps include a total of 30,354 transects spaced at 40 meters along the coast of Massachusetts. Linear regression rates indicate erosion at 68% of the transects, accretion at 30% of the transects, and no net change at 2% of the transect locations.

Both the highest and lowest linear regression rates, 15.2 meters per year of accretion and 12.9 meters per year of erosion, occur on Nauset Beach in Chatham, where the barrier island shoreline faces the open Atlantic Ocean.

Several coastal areas have nearly the same number of accreting areas as eroding areas, such as Hingham (47% eroding, 48% accreting) and Hull (50% eroding, 46% accreting). Both of these areas are located within 40 kilometres of Boston, are highly populated, and have a number of seawalls holding the shoreline in place.

Eroding transect locations predominate in Plum Island (74% eroding, 24% accreting), Truro (83% eroding, 15% accreting), Wellfleet (81% eroding, 18% accreting), and Barnstable (73% eroding, 25% accreting). All of these areas have exposed sandy beaches.

The number of accreting transect locations exceeds the number of eroding transect locations in Harwich (36% eroding, 63% accreting), which is shielded by Monomoy Island to the southeast, and Manchester (40% eroding, 53% accreting), which is characterized by rocky cliffs interspersed with stretches of sandy beaches. The maps and database serve many purposes and interests. For example, state and local government agencies use the historical rate-of-change data to manage and regulate coastal development. Developers and potential homebuyers can examine the maps to determine the stability of beachfront property. Residents can view post-storm erosion in the context of long-term shoreline variability. This information can also aid in evaluating past management efforts, such as the emplacement of seawalls, jetties, and groins that have influenced shoreline change.

MCZM distributes these maps, along with the database and a manual describing how to use the information, to regional government offices and coastal communities. These products will also be made available to the public on the Internet.

16.4 SUMMARY

This project addresses the increasing need to measure and document coastal change by combining established shoreline evaluation techniques with improved GIS applications. The resulting maps and database comprise a current and accessible source of information on coastal change in Massachusetts, including historical shorelines and shoreline change statistics. Coastal managers in Massachusetts can continue to benefit from the database by updating the database with new shorelines and rerunning the suite of AMLs and C programs that calculate change and produce maps.

Academics, consultants, citizen groups, developers, and government agencies can use the map and data products described in this paper for a variety of purposes and interests. The database and maps have the potential to improve decisions concerning coastal management, residential and commercial development, and coastal research by making shoreline data comprehensible and available to a wide audience. The maps displaying shorelines and rates of change are visually appealing and quickly understood, particularly when overlaid on orthophotographs. More detailed information is readily available in an interactive database. If further modifications are desired, new modules can be incorporated into the current program and database structure. These methods and products are an excellent resource for anyone interested in coastal change, and they serve as a solid foundation for future applications.

16.5 ACKNOWLEDGEMENTS

The Massachusetts Shoreline Change Project was funded by the Massachusetts Office of Coastal Zone Management pursuant to a grant from the National Oceanic and Atmospheric Administration, Award NA970Z0165. Susan Snow-Cotter, Steve Mague, Diane Carle, and Rebecca Haney of CZM offered many valuable suggestions for the development of the maps and users guide. Bill Danforth of the USGS provided computer support. Technical reviews by Laura Moore and Larry Poppe improved the clarity and content of the paper.

16.6 REFERENCES

Benoit, J. R., ed., 1989, *Massachusetts Shoreline Change Project.* (Boston: Massachusetts Office of Coastal Zone Management), pp. 19, appendices.

Clow, J.B., and Leatherman, S.P., 1984, Metric mapping: An automated technique of shoreline mapping. In *Proceedings, 44th American Congress on Surveying and Mapping*, American Society of Photogrammetry, pp. 309-318.

Danforth, W.W., and Thieler, E.R., 1992, Digital Shoreline Analysis System (DSAS) User's Guide, Version 1.0. *U.S. Geological Survey Open-File Report No. 92-355.* (Reston, Virginia: U.S. Geological Survey), pp. 42.

Dolan, R., Fenster, M.S., and Holme, S.J., 1991, Temporal analysis of shoreline recession and accretion. *Journal of Coastal Research*, **7** (3), pp. 723-744.

Foster, E.R., and Savage, R.J., 1989, Methods of historical shoreline analysis. In *Coastal Zone '89, Proceedings of the Sixth Symposium on Coastal and Ocean Management.* (New York: ASCE), pp. 4420-4433.

O'Connell, J.F., 1997, Historic shoreline change mapping and analysis along the Massachusetts shore. In *Coastal Zone '97, Proceedings of the Tenth Symposium on Coastal and Ocean Management.* (New York: ASCE).

Thieler, E.R. and Danforth, W.W., 1994, Historical shoreline mapping (II): application of the Digital Shoreline Mapping and Analysis Systems (DSMS/DSAS) to shoreline change mapping in Puerto Rico. *Journal of Coastal Research*, **10**, pp. 600-620.

Thieler, E.R., O'Connell, J.F., and Schupp, C.A., 2001, *The Massachusetts Shoreline Change Project: Technical Report 1800s to 1994.* U.S. Geological Survey Administrative Report. (Woods Hole, Massachusetts: U.S. Geological Survey), pp. 36.

U.S. Bureau of the Budget, 1947, National Map Accuracy Standards. (Washington: U.S. Government Printing Office).

Van Dusen, C., 1996, *Vector based shoreline analysis*. Unpublished report. (Boston: Applied Geographics, Inc.).

CHAPTER SEVENTEEN

GIS for Assessing Land-Based Activities that Pollute Coastal Environments

J.I. Euán-Avila, M.A. Liceaga-Correa, and H. Rodríguez-Sánchez

17.1 INTRODUCTION

According to Heathcote (1998), the development of workable management options (structural and non-structural actions) in a watershed requires the identification of all point and non-point sources of pollution. Discharge of effluents from industrial, urban and sewage treatment plants where a pipe or diffuser outfalls into a water body are called "point sources" of contamination. Another type of source, called a "non-point source" (NPS), is described as the diffuse drainage of rainwater from urban, industrial and agricultural lands that can introduce nutrients, pesticides, and metals to water bodies. Non-point sources are some of the more serious forms of pollution and the effects are often less obvious than those of point sources (Abel, 1998). Clapman *et al.* (1998) indicated that agricultural impacts on ground and surface water quality are more significant than other land use impacts because of their large aerial extent compared to other human land uses. Lack of information related to agricultural practices may lead to overuse of fertilizers and herbicides, and deforestation with serious impacts on soil erosion and water quality.

The Yucatan Peninsula is blessed with large freshwater reserves, bays and coastal lagoons and an exclusive economic zone of 200,000 km^2. However, the maintenance of the water quality seems to be an enormous challenge for the State and other interested groups considering the unfulfilled basic needs of a large part of the population, rates of population growth, immigration, and lack of an integrated approach to the management of the resources and coastal areas. In June 1996 a massive kill of 20,000 fish of the species *Arius felis* was reported in the Bay of Chetumal, State of Quintana Roo (SEMARNAP, 1996). Studies reported harm to their organs and accumulation of PCBs, organochlorine insecticides and polyaromatic hydrocarbon (Noreña-Barroso, 1998). Concerns exist that increasing loss of water quality may have adverse effects on mangroves and coral reefs in an area where the second most important barrier reef, the Mesoamerican reef, and the sanctuary of the manatee *Trichechus manatus* are drawing international attention.

0-41531-972-2/04/$0.00+$1.50

Agricultural activities introduce diffuse pollution to watercourses, aquifers, lagoons and estuaries in the form of sediments, nutrients, pesticides, viruses, salt and other toxins which affect aquatic organisms. Pollution cause-and-effect relationships are complex and the need to find practical tools to generate useful information for decision-making may be addressed with models that introduce expert knowledge. Within this framework, this exercise attempts to rank the agricultural lands according to several factors that may contribute to water contamination on the Mexican side of the Rio Hondo watershed near the Othon P. Blanco municipality border with Belize.

17.2 GEOGRAPHICAL SETTING

17.2.1 The Yucatan Peninsula

The Yucatan Peninsula in México has mainly sub-surficial water dynamics driven by its geological and topographic nature: there are few rivers or lakes in the area. Exceptions are the Candelaria and Champotón rivers in the zone of the Términos Lagoon in Campeche, and the Hondo River in the State of Quintana Roo near the border with Belize. Yucatán has no rivers; however underground discharge in the coastal zone has been estimated at 9.7 million m^3 per year (CNA, 1998). The main features of the coastal areas are bays and lagoons, which are distributed throughout the three states: Términos Lagoon in Campeche; Celestún, Dzilam and Rio Lagartos in Yucatán; and Chetumal, Ascensión and Spiritu Santo bays in Quintana Roo. Surrounding these water bodies, other wetlands cover an area of approximately 8000 km^2, thus forming an important part of the coastal ecosystem (CNA-UNU/RIAMAS, 2000).

Due to its geographical location between the Caribbean Sea and the Gulf of Mexico, the region is influenced by severe hydrometeorological phenomena. The climate of the region is semi-arid in the coastal zone of the north part of the Peninsula and warm with variation of dry to humid in the rest of the peninsula. The mean annual temperature is 26ºC. There are two main seasons in the regional climate: the "rainy season," including extreme phenomena such as hurricanes and tropical storms from May to October; and the "winds of the north" season, from November to April. The region receives abundant but uneven rainfall with mean annual precipitation ranging from 1600 mm in the southeast to 500 mm in the north. Mean annual evaporation is around 1.78 mm (CNA-NU/RIAMAS, 2000).

Human activities in the Yucatán Peninsula are related to agriculture, livestock production, tourism, fishing, oil production, and transportation, and recently to a large number of *maquiladoras*. Fertilizers, pesticides and metal residues have been found in coastal waters and the aquifer. These negative effects on the environment have been reported in several locations (Pacheco and Cabrera, 1996, Benitez and Bárcenas, 1996, Ortiz and Sáenz, 1997, Noreña-Barroso *et al.*, 1998, CAN, 1998, Herrera-Silveira *et al.*, 1998). Coastal resources in the Yucatan Peninsula provide increasing opportunities for economic development in a large number of traditional fishing communities. Lack of awareness of the deleterious effects of land-based activities on water quality may lead to a reduction in biodiversity and esthetical quality of the landscape, thus putting at risk the continuity of fishing, tourism, and recreational activities.

17.2.2 The Municipality of Othon P. Blanco

Othon P. Blanco is a municipality located in the State of Quintana Roo in the southeast part of the Yucatan Peninsula (Figure 17.1). It has an area of 619,799 hectares, 161,226 (26%) of which are occupied by agricultural use, 129,396 (20.9%) by natural and cultivated grass, 325,155 (52.5%) by forest and 4,021 (0.6%) by other uses.

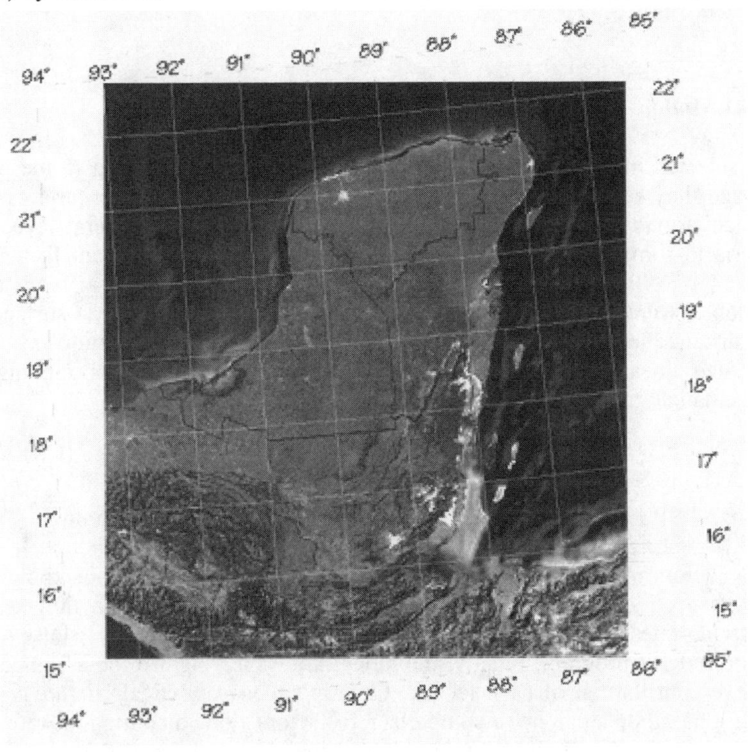

Figure 17.1 Study area in the Yucatan peninsula, Mexico. Municipality of Othon P. Blanco in the state of Quintana Roo. Main hydrological features are the Hondo River and the Bay of Chetumal.

The population in 1990 was 172,563 with 53.5% less than 19 years of age. The municipality has 437 rural and urban centers with 45% of the population living in 436 towns with less than 5,000 inhabitants. Close to 50% of the population are immigrants. Of the population 15 years or older, 17.4% completed elementary school (INEGI, 1991a). Main crops by cultivated area include: corn (19,196 hectares), sugar cane (16,000), chili (5,151), and beans (1, 391) (INEGI, 1993). Production in 1991 was estimated at 12,000 tonnes for corn, 785,000 for sugar cane, 26,000 for chili, and 84 for beans. Other crops cultivated in the area are orange, coconut, and banana (INEGI, 1994).

Some of these agricultural lands can be found in the watershed of Chetumal Bay. The Bay is 67 km in length and 20 km wide, and receives freshwater from the Hondo River and Guerrero Lagoon, causing it to exhibit estuarine characteristics. Man-made channels and other tributaries close to agricultural lands are linked to the Hondo River. The most dominant soil types, Rendzinas and Litosoles, cover 70% of the Peninsula and 85% the study area (INEGI, 1985).

17.3 MODEL AND DATA LAYERS

17.3.1 Model

The selected model attempts to rank agricultural lands according to the potential menace they represent to water quality. The model is a multi-criteria evaluation method provided by the IDRISI software, which combines several layers as the criteria to form an index of evaluation (Eastman, 1999). Relevant factors in the process of NPS pollution assessment (being those for which an estimate was available) were: amount of agrochemical inputs, slope, proximity to surface water, and distance to aquifer. Given the natural continuity in factors, ratio layers and a weighted linear combination of them can provide an index for ranking. Their mathematical representation is as follows:

$$S = (\Sigma\, w_i * x_i) * \Pi\, c_j \qquad (17.1)$$

where s= NPS index, w_i= weighting factor i, x_i= factor i, c_j= constraint j

Factors must be normalized according to the accuracy of our knowledge of their respective ranges of impact, as well as the ways in which they behave at different scales. The procedure used, which is based on fuzzy sets, is also provided by the IDRISI program. Finally, the model allows a weight to be assigned for the relative contribution of each factor. Criteria called Analytical Hierarchy Process (AHP), based on a pair-wise comparison of factors along a continuous rating scale, can derive weighs by calculating the principal eigenvector of the created matrix (Eastman, 1999).

17.3.2 Data layers

17.3.2.1 Geographic location of agricultural lands

A good estimation of the location of agricultural land is a basic prerequisite for assessing NPS pollution. A two-band WIFS image with 180 x 180 m spatial resolution acquired in 1999 was used to estimate the location and extent of the agricultural lands in the Othon P. Blanco municipality (Figure 17.2a). A supervised classification was conducted to estimate the total cultivated land in the area. Census data from the AGROS system elaborated by the INEGI provided crop information for areas called Basic Geo-statistical Area (AGEB in Spanish) (INEGI, 1996). Each AGEB is a well defined polygonal (vector) area with a link to a data base (Figure 17.2b). These two layers were used to estimate the location of cultivated areas by AGEB.

17.3.2.2 Agrochemical practices

The quantification of chemicals used per unit area and frequency of use was another factor in the analysis. A survey was conducted in September 1999 for the purpose of estimating the quantities of fertilizer and pesticides used per crop each year. A total of 97 farmers were interviewed in five agricultural towns: 1) Nicolás Bravo, 2) Palmar, 3) Pucté, 4) Sergio Butron, and 5) Morocoy. The questionnaire was divided into three sections to learn about fertilizer, insecticide, and herbicide practices per crop. Each section focuses on names, dosage, and frequency of applications of agrochemicals by crop per hectare per year. Substances were normalized according to thresholds determined through consultation with experts in the field.

17.3.2.3 Digital elevation model and slope

Slope is a well known factor that favours the movement of the substances on the terrain surface. Slope and aspect are the main factors that determine velocity and direction of the overland flow during storms. In areas with long slopes, the capacity of vegetation to reduce erosion is diminished. Slope was computed from a DEM of the area using the TNT software (Figure 17.2c). The DEM was produced by INEGI in a scale of 1: 250,000. Two archives, E1604 and E1607, were mosaiced to cover the studied area.

17.3.2.4 Proximity to surface water

Distance from agricultural activities to water bodies is also a factor that may facilitate contaminants reaching water courses, ponds, or estuaries. Proximity maps were constructed based on water features identified on band 2 of the WIFS image as well as those features found on topographic maps E16-4-7 at a scale of 1:250,000 produced by INEGI. These two layers were combined to generate a reference object for computing proximity to surface water (Figure 17.2d).

17.3.2.5 Proximity to groundwater

In karst formations (carbonated rocks) such as the Yucatan Peninsula in Mexico, precipitation tends to infiltrate rapidly because of the high number of fractures and solution cavities in the massif. Depth to the aquifer was estimated from a regression model using 15 well depths provided by the Regional Office of the *Comision Nacional del Agua* (CNA) and their corresponding elevation in the DEM (Figure 17.2e).

17.4 RESULTS

17.4.1 Agricultural lands

Agricultural areas detected from the satellite image were defined based upon the 1991 census. These areas were mainly devoted to growing sugar cane, corn, beans and *jalapeño* chili. The statistical data from AGEBS was proportionally assigned to current agricultural lands for a better spatial location of the cultivated areas (Figure 17.2b). As an example, Figure 17.2b shows, by AGEBS, the percentage of the total cultivated surface with sugar cane in 1991. The percentage for corn, beans and *jalapeño* chili was related to the area in a similar fashion.

17.4.2 Agricultural practices

Data from the five sampled regions indicated that farmers are working those lands for 11 years on the average (the interval is from 1 to 30 years with a standard deviation of 7 years, out of 93 valid cases). The mean value of the farmers' age was 42 (the interval was from 18 to 74, and the standard deviation was 12). The number of years each farmer had spent at school had a mean value of 4 (from 0 and 14, with a standard deviation of 3). These farmers reported applying agrochemicals to their primary crops (sugar cane, corn, *jalapeño* chili, and beans) during cultivation. Sugar cane is one of the main crops in the area and it received a large number of agrochemicals (more than 25 commercial products were identified). Given the large number of products, only those with a high frequency of use in each category (fertilizers, insecticides and herbicides) were selected for analysis. From these selected products, active substances were estimated giving the total applied quantity per hectare per year (an example is provided in table 17.1 for sugar cane). A layer with total phosphorous used in one year in kg/ha was calculated from four top selected crops with values of 70.25 kg/ha for sugar cane, 18.8 kg/ha corn, 40.19 kg/ha *jalapeño* chili, and 15.21 kg/ha beans each year. Similar computations were used to estimate nitrogen in the fertilizer category. In the case of herbicides substances such as 2,4D, paraquat, and ametryne were estimated. Finally, estimated substances in the insecticides category were metamidophos, chloropiriphos, and monocrotophos.

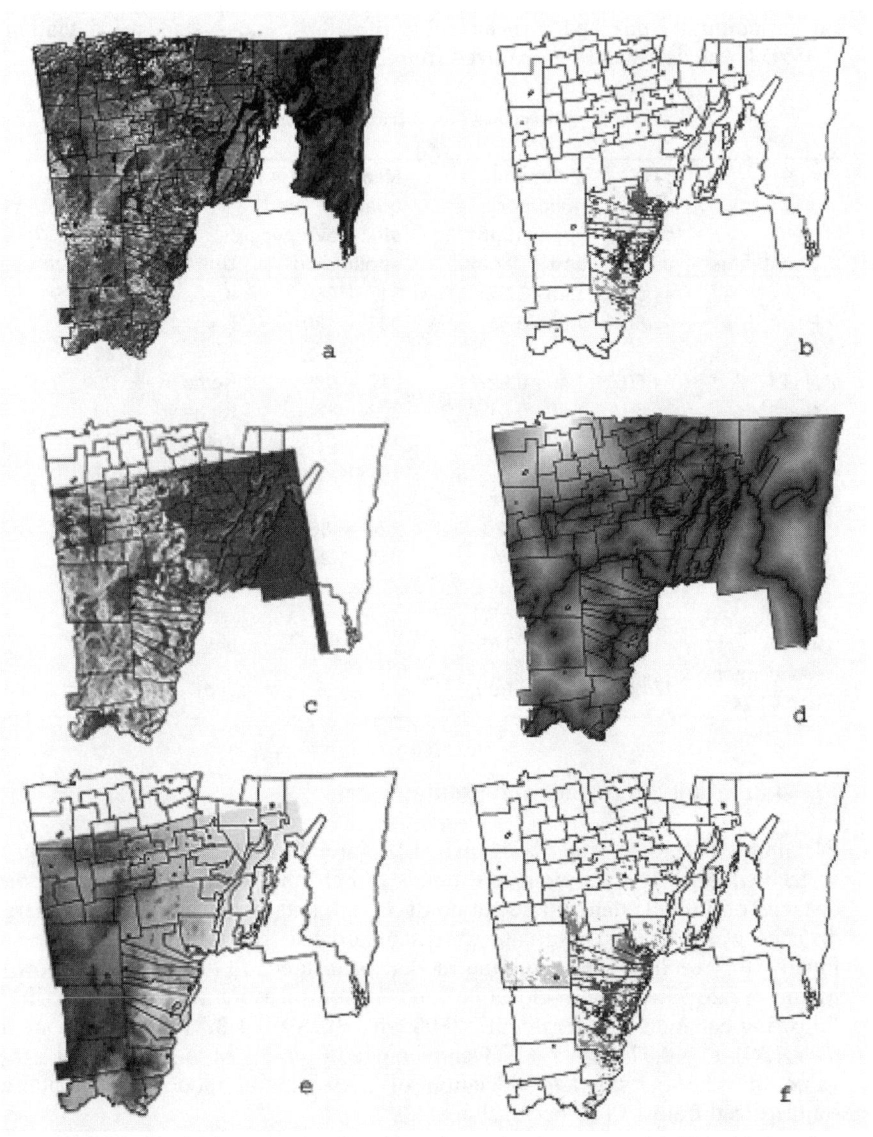

Figure 17.2 (a) WIFS image band 2, (b) cultivated areas for sugar cane and AGEBs, (c) slope from DEM, (d) proximity to surface water, (e) distance to aquifer, and (f) rank of nutrients (P and N).

17.4.3 Slope

Figure 17.2c shows the magnitude in percentage of slopes calculated from the DEM. Slopes in the area range from 0% to 48% with a mean value of 1.3% and a standard deviation of 2.5%. Elevation in the area ranges from 0 to 300 m. Slopes on agricultural lands had a mean value of 1.84% and a standard deviation of 2.98%. These data were standardized from 2 to 10% increasing.

Table 17.1 Agrochemical substances used in sugar cane cultivation.

Substance	% of users	Mean annual number of applications and std. dev.	Mean quantity and std. dev. per application	Unit	Total amount of active substance
PK al 17%	44/60	1.40 - 0.92	317 - 089	Kg/ha	55.25
PK al 15%	14/60	1.30 - 0.61	332 - 130	Kg/ha	15.00
					70.25
N al 17 %	44/60	1.40 - 0.92	317 - 089	Kg/ha	55.25
N al 15 %	14/60	1.30 - 0.61	332 - 130	Kg/ha	15.00
N al 46 %	34/60	1.20 - 0.6	153 - 071	Kg/ha	47.84
					118.1
Ametryne 39.2%	14/60	1.10 - 0.28	2.8 - 0.86	l/ha	0.29
Ametryne 25%	24/60	1.50 - 0.77	3.9 - 1.2	l/ha	0.387
					0.677
2,4-D 49.4%	31/60	1.20 - 0.66	2.3 - 1.1	l/ha	0.73
Monocrotophos 44.24%	17/60	1.90 - 1.00	1.9 - 2.6	l/ha	0.45

17.4.4 Proximity to surface and ground water

Proximity to surface water is shown in Figure 17.2d. Features included in the surface water category were rivers, ponds and channels. Results indicated a mean distance of 3.6 km, standard deviation of 3.2 km and mode of 0.9 km. The largest distance was 17 km. These data were standardized from .5 to 5 km decreasing. Finally, the result of the regression model for distance to the aquifer is shown in Figure 17.2e. Regressed elevation and depth measurements resulted in the following equation: aquifer depth = $0.0029h^2 + 0.2891h + 3.8161$ (h = elevation in the DEM), $r^2 = 0.97$ for n = 15. Depths ranged from 3.8 m to 211 m with a mean value of 42.5 m, standard deviation of 37.9 m and mode 3 m. Data was standardized from 1 to 15 decreasing.

17.4.5 Model results

Distribution of the S index for fertilizers, herbicides, and insecticides was calculated as a potential threat to water quality. An example, based upon nutrient load, is illustrated in Figure 17.2f. Parameters to compute the S index for standardized factors were selected from the observed data range of factors in the area. Equal weights were assigned to factors. With these parameters, the NPS index for fertilizers (phosphorous and nitrogen) had values from 1 to 204, a mean value of 82 and a standard deviation of 44. Higher values of the index were found in the areas of Pucté and Palmar where sugar cane is the main crop. In the vicinity of Morocoy and Sergio Butron medium values were recorded, and the smallest values could be found in the area of Nicolas Bravo. The NPS index for herbicides (2-4D, paraquat, and ametrine) had values from 1 to 155, with a mean of 50 and a standard deviation of 32. Large tracts of Pucté and Palmar displayed higher values, while Sergio Butron displayed medium values and Morocoy and Nicolas Bravo the smallest. Finally, for insecticides (monocrotofos, chloropyrifos, and metamidofos) the NPS index ranged from 1 to 124, with a mean value of 40 and a standard deviation of 23. In this case, large areas of Pucté and Palmar displayed higher values, Sergio Butron and Morocoy displayed medium values, and Nicolas Bravo showed the smallest. The analysis suggest that Pucte and Palmar are major potential areas of threat to water quality as a result of nutrient, herbicide and insecticide inputs, while the importance of Morocoy, Sergio Butron and Nicolas Bravo in this respect changes according to the type of agrochemical under analysis.

17.5 CONCLUSIONS

An exercise in the use of GIS tools was conducted to characterize agricultural NPS pollution in the Yucatan Peninsula. Standard GIS processes and digital data from satellite, maps, and other digital products available in Mexico allowed ranking of agricultural lands in the Hondo river watershed according to the threat posed to water quality by the use of agrochemicals on these lands. Standard procedures available in the TNTmaps and IDRISI software packages were used to feed and run a multi-criteria model for decision-making. Three indices suggested that two out of the five agricultural study areas, where sugar cane is the main crop, have a large potential and non-homogeneous threat to water quality as indicated by the NPS index. The potential contribution of the other agricultural areas varies when examined by category (nutrients, herbicides, and insecticides). Increasing water contamination and limited resources in developing countries can be better allocated when knowledge of the co-occurrence of practices and terrain features is integrated into models for the assessment of the potential effects of human activities on water resources. As new knowledge is integrated these scenarios can be fine-tuned with more precise data and expert knowledge, providing improved spatial information to assist decision makers in the selection of appropriated management actions in the arena of NPS pollution.

17.6 ACKNOWLEDGMENTS

We thank the following people for their many contributions: J. Acosta, H. Hernández, G. Mexicano, Ricardo Rodríguez, Dr. Jorge Alvarado, P. I. Caballero, and students from the Centro de Estudios Tecnológicos del Mar (CETMAR). This research has been supported in part by Secretaría del Medio Ambiente Recursos Naturales y Pesca (SEMARNAP).

17.7 REFERENCES

Abel, P.D., 1998, *Water Pollution Biology.* (London: Taylor & Francis).

Benites, J.A. and Bárcenas, C., 1996, Sistemas fluvio-lagunares de la Laguana de Términos: hábitats críticos susceptibles a los efectos adversos de los plaguicidas. In *Golfo de México, contaminación e impacto ambiental: diagnóstico y tendencias*, edited by Botello, A.V., Rojas Galaviz, J.L., Benítes, J.A. and Zárate Lomeli, D. *EPOMEX* Serie Científica No. 5.

CNA, 1998, *Diagnóstico para la región XII, Península de Yucatán*, edited by Gerencia Regional de la Península de Yucatán de la Comisión Nacional del Agua.

CNA-UNU/RIAMAS, 2000, *El Proceso Final para el Cuidado y Manejo Responsable del Recurso Agua en la Península de Yucatán*. Reporte final, Comisión Nacional del Agua, pp. 2–6.

Eastman, J.R., 1999, *Guide to GIS and Image Processing, Vol. 2*. (Massachusetts: Clark Labs).

Heathcote, I.W., 1998, *Integrated Watershed Management: Principles and Practice.* (New York, Toronto: Wiley).

Herrera-Silveira, J.A., Ramírez, R.J., and Zaldivar, J.A., 1998, Overview and characterization of the hydrology and primary producer communities of selected coastal lagoons of Yucatán, México. *Aquatic Ecosystem Health and Management,* **1**(3–4), pp. 353–372.

INEGI, 1985, *Carta Edafológica 1:250 000.*

INEGI, 1991, Quintana Roo. Resultados Definitivos. Tabulados Básicos. *XI censo General de Población y Vivienda de 1990*, pp. 1–37.

INEGI, 1993, Othón Pompeyo Blanco, Estado de Quintana Roo. Cuaderno Municipal, Gobierno del Estado de Quintana Roo, edited by INEGI and H. Ayuntamiento Constitucional del Estado.

INEGI, 1994, Quintana Roo. Panorama Agropecuario. *VII censo agropecuario 1991*, pp. 23–33.

INEGI, 1996, AGROS, Información Censal Agropecuaria.

Noreña-Barroso, E., 1998, Contaminantes orgánicos y sus efectos a nivel histológico en bagres *Ariopsis assimilis* de la Bahía de Chetumal, Quintana Roo, México, Tesis de Maestría, CINVESTAV, Unidad Mérida, Yucatán, México.

Noreña-Barroso, E., Zapata-Perez, O., Ceja-Moreno, V., and Gold-Bouchot, G., 1998, Hydrocarbon and organochlorine residue concentrations in sediments from Bay of Chetumal, Mexico, *Bull. Environ. Contam. Toxicol.* **61**, pp. 80–87.

Ortiz, M.C. and Sáenz, J.R., 1997, Detergents and orthophosphates inputs from urban discharges to Chetumal Bay, Quintana Roo, México. *Bull. Environ. Contam. Toxicol.* **59** (03), pp. 486–491.

Pacheco, J. and Cabrera, A., 1996, Efecto del uso de fertilizantes en la calidad del agua subterránea en el Estado de Yucatán. *Ingeniería Hidráulica en México.* **11**(1), pp. 53–60, enero-abril.

SEMARNAP, 1996, Contaminación química en la Bahía de Chetumal, *Boletín Caribe*, edición de julio, Secretaría de Medio Ambiente y Recursos Naturales.

CHAPTER EIGHTEEN

Applying the Geospatial Technologies to Estuary Environments

David R. Green and Stephen D. King

18.1 INTRODUCTION

Around the World the value and vulnerability of estuaries has long been recognised. The National Estuary Study of 1969 in the U.S. is just one such example (http://www.inforain.org/mapsatwork/oregonestuary/Oregonestuary_page 4.htm). In the UK, initiatives by English Nature (EN) and Scottish Natural Heritage (SNH) to develop new approaches to estuary management in England and Wales, and Scotland respectively, led to Estuary Management Plans (English Nature, 1993). In Australia there have also been many similar initiatives to develop sustainable management of estuarine environments increasingly coming under pressure from tourism and industry (NLWRA, 2002) and, more recently, in the UK, from the impact of offshore windfarms e.g. Robin Rigg in the Solway Firth (Anonymous, 2003).

Green (1994; 1995) proposed the idea of GIS-based estuary information systems in the UK as a means by which it would be possible to collect, store, analyse and display spatial data and information to aid in estuary management. At the time, GIS was just beginning to develop into a practical tool for environmental applications in coastal zone management. In the intervening years, a great deal of interest has been shown in the development of GIS and the related geospatial technologies to aid in environmental monitoring, mapping, modelling, and management. Rapid developments in information technology (IT), including the Internet and the related technologies, have also led to the more widespread use of geospatial data and information for environmental applications by coastal managers and practitioners. Such developments have been responsible for providing the basis for access to data and information for management and public participation exercises. Increasingly, decision support systems (DSS) and information systems are also being used to support data and information requirements for coastal management.

With the continuing evolution of Information Technology (IT), the collection of, and access to, both data and information have rapidly been extended to the use of mobile technologies. These include Personal Digital Assistants (PDAs), Wireless Application Protocol (WAP)-enabled mobile phones, Global Positioning Systems (GPS), digital still and video cameras, and mobile GIS products such as ESRI's ArcPad, PocketGIS (http://www.posres.com), HandyGIS

0-41531-972-2/04/$0.00+$1.50

(http://www.handgis.com) and FastMap (http://www.surveysupplies.com). Field data collection and access to information, via remote uploading and downloading, sending email attachments (e.g. photographs taken with a digital still camera or camera accessory for a mobile phone), and accessing the Internet using wireless technology are now all providing new opportunities to collect, work and interact with spatial data 'on-the-fly' (e.g. Vivoni and Camilli, 2003).

Alongside the hardware and software developments there have also been a number of new and important airborne and satellite-borne sensors. These have provided new sources of finer spectral and spatial resolution data to assist in the monitoring of coastal and estuarine environments e.g. CASI (Compact Airborne Spectrographic Imager), whilst LiDAR (Light Detection and Ranging) data provides unique high-resolution height data. The launch of ENVISAT in 2002 by the European Space Agency (ESA) for dedicated environmental monitoring, and European projects such as COASTWATCH (http://www.coastwatch.info) are also increasing awareness of the role of EO data for coastal monitoring in the context of GMES (Global Monitoring of Environment and Security).

This chapter presents an overview of some of the recent developments in the application of the geospatial technologies to estuary environments as the basis to support growing requirements for data and information in the context of environmental monitoring, mapping, and management. The role of some of the different geospatial technologies for environmental data collection, processing and access to information is examined. To conclude, some future developments are discussed. The chapter is illustrated using a number of examples drawn from Europe (including the UK), North America, and Australia.

18. 2 THE ESTUARY ENVIRONMENT

Estuaries are very well studied features of our coastal environment. A great deal of research has been undertaken into the physical processes operating in an estuary including sediment movement, water circulation, pollution and erosion (e.g. Stapleton and Pethick, 1996; Stove, 1978; Townend, 2002). The use of modelling techniques has been widely applied in an attempt to increase our knowledge and understanding about the processes that are active (e.g. Hinwood and McLean, 2002). Ecological studies of estuarine environments have included investigations into coastal habitat, macro-algal weedmats, and bird population distribution (e.g. Jernakoff *et al.*, 1996; Young *et al.*, 2000; Lewis and Kelly, 2001; Ripley *et al.*, 2002).

Estuaries are complex and highly productive ecosystems (among the most productive on Earth) supporting a wide range of habitats and species, extending from the river's upper tidal limit to the sea, and provide a constantly changing environment where sea and fresh water mix. They provide:

- Habitat (live and feed)
- Nursery Areas (reproduce and spawning)
- Productivity (abundance and biodiversity)
- Water Filtration (fresh and salt marshes)
- Flood Control (buffer to flood waters, dissipate storm surges)

- Erosion Prevention and Stabilisation (grasses and plants)
- Cultural Benefits (recreation, scientific knowledge, education, aesthetic value)
- Economic Benefits (natural resources used for recreation, industry, fishing, tourism)

(http://inlet.geol.sc.edu/nerrsintro/nerrsintro.html and http://www.epa.gov/owow/estuaries/about1.htm)

Many different habitat types, for example, are found in and around estuaries (http://www.epa.gov/owow/estuaries/about1.htm) including:

- Shallow open water
- Freshwater and salt marsh
- Sandy beaches
- Mud and sand flats
- Rocky shores
- Oyster reefs
- Mangrove forests
- River deltas
- Tidal pools
- Sea grass and kelp beds
- Wooded swamps

These are home to an abundance and diversity of wildlife including:

- Shore birds
- Fish
- Crabs and lobsters
- Marine mammals
- Clams
- Shellfish
- Marine worms
- Sea birds

(http://www.epa.gov/owow/estuaries/about1.htm)

Estuaries are also multi-value resources to society (NWLRA, 2002). They provide a whole suite of human resources, benefits and services (http://www.epa.gov/owow/estuaries/about1.htm). Such environments support commercial, traditional, and recreational fisheries, tourism, and wilderness experiences. They have also long been a source of inspiration for the human mind through aesthetics, art, and poetry. Since early times estuaries have been places for human settlement, providing sheltered harbours, transport and trade routes and natural resources for industry (http://www.iwight.com). Some are still relatively sparsely developed whilst others have become densely populated areas.

Not surprisingly the wide range of potential uses results in many conflicts, and therefore necessitates a management strategy. Each estuary has its own distinct characteristics, susceptibility to impacts, and management requirements. An increasing number of national and international designations, policy and bylaws also make management of an estuary environment potentially very complicated. Worldwide there is now a wide range of organisations with statutory responsibilities on land, sea, and across the intertidal zone. In addition, there are many coastal and estuary management plans (http://www.iow.gov.uk) designed to provide a more holistic approach to sustainable management of coastal areas.

As local, national and international interest in estuaries has grown, so too have the sources of information available to managers and the public. Besides documents such as estuary management plans, there are now many local community groups who seek to involve and inform the public about estuaries. In the UK, for example, there are a number of organisations who look after the interests of the local estuary communities; for example, the Thames Estuary Partnership, the Solway Firth Partnership, the Moray Firth Partnership, and the Forth Estuary Forum to mention but a few. Besides providing documentation in the form of management plans, papers and publicity, these groups are also increasingly actively involved with local communities through the development of websites, online mapping and various local activities e.g. litter monitoring, habitat surveys, field visits and seminars.

In other parts of the world, for example the US and Australia, there are similar approaches to estuary management. For example, in Australia: the site http://www.dlwc.nsw.gov.au/care/water/estuaries/Inventory/Index_Geogr.html, and in the USA the sites http://www.spn.usace.army.mil/bmvc/baylink.html and http://www.estuaries.gov/about/programs.html provide lists of national programs and organisations involved in estuaries and resources. A number of recent projects have also focused on the status of Australian estuaries. For example, Australia's Near Pristine Estuaries: Assets Worth Protecting (NLWRA, 2002) has made a preliminary classification of all of Australia's estuaries based on their 'condition.'

18.3 ESTUARY MANAGEMENT

Estuaries are clearly very complex, interesting, and dynamic components of coastal zones around the World. They provide an interface between freshwater and saltwater environments, one that is constantly changing. They are also fragile environments and easily affected by human activities whether it be settlement, industry or tourism. Changes, however small, can be very harmful to the survival of an estuary (http://estuaries.gov/about/aboutestuaries.html).

With growing numbers of people settling and utilising estuaries, there are many activities that are now considered to endanger the survival of estuaries; such as dredging, the infill of flats and marshes, pollution, reconstruction of shorelines for housing, transportation, and agricultural needs. Also, marinas and tidal barrages can lead to dramatic changes in water circulation, erosion, and sediment movement. The effects of changes that have taken place have been unsafe drinking water, closure of beaches and shellfish beds, the development of harmful algal blooms, unproductive fisheries, loss of habitat and wildlife, and fish kills, as well as a host of other human health and natural resource problems (http://www.epa.gov/owow/ estuaries/about1.htm). Addressing such problems involves the use of planning, protection measures, and the implementation of management strategies. The protection of estuaries from detrimental change is considered vital in order to preserve natural 'beauty and bounty,' as well as to sustain livelihoods that depend upon fishing and tourism. The designation of protective zones, use of patrols and enforcement can be used, whilst management, for example, can help to restore estuaries (http://www.estuaries.org).

But estuaries are often difficult areas to manage successfully because of all the conflicting interests that exist. Whilst local partnerships between communities, industry and government can be, and have been, very helpful in addressing some of the problems facing estuarine environments (for example, the Scottish Firths initiative e.g. the Forth Estuary Forum) the many competing interests are often very difficult to accommodate satisfactorily. Ideally for management to be successful requires the identification of a lead organisation at a higher level. National policy in Australia, for example, is still deemed essential despite the local and regional initiatives. But, the piecemeal approach to coastal management that seems to have developed around the World has also led to many problems. As noted by Brown (1995), "the current coastal management system is the cause of the failure to redress the problem of the coastline. This management system is characterised by fragmentation of responsibility between spheres of governments, among different professional areas of expertise, and across different social, economic and environmental interests. It is indeed difficult for the existing systems of coastal management to become part of the solution to coastal degradation while it remains part of the problem."

18.4 GEOGRAPHY AND THE GEOSPATIAL TECHNOLOGIES

Geospatial technology has the potential to assist in the future management of estuarine environments in a number of ways. Brown (personal communication, 2003) considers GIS, for example, to have multiple roles including record keeping, monitoring change, advising scientists, and most crucially (and currently the least developed) for demonstrating relationships between biophysical systems and socio-political options. The rationale for using geospatial technologies in a coastal environment is also clearly highlighted by Fabbri (1998) in the following quote:

"Given the complexities of coastal systems and the multidisciplinarity required for sustainable coastal development, computerized systems are necessary for the integration and distribution of vast amounts data and expert knowledge. They are also vital for performing analyses to aid decision makers in their difficult task of proving optimal and compromise coastal management solutions"

In this chapter, 'geospatial technologies' are considered to include remote sensing (aerial photography, airborne and satellite data and imagery), Geographical Information Systems (GIS), Global Positioning Systems (GPS), mobile computing e.g. portable computers and Personal Digital Assistants (PDAs), WAP-enabled mobile telephones, digital still and video cameras, portable data storage and compression, and the Internet. Together these technologies are all rapidly becoming increasingly useful tools for a wide range of environmental projects that have a requirement for data collection and access to digital spatial data and information. These technologies, comprising both computer and other hardware and software, enable us to undertake:

- Data collection
- Data processing
- Data analysis
- Modelling and simulation
- Visualisation
- Communication and networking
- Provision of distributed data services
- Remote data access

Remote sensing (airborne and satellite sensors), for example, offers data collection and monitoring capabilities for estuarine environments at a number of different spatial, spectral, and temporal resolutions. This ranges from simple low-cost data acquisition platforms such as kites and balloons, model aircraft, and helicopters (Green *et al.*, 1998), and microlights for small-scale project work using 35mm cameras, digital cameras, and video, to the use of CASI and LIDAR, and satellite-based sensors on platforms such as IKONOS, ERS (European Remote Sensing Satellite), ENVISAT (Environment Satellite) and RADARSAT. Taken together these offer a powerful source of environmental data and information about estuary environments providing opportunities to derive up-to-date information on, for example, habitat type, algal weedmats and algal blooms, water depth, and sediment movement.

Rapidly decreasing costs of both hardware and software, together with improved usability and functionality, now offer end-users a wide range of tools that, when integrated, provide a practical framework for both acquiring and accessing data, storage, processing, analysis, modelling, and visual display. Recent developments in user-interfaces to both hardware and software have also made the geospatial technologies much easier to use than in the past. Likewise, developments in infrared (IR) and Bluetooth technologies have revolutionised communication links, facilitating the movement of large datasets from one location to another e.g. local to remote and vice versa. All of these developments now provide significant opportunities to interact with geospatial data in a laboratory and a field environment, the latter providing a means by which it is possible to access data and information 'on the fly.'

As these technologies have become cheaper and more usable, so they have become more accessible to a wider and more diverse end-user community including: academics who want to collect data in the field as part of a research project; commercial users working on contracts; ecologists, botanists, and biogeographers who need to gather data at a field study site; students collecting data as part of a dissertation; and schoolteachers and pupils wishing to undertake practical fieldwork as part of a curriculum requirement, class, or to aid in the development and creation of a virtual field course. All of these different categories of people are potential contributors to the well-being of an estuary, the very basis of the UK estuary fora.

Geospatial technology is rapidly becoming all-pervasive and there is growing interest in the potential that it can offer in terms of mobile access to data and information. Already there is much 'geospatial' data and information available over the Internet, and together with developments in interface technology and the capability for greater end-user interactivity, the Internet is now providing access to powerful new tools that allow the delivery of online maps, animation, and virtual reality, not only on the desktop but now also in the field. For example, a combination of a PDA and a mobile phone will allow remote access to the Internet. Map data collected in the field using a GPS can be captured on a PDA and processed using software such as PocketGIS. Maps can be downloaded locally to a PDA, updated in the field and, with the aid of a mobile phone, uploaded into a remote database for storage, querying, retrieval, analysis, and display.

18.5 EXAMPLE GIS APPLICATIONS IN ESTUARY MANAGEMENT

The following are some examples of the different ways in which the geospatial technologies can and are currently being used in relation to an estuary environment:

18.5.1 Artificial Reef Siting

Artificial reefs have become increasingly popular as the means to attract fish populations for a variety of purposes, one of which is commercial fishing sport. These usually take the form of large metal structures, e.g. old boat hulls or drilling rigs, that are sunk thereby forming an 'artificial reef,' an ideal fish habitat. Where such structures are to be sited in shallow waters, close to the coast, and in areas in which there are many other potential users of the coastal area or zone, care must be taken in the siting process. The process of siting can be undertaken manually using relevant geographical information and siting criteria. An alternative approach is to use GIS to achieve the same task, with the added benefit of being able to undertake the process more quickly and efficiently, using more geospatial information, and the bonus of being able to generate a number of different siting scenarios. This type of siting exercise is a relatively simple example of using GIS in an analytical capacity. It involves the creation of an environmental database, comprising a number of raster and vector datasets, the selection of suitable environmental datasets according to certain criteria, the use of GIS functionality e.g. buffering, and the use of multi-criteria overlay analysis to create a composite in which only those areas that fulfil all the siting criteria are shown, those where artificial reefs might be located. An example of using this approach is a study in Moray Firth in the northeast of Scotland (Green and Ray, 2002; Wright *et al.*, 1998).

18.5.2 Habitat Management

The Nature-GIS (http://www.gisig.it/nature-gis/) project began in 2002 to examine the different ways in which geographic data and information are currently being used for protected areas management. Protected Areas Management includes Marine Protected Areas (http://www.ukmarinesac.org.uk) within estuaries.

Involving a total of nineteen European partners, the project also has as one of its aims the development of a protected areas geospatial data model (Green *et al.*, 2003). The data model aims to help standardise the way in which protected areas data is collected and stored for display and analysis with geographic data-handling tools, and subsequently used in protected areas management.

18.5.3 Erosion

Coastal erosion is of growing concern around the World. A European project, EUROSION (http://www.eurosion.org), has recently begun to provide documentation and information about coastal erosion along the coastlines of Europe. One particular area of concern is erosion in estuarine environments. Typically estuaries are sheltered areas of the coast protected from the force of the sea. However, the process of erosion is still active within such an environment. Consideration is being given to coastal erosion management, including some cases of estuaries in the Netherlands and Portugal. (Further details can be found at http://www.eurosion.org/project/reports.html.)

18.5.4 The Siting and Management of Yacht Marinas

In the UK, yacht and boat marinas have become increasingly popular coastal developments to cater for the growing leisure and tourism industry (Sidaway, 1991). Many of the estuaries around the UK coast are now home to both small- and large-scale marina developments. GIS can be used in a similar way to the artificial reef-siting example discussed above, to assist in the siting of a new marina. The GIS utilises a number of geospatial datasets and marina siting criteria to undertake an environmental impact assessment (EIA) and subsequently to isolate the ideal or optimum geographical location(s) for a marina. GIS can also be used as an effective data management tool for the marina and its assets. Already there are a number of commercial marina management tools available e.g. The Marina Program (http://www.ccmarina.com/index.htm). Although many largely are extended mapping toolboxes, the addition of a database in which to store the attributes describing the marina infrastructure and the location and characteristics of the boats in the marina provides an example of the functionality of the GIS toolbox. Mobile GIS hardware (e.g. Compaq iPAQ) and software (e.g. PocketGIS) and Global Positioning Systems (GPS) also provide considerable potential to enhance a marina database with information about the features that are not available from traditional data sources such as maps e.g. boat locations. Online GIS and mapping tools can provide the means to develop interactive retrieval and display systems for marina users, even extending to public information systems.

18.5.5 Weedmat Mapping

The problem of weedmats is common to many estuarine environments throughout the World. In estuary catchments that are predominantly agricultural, and where there are areas of shelter, nutrient enrichment, and the growth of macro-algal weedmats may be common threats. There are also many other factors besides pollution that can affect the location and spatial distribution of weedmats, including sediment type and distribution as well as water flow. Macroalgae such as *Enteromorpha*, *Ulva* and *Chaetomorpha* typically form vast 'mats' that lie on the surface of estuarine mud- and sand-flats. These mats can either be very fine coverings (thin) or several centimetres deep (thick).

Remote sensing, most often aerial photography (panchromatic, colour, and colour infrared (CIR)) because of its spatial and temporal resolution, and more recently hyperspectral data (because of its finer spectral resolution), has been used to monitor and map weedmats with varying degrees of success. An example of this is the recent work undertaken on the Ythan Estuary in the north east of Scotland to establish a practical methodology for deriving weedmat area over time (Green and King, 2002; Orr, 2003).GIS can be used to undertake simple temporal differencing using overlay techniques to estimate change in weedmat area and distribution over time, as well as to establish area estimates for weedmat coverage using additional functionality contained in ArcView extensions such as Xtools (http://www.odf.state.or.us/DIVISIONS/management/state-forest/XTools.asp). In the longer term, the data may form the basis of an Ythan Estuary GIS (desktop and online using ArcIMS), which seeks to establish an environmental database comprising remotely sensed imagery, Ordnance Survey (OS) digital map data, digital hydrographic bathymetry from the United Kingdom Hydrographic Office (UKHO) and other digital datasets e.g. bottom sediment distribution, bird count data and distribution, and water quality data.

18.5.6 Search and Rescue (SAR)

In coastal search and rescue operations access to geographical data and information is vital. As well as occurring along the coast and out to sea SAR operations also occur in estuarine environments.

Information may be required during both the search and the rescue stages. Whilst some relevant data and information may be available e.g. in the form of maps and charts, much of the baseline data and information is currently not available at a scale that is useful to a SAR operation. Increasingly it has been recognised that such information should ideally (a) be available in a digital format, (b) take advantage of the digital communications networks, (c) embrace the Internet and online Geographical Information Systems (GIS) technology and (d) comprise sources of information at many different scales, the most important of which is local knowledge (Figure 18.1).

The acquisition of local knowledge and information can be greatly enhanced through the use of the mobile geospatial technologies that are now available, integration of which can be developed into a decision support system framework for coastal and marine search and rescue operations. For example, a wide range of mobile technologies can be brought together to provide a basis for gathering, processing, displaying, and communicating geospatial data and information in real-time. Hardware in the overall IT equation might include a PDA with PocketGIS for detailed mapping of an access footpath to a beach; a mobile phone to upload the datafile to an online GIS e.g. ArcIMS; an on-shipboard PC with a wireless access to the Internet, accessing the ArcIMS-based online decision support system; a digital camera that sends by email on a mobile phone a picture of the coast from the seaward side to the local Coastguard either onboard a ship or to a rescue team on the cliff top; sonar information that can be uploaded to an ArcIMS system and combined with e.g. an Ordnance Survey digital coastline; a video clip or panoramic photograph of a section of the cliffline that can be zipped and sent as an attachment by email. Additional components of the system may also include digital or paper fax machines, which could, for example, be used to communicate an annotated sketch map or photograph to someone on land, on a lifeboat or inflatable, or at an RNLI (Royal National Lifeboat Institution UK) or Coastguard station.

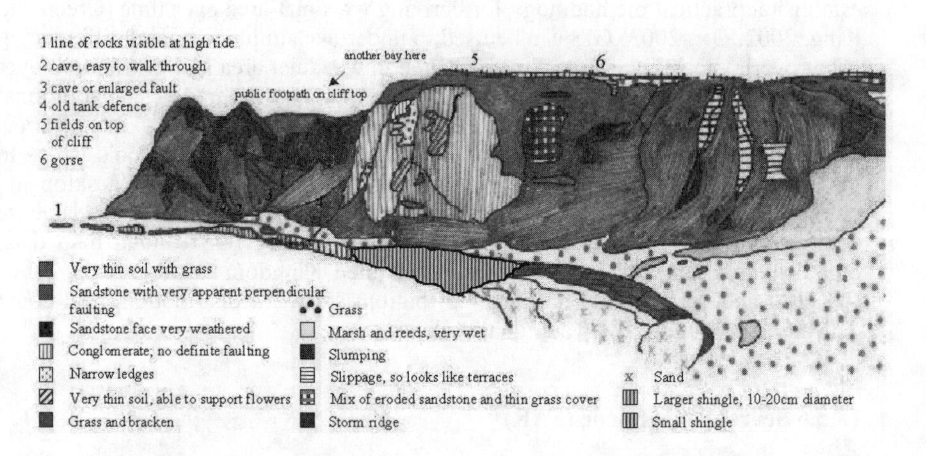

Figure 18.1 A field sketch of a section of coastline (courtesy of Joanna McDonald, MRI)

18.5.7 Participatory Planning

One outcome of the work reported earlier in this chapter about weedmat monitoring and mapping was to input the resulting information into a GIS, as a means to deliver information to a wider audience in the context of environmental awareness and participatory planning. Whilst geographic datasets can be viewed in a desktop GIS environment, not everyone has access to GIS software. An alternative is to place the datasets on CD to disseminate the information to the end-user. A more effective approach is to distribute the datasets with a GIS viewer such as ESRI UK's MapExplorer (http://www.esriuk.com). MapExplorer is freeware that allows the user to access GIS datasets for viewing and output, as well as providing limited GIS functionality such as zoom, pan, overlay, and measurement tools. MapExplorer is a 'cut down' GIS package with navigational functionality allowing an end-user to open and explore the information. GIS tools have also been developed for the Internet. The Internet is an ideal medium in which to deliver maps and images because it is a familiar interface to many people. ERSI's ArcExplorer (http://www.esri.com/arcexplorer), for example, can be used to connect to an Internet webserver. Online mapping and GIS software provides an ideal mechanism to involve the public in the environmental planning process through simple interactive interfaces to visualise geospatial data and information in the form of maps and imagery.

18.5.8 Modelling

Developing an improved knowledge and understanding of the complexity of an estuary can often be aided through the use of modelling techniques and tools. Whilst often abstract, such models nevertheless help to provide insight into the physical and biological processes active in such an environment. One such model is the Forth Estuary Modelling System (FEMS). FEMS was developed as part of a Forth Estuary Forum (FEF) Flagship project. The aim was to take an ecological model and present this to the wider community in a form that would help to educate people of all ages about ecological processes active in the Forth Estuary. It comprised a Pelagic Submodel; Benthic Submodel; Fish Submodel; Sea Birds/Mammals Submodel; and a Human Influences Submodel. The model was written using an interactive modelling toolbox, STELLA. Initially distributed on a CD-ROM complete with HTML documents, FEMS was designed to be accessed as an educational tool from the Forth Estuary Forum website, facilitating updates as the model evolved (Green and King, 1999).

18.5.9 Education and Training

One application of GIS technology, either as a desktop or web-based system, is to provide the basis for helping to aid in education and training programmes about estuaries and estuary management. Access to data and information stored in a GIS using standard information retrieval and query tools is a simple way for people to retrieve, display, and communicate geographical information, most often in the form of an interactive map. Simple querying of a spatial database and visualisation in the form of a chart or a map can often help to raise awareness and educate people. Additionally, provision of simple end-user interfaces to the information e.g. using ArcView or ArcIMS on the Internet increases the power of the 'map' to communicate spatial information. Toggling map layers on or off can help to communicate spatial patterns and reveal spatial relationships. GIS mapping can also provide a simple data exploration tool.

GIS can also be used as an effective training environment to learn about estuarine environments. The GIS toolbox provides the means to integrate geographical datasets, to examine patterns, explore data, establish relationships, as well as to visualise data and information in a variety of different ways e.g. 2D, and 3D. Much of the output resulting from simple database queries, statistical summaries, analysis, and visualisation exercises provides a very effective form of communication that can be used as input to documents, reports, slide shows and presentations.

Many of the estuary websites now becoming available are primarily designed to provide access to both data and information. They usually comprise webpages, downloadable documents, datasets, software and sometimes even models. These help to maximise access to information for a wider community. In some cases, these websites are taking advantage of many of the software tools that accompany the Internet including, for example, online mapping and GIS. Although most online mapping systems currently have relatively limited GIS functionality, they nevertheless provide significant potential to allow end-users to view information in the form of text and images. Some also include access to remotely sensed imagery (airborne and satellite) that can be viewed as 'quicklooks' or in some cases takes advantage of image compression software such as Lizardtech's MrSID, or ERMapper's ECW format, to deliver high-resolution images rapidly across the Internet.

18.6 INTERNET-BASED INFORMATION SYSTEMS

18.6.1 GDSPDS - Solway Firth

In 1999, a research project began at the University of Aberdeen to develop an Internet-based coastal information system. The aim of the project is to prove the concept of the need for and the capability to deliver geospatial information (digital images, documents, maps and remotely sensed imagery) to the coastal manager via the Internet, using the Solway Firth as a case study area. The Geo-information Decision Support Processing and Dissemination System (GDSPDS) is capable of serving GIS data, digital photography and video as well as remotely sensed imagery via the Internet.

The Solway Firth GDSPDS is built around Windows NT Server 4.0 using Internet Information Server 4.0 as the web server. In addition, ESRI ArcIMS 4.0 is used to deliver online mapping capabilities, and ER Mapper's Image Web Server (IWS) version 1.7 and LizardTech's Content Server are used to deliver remotely sensed imagery. The imagery is compressed using either ER Mapper's Enhanced Compressed Wavelet (ECW) technology or LizardTech's MrSID Geospatial Encoder (v1.4). Both of these compression technologies allow remotely sensed imagery to be both stored more easily (for example, a colour air photo mosaic of 178.7Mb can be compressed to 3.2Mb in ECW (30:1) and 3.11Mb in MrSID format (30:1)), and also served rapidly across the Internet to any PC with a web browser. In the case of ECW files, a relatively small plug-in must be installed to view the imagery. MrSID imagery can be delivered either via a plugin viewer or as a series of JPEG images. The imagery is stored in an MS Access database which is searchable using a web-based form. A number of basic search queries have been developed to try and help guide the user to the correct imagery for their specific management problem or application.

As well as being accessible on a desktop computer, the GDSPDS can also be accessed via a PDA connected to the Internet using a mobile phone. Remotely sensed images can be viewed in the web browser on the PDA, while ESRI's ArcPad software can be used to access and retrieve maps from the ArcIMS map service. This enables a coastal manager to go into the field and access geospatial information to use for navigation and for carrying out temporal comparisons as well as completing updating tasks.

The GDSPDS is a means of more easily disseminating geospatial information to a wider audience without the need for each user to have expensive GIS software on their own computers. It provides support for data collection and updating, decision-making and report writing as well as greater interaction between coastal managers and users (Figure 18.2).

Figure 18.2 The Solway Firth GDSPDS. Viewing colour aerial photography via the Internet on a PDA (Aerial Photography Copyright © English Nature 1997)

18.6.2 Interactive Mapping using the Oregon Estuary Plan Book

Inforain has developed the Oregon Estuary Plan Book (http://www.inforain.org/interactivemapping/). Interactive maps from the Oregon Estuary Plan Book have been designed to help planners and decision makers in the evaluation of estuaries. The maps comprise several data layers allowing users to explore specific management units and habitat types. There is a base map layer together with other layers comprising information on dredge material disposal sites, estuary habitat, mitigation and special mitigation sites, estuary management units, shoreline management units, and areas of significant habitat (http://www.inforain.org/dlcd.asp). Datasets for download are stored in a compressed format. Provision of an image record is especially useful for interpretation and communication to the layperson, either in a raw or processed format.

18.6.3 FREMP GIS: An Estuary Management Tool

FREMP (Fraser River Estuary Management Program) uses a GIS as a tool to promote integrated decision-making in the estuary (North and Beckmann, 1999). It uses orthophotos as a base layer and a number of thematic data layers including habitat inventory to assist in environmental recommendations. Initially FREMP did not use GIS software, but instead utilised aerial photographs and digital databases. Adoption of GIS, however, improved spatial data organisation and integration and provided wide access to data and information. It also provided a communication tool. Datasets comprised habitat inventory and shoreline classification maps, digital photographs, development projects, basemaps from FREMP and other external sources. In addition, metadata was provided. The interface to FREMP is an extension of the standard ArcView software but designed to be more user-friendly and including added functionality.

18.6.4 Virtual Estuary

This is a comprehensive online system for retrieving, analysing and integrating data and information related specifically to the Hudson Estuary in the USA. It has been targeted at a wide range of end-users from the scientist, to the decision-makers, and the general public. A joint project of Columbia University, COAST (Clean Ocean and Shore Trust) and a range of other institutions and agencies with a direct interest in the estuary. An online mapping system (using ESRI's ArcIMS) is also part of the Virtual Estuary project and is designed allow the creation of 'on-the-fly' maps from the estuary databases (http://www.nynjcoast.org/Hudson_Estuary/hudson_estuary_.html).

18.6.5 OZEstuaries

OZEstuaries (http://www.ozestuaries.org) is an Australia-wide database that enables managers and stakeholders to develop a greater understanding of their own estuary, and where it fits into the national picture. It facilitates comparison, provides information on estuarine form and function, identifies potential threats, provides information on environmental indicators, delivers the National Land and Water Resources Audit (NLWRA Estuary Assessment, 2000) condition assessment, and provides a single point of entry for information on Australia's estuaries. This will be enhanced over time with the help of scientists, managers and stakeholders, with the provision of additional data, the addition of a GIS front end, estuary models and feedback from end-users.

18.6.6 PIVOT

In the USA, the National Estuary Program of the Environmental Protection Agency links community to agencies. PIVOT (Performance Indicators Visualization and Outreach Tool) provides an interactive graphics and maps toolbox to help users better understand the issues and visually track the National Estuary Program's progress toward achieving its habitat restoration goals. Provision of a map that combines spatial relationships of factors to priority issues and the management actions designed to address those issues is deemed to be a powerful tool as it helps to raise questions about the effectiveness of management actions (http://www.epa.gov/owow/estuaries/pivot/overview/intro.htm).

18.6.7 DEMIS

DEMIS (Dynamic Estuary Management Information System) is part of the Oregon Coastal Management Program and is designed to provide a useful information depot for both traditional and digital information for the estuaries of Oregon (http://www.lcd.state.or.us/coast/demis/demisfaq.htm). It provides access to data, GIS layers, metadata, estuary plan book information, as well as a CD. DEMIS has arisen in response to the problems associated with the current fragmentation of information. It has been designed to help improve access to information; establish a GIS (aggregate, standardise, distribute digital data); accumulate and analyse data for estuary restoration; incorporate new data into an information management system; and to maintain timely and efficient updates.

18.7 SUMMARY AND CONCLUSIONS

Powerful information technology tools now support the development, operation and use of estuary information systems of various different types, providing considerable potential to aid in estuary management. Information technology has greatly enhanced the data-into-information pathway, providing the means and the tools to collect more data, to process it more rapidly, and to communicate the results of analyses more effectively to a wider community. In recent years, rapid developments in the geospatial technologies have greatly enhanced the collection of spatial data, information processing, modelling, and visualisation of information in new, interesting, and informative ways. The capability to gather higher resolution data from remote sensing, to process, integrate and visualise environmental data within a GIS, and the use of GIS as a data handling tool for spatial analysis and modelling has provided the basis for studying the environmental processes operating in an estuary. Online GIS and decision support tools are also empowering the policy and decision-maker, as well as the coastal manager and practitioner.

In the future the practical application of the geospatial technologies will grow considerably providing tools that will aid the coastal manager and practitioner in the workplace. Some of the important issues that will govern the greater use of this technology in the future, however, will be end-usability of the hardware and software, integration, and functionality. Moreover, data availability, data standards, data costs and copyright will also largely determine how much of the potential of geospatial technology can be practically implemented and realised in the workplace.

18.8 REFERENCES

Anonymous, 2003, Wind Turbines – I Can Hardly See Them! *Times and Star*, Friday March 21st 2003.

Brown, V.A., 1995, *Turning the Tide: Integrated Local Area Management for Australia's Coastal Zone*, Department of Environment, Sport and Territories, Canberra (second printing).

Brown, V.A., 2003, Personal Communication.

English Nature, 1993, *Estuary Management Plans - A Co-Ordinator's Guide. Campaign for a Living Coast.*

Fabbri, K.P., 1998, A Methodology for Supporting Decision-making in Integrated Coastal Zone Management. *Ocean and Coastal Management*, **39**, pp. 51-62.

Green, D.R., 1994, Using GIS to Construct and Update and Estuary Information System. *Proceedings of Management Techniques in the Coastal Zone. University of Portsmouth*. 24th-25th October, pp. 129-162.

Green, D.R., 1995, User-Access to Information: A Priority for Estuary Information Systems. *Proceedings of CoastGIS'95. International Symposium on GIS and Computer Mapping for Coastal Zone Management*. University College Cork, Ireland. February 3rd-5th, pp. 35-60.

Green, D.R. King, S.D., and Morton, D.C., 1998, Small-Scale Airborne Data Acquisition Systems to Monitor and Map the Coastal Environment (D.C. Morton and S.D. King) *Paper/Poster Paper for the Marine and Coastal Environments Conference* - San Diego, U.S. (October 1998). (Paper published in Proceedings. Volume 1: 439-449).

Green, D.R. and King, S.D., 1999, *Forth Estuary Modelling System (FEMS)*. CD-ROM.

Green, D.R. and King, S.D., 2000, *Practical Use of Remotely Sensed Data and Imagery for Biological and Ecological Habitat Monitoring in the Coastal Zone.* Report/Monograph, (JNCC/English Nature).

Green, D. R. and King, S.D., 2003, Applying Geospatial Technologies to Weed Mat Monitoring and Mapping: The Ythan Estuary, NE Scotland. Special Session. *Proceedings of the Association of American Geographers*. March 5-9, New Orleans, Louisiana, USA.

Green, D.R., King, S.D, Annoni, A., Margoulies, S., and Humblet, J-P., 2003, Nature-GIS: Developing a Pan-European Geospatial Data Model for Protected Areas (PEPAN). Paper presented to 6th AGILE Conference on Geographic Information Science, 24-26 April, Lyon, France.

Green, D.R. and Ray, S.T., 2002, Using GIS for siting artificial reefs – Data issues, problems and solutions: 'Real World' to 'Real World', *Journal of Coastal Conservation*, **8** (1), pp. 7-16.

Hinwood, J.B. and McLean, E.J., 2002, Modelling the Evolution of an Estuary. In, *Proceedings of Littoral 2002 – The Changing Coast,* edited by Veloso Gomes,

F., Taveira Pinto, F., and Das Neves, L., EUROCOAST/EUCC. Porto, Portugal. September 22nd -26th. pp. 245-251.

Jernakoff, P., Hicl, P., Ong, C., Hosja, W., and Grigo, S., 1996, Remote Sensing of Algal Blooms in the Swan River. *CSIRO Marine Laboratories Report 226. CSIRO Blue-Green Algal Multi-Divisional Program.* CSIRO, Australia, March 1996.

Lewis, L.J. and Kelly, T.K., 2001, A Short-Term Study of the Effects of Algal Mats on the Distribution and Behavioural Ecology of Estuarine Birds. *Bird Study.* **48**, pp. 354-360.

North, S.M. and Beckmann, L., 1999, FREMP GIS: An Estuary Management Tool. *Proceedings of GIS'99: GeoSolutions in a Coastal World - Integrating Our World.* Vancouver. Thirteenth Annual Conference on Geographic Information Systems. March 1st-14th, pp. 125-129.

NLWRA, 2002, Australia's Near Pristine Estuaries - Assets Worth Protecting. http://www.nlwra.gov.au.

Orr, G., 2003. A Study of the Ythan Estuary. *Unpublished Undergraduate Dissertation.* University of Dundee, Scotland, UK.

Ripley, H., Jones, W., and Scheibling, R., 2002, Mapping Invasive Species: Assessing the Impact of Species Invasions on Coastal Ecosystems Using Remote Sensing. *Backscatter.* Fall/Winter 2000. pp. 8-12.

Sidaway, R., 1991, *A Review of Marina Developments in Southern England. Report.* Royal Society for the Protection of Birds (RSPB) and the World Wildlife Fund UK (WWF-UK).

Stapleton, C. and Pethick, J., 1996, Coastal Processes and Management of Scottish Estuaries. III: The Dee, Don and Ythan Estuaries. *Scottish Natural Heritage Review*, **52**.

Stove, G.C., 1978, The Hydrography, Circulation and Sediment Movements of the Ythan Estuary. *Unpublished PhD Thesis*. University of Aberdeen.

Townend, I., 2002. Identifying Change in Estuaries. In, Veloso Gomes, F., Taveira Pinto, F., and Das Neves, L. (Eds). *Proceedings of Littoral 2002 – The Changing Coast.* EUROCOAST/EUCC. Porto, Portugal. September 22nd -26th. pp. 235-243.

Vivoni, E. R., and Camilli, R., 2003, Real-Time Streaming of Environmental Field Data. *Computers and Geosciences*, **29,** pp. 457-468.

Wright, R., Ray, S., Green, D. R. and Wood, M., 1998, Development of a GIS of the Moray Firth (Scotland, UK) and its application in environmental management (site selection for an artificial reef), *The Science of the Total Environment*, **223** (1), pp. 65-76.

Young, D.R., Cline, S.P., Specht, D.T., Clinton, P.J., Robbins, B.D., and Lamberson, J.O., 2000, Mapping Spatial/Temporal Distributions of Green Macroalgae in a Pacific Northwest Estuary Via Small Format Color Infrared Aerial Photography. *Proceedings of the Sixth International Conference on Remote Sensing for Marine and Coastal Environments*, Charleston, South Carolina, USA. 1st -3rd May, Veridian ERIM International, Ann Arbor, Michigan, pp. II-285-II-291.

CHAPTER NINETEEN

A Territorial Information System (TIS) for the Management of the Seine Estuary – Environmental and Management Applications

Jean-Côme Bourcier

19.1 INTRODUCTION

Located at the geographic, sociological, economic, and environmental convergence of multiple interest groups, the estuary of the Seine (Normandy, France – Figure 19.1) presents itself as a coastal zone subject to divergent, if not contradictory, anthropogenic pressures. In fact, its position vis-à-vis its ecosystems gives it rare and specific functions and its location within geographic space confers strategic advantage. The urbanisation and level of primary industry of the region must develop in directions that respect an ecological heritage, and an exceptional landscape, that is subject to the attention of local communities as well as local, state and European levels of administration. Furthermore, the existence of various pollutants, risks from technological hazards, the exploitation of natural resources, an upstream/downstream divide, a disparity between left and right banks, and multiple activities, etc., are just some of the key issues in this unique space.

In this context, geographical information systems (GIS) present a wealth of applications as tools capable of collecting, harmonising and analysing data about the estuary coastline in order to monitor the health of the region and provide summary information as an aid to decision-making. Harnessing these technological, methodological, and operational tools enables coastal management problems to be defined and resolved in an holistic and consensual manner. They could also lead to a new approach to management and a remediation, by providing a better understanding of coastal processes, functioning and change; an approach more attuned to the natural and human complexities of a geographic space that encompasses all levels of organisation, and which has set sustainable development as primary goal.

These distinguishing features justify the establishment and use of a GIS specific to the Seine estuary within the CIRTAI laboratory at the University of

0-41531-972-2/04/$0.00+$1.50

Le Havre (UMR 6063 CNRS – Research's group CIRTAI). More specifically, it consists of a territorial information system (TIS) tied to a multidisciplinary and multi-thematic, geographically referenced database designed to improve understanding of the processes and functioning of the estuary in order to support operational and participatory management practices. Within the range of spatially referenced information systems, (Prélaz-Droux, 1995), this territorial information system sits mid-way between the two end-points of land information system and geographical information system as defined by the author: it exceeds the institutional limits of a land information system, and applies the potential of a GIS in its ability to support development of land and environment in a complex geographic space that requires management at multiple scales and involving multiple actors.

The aim of this article is to demonstrate a particular application of the TIS, as an observatory of the natural heritage of the Seine estuary. Within this application, different levels of information have been integrated and analysed in order to provide qualitative and quantitative assessments of this heritage by means of territorial diagnostics.

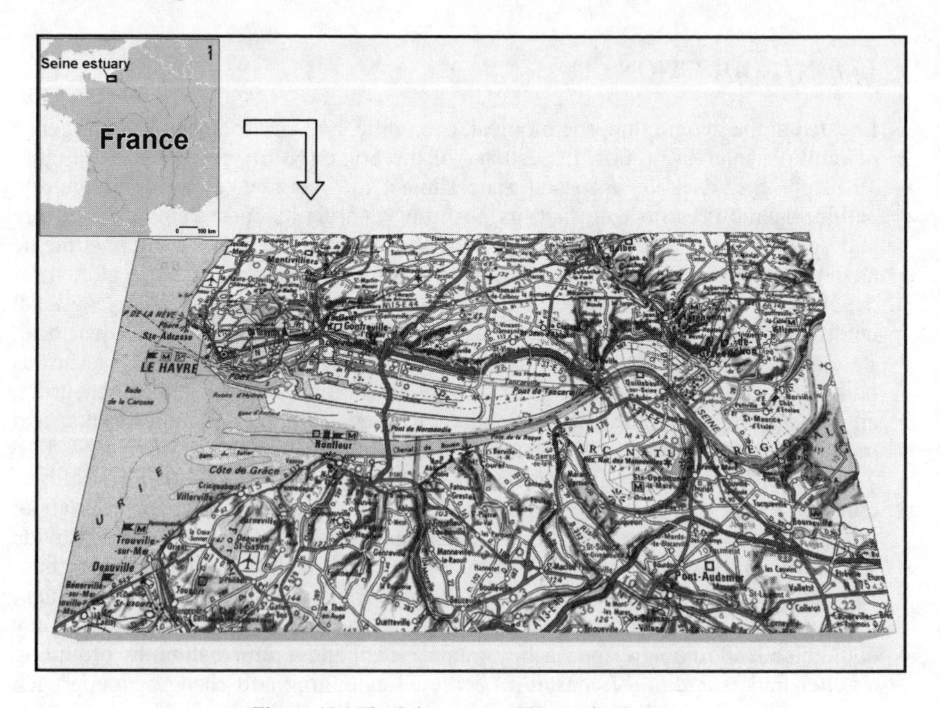

Figure 19.1 The Seine estuary (Normandy, France).

19.2 ON THE NEED FOR AN OBSERVATORY OF THE NATURAL HERITAGE IN THE SEINE ESTUARY

As a meeting point for terrestrial, fluvial and marine environments, the Seine estuary integrates a number of biologically productive and varied ecosystems. It is an integral element in the routeway for many species of migrating birds and provides shelter and nutrients for fisheries and shellfish farming as well as for certain invasive species.

The estuary basin is characterised by the juxtaposition of alluvial plains and calcareous plateaux, giving rise to an extraordinarily diverse range of humid-zone landscape types that act as sponges and filter-cleaners for the environment.

Nowadays, the natural heritage of this estuary is subject to the impacts of numerous activities and processes at the local scale (industrialisation, urbanisation, agriculture, resource exploitation) as well as nationally (it is the outlet of a drainage basin within whose borders is found 30% of the French population and 40% of the national economic activity). This convergence of interests, and the conflicts that arise, raises the problem of how to conserve natural spaces whose resources are limited and whose equilibrium is fragile.

Management of this territory must therefore ensure balanced development, taking into account both scientific data and socio-economic factors (Chappuis, 1993), at a scale that is essentially local but which also has national and international elements (Bailly, 1995). Every stage in the process involves nature and the landscape, and it is often necessary to draw on a complete inventory of the natural and landscape heritage in order to steer decision-making in appropriate directions whenever choices present themselves.

In order to address these requirements, the Seine Estuary TIS acts as an observatory of natural heritage, and an intelligent tool for providing territorial and environmental indicators needed to manage this unique coastal system. GIS offers technologies with potential for better planning, management and analysis of our environment (Bryant, 1993); for tackling the integrated, holistic and rational management of space; for pooling necessary information; and for overcoming institutional limitations, to allow all levels of environmental organisation (hydrological systems, landscape systems, drainage basins, etc.) to be taken equally into account.

The objective is to make available to environmental managers and decision-makers a suite of key indicators that will allow better understanding of disparities between left and right banks of the river, conditions of flood and low water, competition and compatibilities of estuarine biota, etc., and ultimately allow the delimitation of zones of strong environmental vulnerability. This will also help bridge the gaps between scientific research and real-world practice and, more precisely, between those scientists, technicians, professionals, and politicians who are involved in strategic and political decision-making.

The territorial information system for the Seine estuary aims to bring the many facets of the same geographic reality into a common framework. Thus organised and structured, all available data for a given location may be integrated as part of a single analysis, in order to better define environmental objectives, and to provide support for decisions relating to territorial planning and management within the estuary.

19.3 FROM DATA COLLECTION TO DEVELOPMENT OF ENVIRONMENTAL DIAGNOSTICS

This TIS for the Seine estuary exploits a multi-source and multi-thematic geographically referenced database, presently consisting of more than 80 layers of data in vector format, standardised on the Lambert conformal conic projection and georeferenced on the Lambert zone 1 coordinate system (the local system for the north of France). Data in raster format are used to provide rapid enrichment and updating of the database, according to immediate needs. These data layers are all subject to a number of quality-control parameters (Laurini and Milleret-Raffort, 1993) including accuracy, precision, and timeliness (Rouet, 1991), and each layer is the subject of a file of metadata (Bourcier *et al.*, 1998) to ensure that it is used appropriately and optimally. A dynamic link has been developed, so that the metadata may be consulted directly via the user interface of the GIS software.

The operational use of the TIS requires, *inter alia*, data to be updated regularly, since they describe a territory that is subject to strong spatio-temporal evolution.

Data relating to the natural heritage of the Seine Estuary are collected on a regular basis by various groups: the regional Ministry of Environment for Upper and Lower Normandy, the Department of Hunting, an organisation for the conservation of wetlands, the Regional Nature Park of Brittany, Nature Park, etc. This demands an effective partnership, in order to respond to the specified quality criteria. Initially, exchange of these data was via hard-copy paper media (maps, cadastral plans, etc.), which implied heavy, time-consuming digitisation overheads. More recently, all the data-providing partners have opted for the advantages of digital technologies (GIS platforms, CAD, computer-assisted cartography), which has enabled much more efficient implementation of the system.

An environmental geodatabase is by definition multidisciplinary and multi-territorial; its quality and relevance depend on, among other factors, the abilities of the GIS team to manage interdisciplinary conflict, institutional compartmentalisation and protectionism, hierarchical management "pyramids" and cultures of regulation, etc.; and especially possessive attitudes and the equation of information with power, leading to a reluctance among some players to share information with others.

19.3.1 Environmental heritage diagnostics in the Seine Estuary

As has been discussed above, all management practices have environmental implications, and therefore require access to a complete inventory of the natural heritage. This essential knowledge of the natural resources of a region needs no further justification.

The natural heritage value of the Seine estuary is recognised through the implementation of a number of statutory instruments of national or European origin: the Ecological Fauna and Flora Inventory ("ZNIEFF"), Nature Reserve, Special Areas of Conservation (SACs), Biotope conservation, etc. To these are added other protective measures, such as the purchase of land.

It seems evident that this accumulation of regulations makes the direction of management projects even more difficult, especially since many of these provisions lack clarity, are obsolete, are contradictory and of uncertain relevance, or are imprecise and subject to multiple, subjective interpretations.

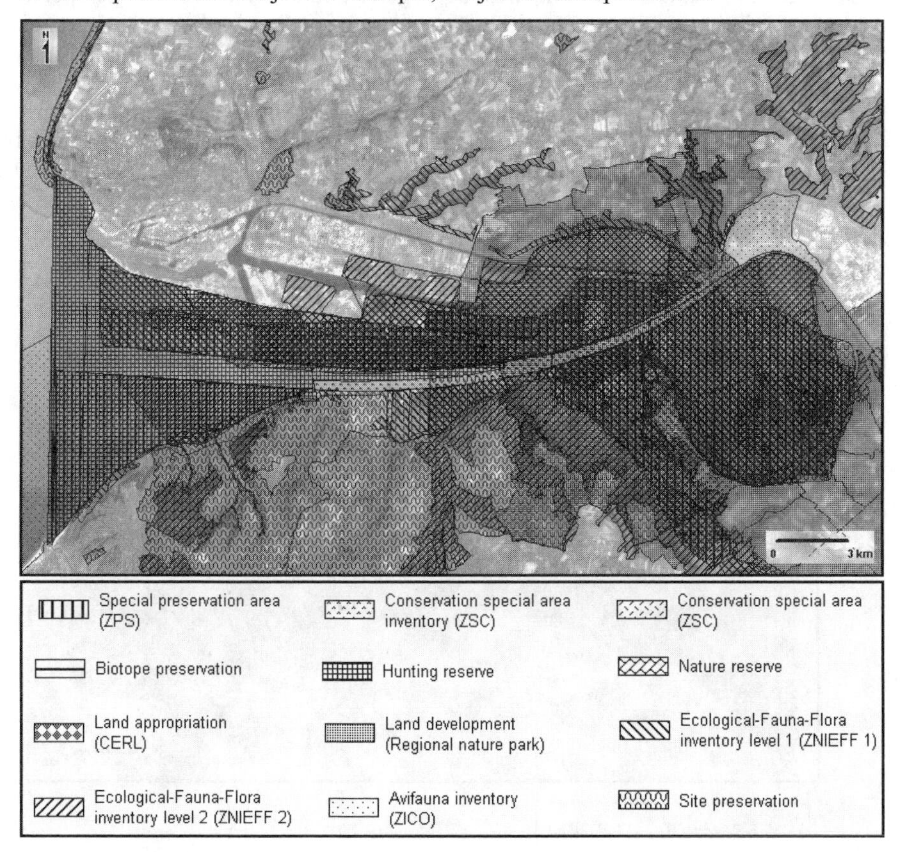

Figure 19.2 Spatial concentration of environmental conservation in the Seine estuary.

Superimposition of all these layers demonstrates visibly the multiplicity of initiatives and regulatory measures that currently apply to this territory. We may also note several important geographical overlaps between these various measures, rendering their representation in cartographic form particularly difficult.

Overall, such a representation implicitly demonstrates the high heritage value of this estuary, but the resulting legislative overload demonstrates clearly the need for harmonization of French Environmental law. French estuaries present a major challenge for the law, which must find ways to address their hybrid character and their multifunctional nature. Analysis of the legal treatment accorded these spaces demonstrates the weak acknowledgement in national law of their distinctive nature and, indeed, a strong tendency to promote their development, despite the accumulated knowledge of their ecological interest and their vulnerability (Auger and Verrel, 1998).

In terms of spatial analysis, it is necessary to go beyond simple superimposition of data layers, to examine the spatial distribution of particular entities and to synthesise the information they present. The aim is to create a new layer of information that compares, aggregates and analyses various perspectives on the region, so as to present alternative views of reality, for better environmental comprehension, decision-making and management. The complexity of the real world is thus rendered easier to read and communicate.

Thanks to the topological functionality of GIS, an expansion of the various layers of information relating to this natural heritage has enabled a "quantitative synopsis of protection measures" to be drawn up. This spatial analysis facilitates easy identification and classification of zones according to the sum of the protective measures that apply to them (Figure 19.3).

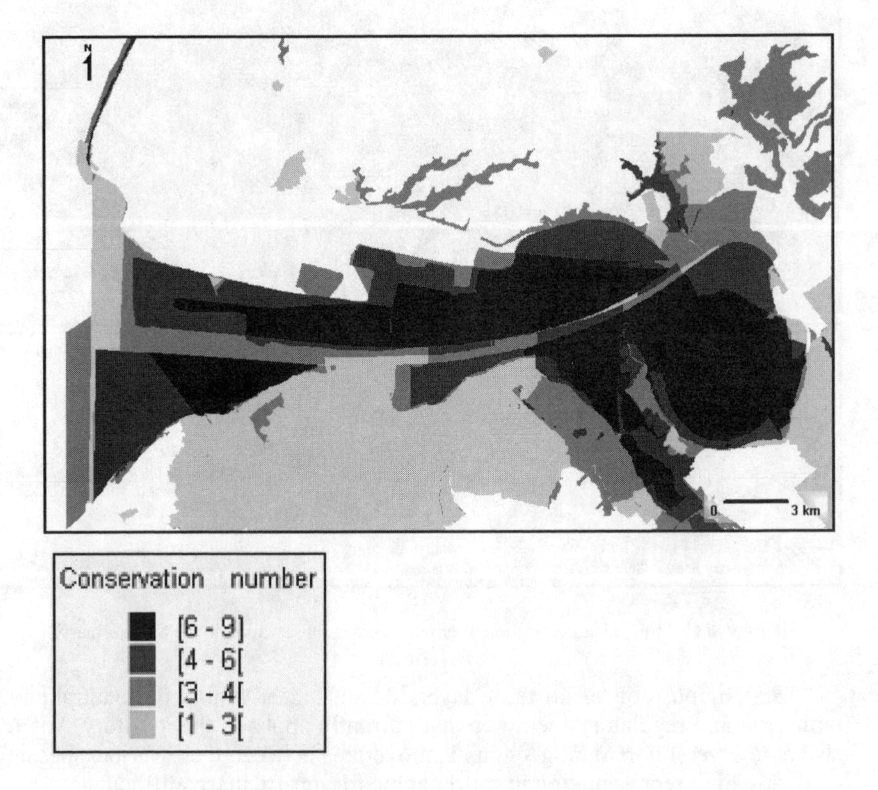

Figure 19.3 Quantitative synopsis of nature conservation in the Seine estuary.

While a diagnostic utility of this type is visibly effective at demonstrating the multiple protective measures, it also underscores the relevance of the known natural heritage of the Seine estuary, and explicitly reveals the conservation importance of the intertidal areas and of the flood plains. These latter are wetlands of acknowledged ecological importance, according to the National Programme of Research on Wetlands (Bourcier *et al.*, 1999). The TIS allows real-time spatial and thematic analyses, such as calculating the type, number and surface area of biotopes for each of the identified sectors, thereby allowing optimal planning and management for these environments.

Still within the perspective of generating useful indicators for stakeholders and decision-makers, the foregoing increase in topological data is further enriched by the addition of information relating to the level of legal protection accorded to these zones. This leads to a "qualitative synopsis of the degree of protection of the natural environment", which identifies and classifies different zones according to a cumulative index of conservation (Figure 19.4).

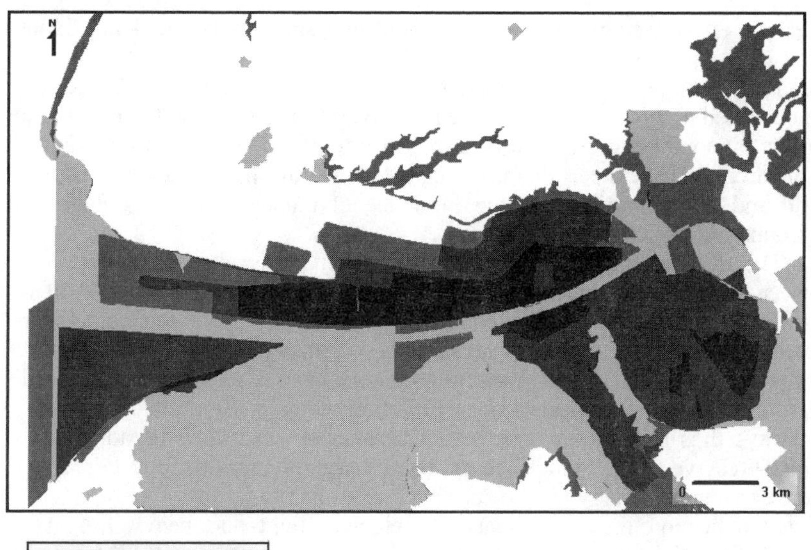

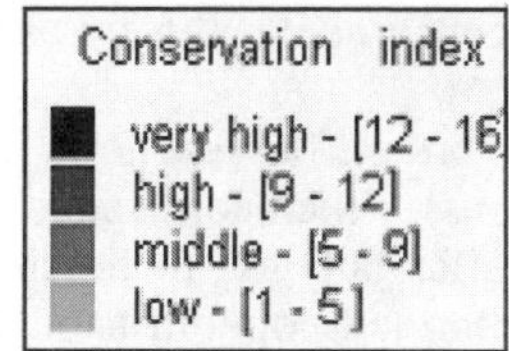

Figure 19.4 Qualitative synopsis of nature conservation in the Seine estuary.

The most protected zones (very high value) are located mainly in the alluvial plains, which underscores the ecological importance of these habitats.

This territorial index usefully complements the preceding map, by conveniently summarising the environmental constraints of the estuarine lands, providing indispensable information when territorial management projects or planning exercises are undertaken.

Achieving an optimal management of the natural heritage, at all levels of organisation, will probably require a greater harmonisation of the many superimposed levels (local to international) of environmental protection, in legal as much as in economic terms.

Additionally, this harmonisation offers the advantage of representing environmental constraints as homogenous zones that transcend administrative boundaries, and which are thereby more appropriate to holistic understanding of the environment from the global to the estuarine scale.

From the perspective of optimal management of the natural environment, would it not be useful to avoid the heavy imposition of all these different protected zone boundaries and, perhaps, rely on European and international commitments (e.g. the Natura 2000 network)? These levels of administration could provide powerful tools for protection through regulatory and financial instruments.

19.3.2 An environmental impact of coastal management in the Seine Estuary

The usefulness of the spatial analytical functions of the TIS may equally be demonstrated in the assessment of the impacts of traffic noise on the bird life of the estuary: by combining different of layers of information which, at first sight, have no apparent link between them, we can demonstrate the complexity of the real world and, more particularly, the influence of human activities on the natural environment (Figure 19.5).

The alluvial plain north of the Seine estuary is also subject to conflicts between natural environment and economic development, the resolution of which calls for sustainable development strategies to be drawn up. In this context, spatial analysis allows vehicular traffic levels to be correlated with the natural habitat of the Spoonbill (*Platalea leucorodia,* a migratory bird, see Figure 19.6), in order to measure the potential impact of noise pollution on the distribution of this bird. The routeways thus identified are of particular interest when they lie adjacent to the Natural Reserve of the Seine Estuary, liable to be directly linked to Port 2000, the future expansion strategy for the current commercial port.

Statistics on the movements of vehicles (light and heavy freight) were obtained from the authorities of the three departments concerned, and were used to calibrate the "road traffic" data layer.

A noise level envelope was calculated, based on traffic levels of 5000 vehicles per day (all vehicles, travelling in both directions), which corresponds to a noise level of 70 dB SPL. The TIS then allows a buffer polygon to be created around selected entities, and this can be intersected with the Spoonbill habitats to create an output layer that precisely identifies those areas where noise disturbance could be a potential problem. By simple querying, it is easy to obtain the location of each area, as well as their spatial attributes (length, width, perimeter, area, etc.).

Figure 19.5 ***Platalea leucorodia*** (Spoonbill: a migratory bird).
(© Cybernat - Le Havre University).

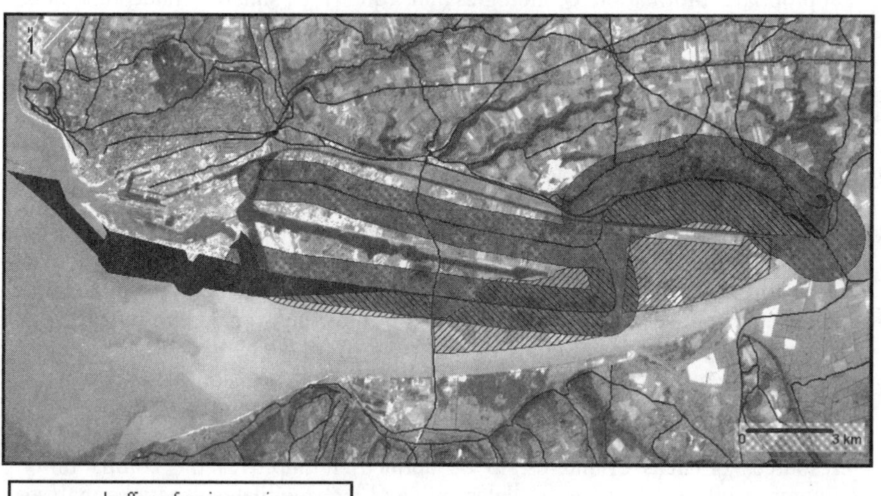

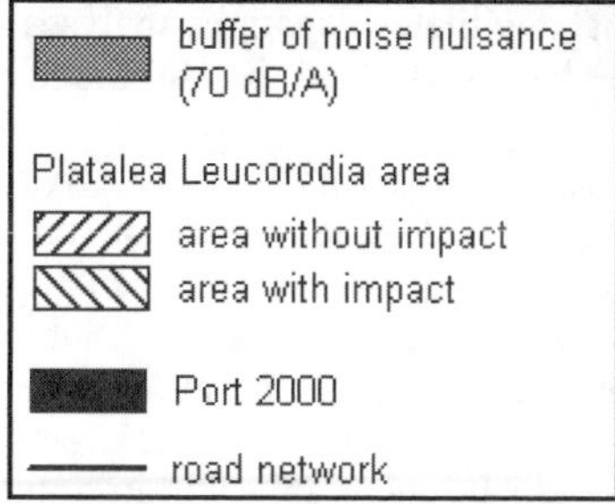

Figure 19.6 Effect of road traffic noise on *Platalea leucorodia* area.

This type of spatial analysis can also be applied to other domains (pollution of air or soil, etc.), by developing specialised algorithms for use in modelling impact analysis or simple diagnosis and visualisation. In the final analysis, the GIS tools have demonstrable relevance to aiding understanding, decision-support and management of environmental resources in a territory such as the Seine Estuary. By means of regional indicators, they can contribute to more holistic approaches to territorial management operations.

19.4 CONCLUSION

The concept of environmental management demands engagement with the interface between society and nature, since the multiple activities (often concurrent and mutually conflicting) that take place within a region are governed by economic and social paradigms.

Even if environmental considerations are nowadays better taken into account during processes of economic and social decision-making, they are still often only brought in towards the end of the process, while the various actors may frequently lack important elements needed to establish priorities and anticipate problems. Current practices and procedures, at different scales, are poor at anticipating environmental consequences of territorial management, due to a lack of appropriate information technologies and, equally, of suitable conceptual tools.

Spatial management of an estuary such as that of the Seine can only be successful and balanced if suitable structures are established to arbitrate effectively and equitably between the many pressures that currently lead to conflict.

Sustainable development is frequently advocated as a solution to these problems, whereby a dynamic balance is sought between economics, ecology and the needs of society, at local levels of organisation but also compatible with national and international levels.

Implementation of sustainable development policies requires a global approach and interdisciplinary vision, so that policies can lead to improved decision-making. In this context, the TIS for the Seine estuary has many potential applications. It serves as a tool capable of collecting, organising, integrating, and analysing information about the environment, from which territorial indicators and syntheses may conveniently be obtained to support decision-making, and to help solve problems of regional management and development.

In particular, it allows the properties and geographic distribution of natural resources or areas of interest to be defined; it improves the compatibility and interoperability of diverse datasets; and it facilitates spatial analyses of environmental impacts that would otherwise be overlooked or discarded for reasons of cost or procedural complexity.

This TIS constitutes a relevant tool because it has as its first objective the representation of a region, in this way improving our understanding and permitting the description and analysis of natural and human phenomena produced by the landscape. Since it lies at the junction of several disciplines, this TIS brings convergence and standardisation to the description and representation of space. It aims to bring coherence to all the many aspects of the same reality and, structured accordingly, data relating to the same location can be integrated within a single analysis, in order to better define environmental objectives and provide support for arguments or reasoning concerning regional development and planning within the estuary. The TIS has thereby enabled information relating to the natural heritage of the Seine estuary to be related, aggregated, and synthesised, in order to create territorial indicators of use for understanding, decision-making and management of the region.

Nonetheless, "mastering GIS, and developing robust methods of analysis, implies the ability on the one hand to assimilate elements of different terminologies, and the concepts associated with them, so as to permit truly multidisciplinary dialogue; while, at the same time, allowing questions to be formalised in terms that are compatible with the various elements of the information system" (Meillet, 1999).

Finally, it remains the case that these techniques carry limitations, even though they allow the integration of multiple sources of data. The user must always understand the consequences of their implementation, and minimise possible errors due to data quality, problems linked to the treatment of data, or due to sensitivities inherent in the selected methodologies (Marois, 1993). In other words, their use within the process of large-scale choice and decision-making requires those concerned to have a high degree of confidence in the relevance and reliability of the data that are exploited. The quality of data and of their interpretation thus prove to be the crucial *sine qua non* of any process of decision-making.

19.5 REFERENCES

Auger, Ch. and Verrel, J.-L., 1998, *Les estuaires français – Évolution naturelle et artificielle*, (Plouzané: IFREMER).

Bailly, A., 1995, *Les concepts de la géographie humaine*, (Paris: Masson).

Bourcier, A., Bourcier, J.-C. and Pouchin, Th., 1999, Appréhension de la cinématique paysagère par télédétection spatiale en estuaire de Seine: contribution au programme national de recherche sur les zones humides. In *La télédétection en francophonie: analyse critique et perspectives*, Collection Actualité Scientifique, (Montréal: Agence Universitaire de la Francophonie), pp. 215–223.

Bourcier, J.-C., Bourcier, A. and Pouchin, Th., 1998, *L'Observatoire des zones humides - Zones humides de l'estuaire et des marais de Seine: structure, fonctionnement et gestion*, (Le Havre: Université du Havre).

Bryant, C.R., 1993, Les SIG au service de l'aménagement et du développement communautaire. *Revue de Géographie Alpine*, **9**, (Grenoble: Université Joseph Fourrier), pp. 63–66.

Chappuis, J.-B., 1993, *Protéger la nature – Guide pratique de la protection de la nature et du paysage*, (Lausanne: Delachaux et Niestlé).

Laurini, R. and Milleret-Raffort, Fr., 1993, *Les bases de données en géomatique*, (Paris: Hermès).

Miellet, Ph., 1999, *SIG pour la gestion et l'aménagement urbain*, (Paris: Ministère de l'Équipement, des Transports et du Logement).

Marois, C., 1993, Environnement et Systèmes d'Information Géographique. In *Revue de Géographie Alpine*, **9**, (Grenoble: Université Joseph Fourrier), pp. 11-14.

Prélaz-Droux, R., 1995, *Système d'information et gestion du territoire: Approche systémique et procédure de réalisation*, (Lausanne: Presses Polytechniques et Universitaires Romandes).

Rouet, P., 1991, *Les données dans les systèmes d'information géographique*, (Paris: Hermès).

CHAPTER TWENTY

Developing an Environmental Oil Spill Sensitivity Atlas for the West Greenland Coastal Zone

Anders Mosbech, David Boertmann, Louise Grøndahl, Frants von Platen, Søren S. Nielsen, Niels Nielsen, Morten Rasch, and Hans Kapel

20.1 INTRODUCTION

Marine oil spill sensitivity mapping has become widespread. The purpose is to provide oil spill response planners and responders with tools to identify resources at risk, establish protection priorities and identify appropriate response and clean-up strategies. GIS is an important tool in the development of oil spill sensitivity maps and can also be used for presentation. Several different principles for information integration have been used in different countries (Anker-Nilssen, 1994; Dickens et al., 1990; Hall et al., 1997; Nansingh & Jurawan, 1999; Moe et al., 2000). An environmental oil spill sensitivity atlas was produced as part of the preparations for exploratory drilling off the West Coast of Greenland (Mosbech et al., 2000). We have adapted a Canadian sensitivity index system integrating physical, biological and human-use information and combined it with elements from a Norwegian sensitivity mapping system.

The Atlas covers the West coast of Greenland between 62° N and 68° N latitude. Although the area under study stretches only 700 km from north to south it encompasses approximately 18,000 km of coastline. It is the most populated area in Greenland with about 35,000 inhabitants living in 4 towns and 6 settlements. It is extremely important for fisheries and it is ecologically highly important for a number of seabird and marine mammal species.

The Atlas is a multidisciplinary GIS project integrating many kinds of scientific data and local knowledge (traditional ecological knowledge). Although studies on geomorphology and coastal spawning areas were initiated for the project it was a major challenge to compile and get the most out of the existing data from many sources. This paper will outline and discuss how the data was integrated, the principles used to identify and prioritise the sensitive areas, the final Atlas product ,and the dialogue during the community consultation.

0-41531-972-2/04/$0.00+$1.50

20.2 METHODS

20.2.1 The Atlas

The following elements are included in the Atlas

- coast types,
- oceanography, ice and climate,
- biological resources (fish, birds etc.),
- fishing and hunting,
- selected areas (e.g. seabird breeding colonies),
- archaeological sites,
- logistics and oil spill response methods.

The Atlas covers the coastline in the scale 1: 250 000 and the offshore area in the scale of 1:3.5 million. In a PDF-version 34 maps cover the coastline and one set of maps shows index values for coastal sensitivity and symbols for the elements of classification (see map and legend example plates 20.1a and 20.1b and colour inserts following page 164). Another set of maps show coast types, logistics, and proposed methods of oil spill response for each area (see legend and map example plates 20.2a and 20.2.b, see colour inserts).

Although a limited number of printed copies of the Atlas have been produced, the main version is published on CD. Since the target group of this atlas is not necessarily familiar with GIS, it was decided to distribute the final product in a form that is simple and quick to use and with an easy-to-use interface. The atlas with all text, maps, tables, and images is distributed on CD and on the Internet as a number of linked PDF-documents, which can be read with the free Acrobat Reader application. This allows a large amount of background information to be included as hypertext and at the same time to have a succinct and operational document. Furthermore, on the CD all geographic data is available for users to pan, zoom and print seamless maps with the free GIS-viewer ArcExplorer.

The Atlas was developed with an ESRI ArcView project and an MS Access database. This database software was chosen because many of the existing data and institutional databases employ MS Access. Spatial information on the model parameters is stored in ArcView whereas all model parameter attribute values are in the Access database. Index calculations are performed with Visual Basic for Applications in the database.

An important component of the Atlas is a sensitivity ranking system, which is used to calculate an index value describing the relative sensitivity of coastal and offshore areas. The sensitivity index value is calculated based on information on resource use (human use), biological occurrences, and physical environment. The sensitivity index system is based on a Canadian system used in Lancaster Sound (Dickins et al., 1990) and modified to the west Greenland fauna and other specific requirements of the Greenland study area. As a supplement to the Canadian ranking system, a number of smaller areas have been selected for priority in case of an oil spill. The selection of these areas is based on the principles from a Norwegian system (Anker-Nilssen, 1994), which gives priority to oil spill-sensitive areas for the purposes of oil spill contingency planning. While the

Canadian index system covers the entire coast in 50-km units, and thus gives a general overview of the sensitivity, the Selected Areas system uses actual borders and thus pinpoints the sensitive areas and leaves the rest unclassified.

20.2.2 The Sensitivity Index Calculation

The shoreline zone in the study area has been divided into 279 shoreline areas, each consisting of approximately 50 km of shoreline or, in the case of archipelagos, groups of islands and skerries having roughly 50 km shoreline. The offshore zone in the study area has been divided into 12 offshore areas (including 1 major fjord). The boundaries of the offshore areas are based on bathymetry and ice conditions during the winter.

The importance of resource use and the abundance of a number of biological occurrences in each of the 279-shoreline and 12 offshore areas were rated on a scale from 0 to 5 by using a number of subindices. Site-specific significant habitats are indicated on each shoreline segment. For example, such sites include important bird colonies and terrestrial haul-out for harbour seals. Photos of the coastal setting for about 50 bird colonies have been included and can be accessed from links.

As a part of the project, classification of the coastline geomorphology has been conducted from aerial photographs, e.g. the occurrence of rocky shores and beaches. An index value (oil residence index) of the self-cleaning ability of the coast after an oil spill has been calculated based on this classification in combination with shoreline exposure to waves and ice. For example, oil on a rocky coast exposed to wave action will be cleaned faster than oil on a beach in a protected lagoon.

An oil spill sensitivity index value has been calculated for each of the 279 shoreline and 12 offshore areas based on:

i) the abundance and sensitivity of selected species (or species groups);
ii) resource use (human use), mainly fishing and hunting;
iii) the potential oil residency on the shoreline (Oil Residency Index) based mainly on wave exposure, substrate and slope of coast;
iv) the presence of towns, settlements and archaeological sites (for shorelines).

The sensitivity index value for each of the 279-shoreline areas and 12 offshore areas is given. All areas are ranked as extreme, high, moderate or low sensitivity areas and a corresponding colour code has been used. The detailed index value calculations for each shoreline and offshore area can be accessed by hyperlinks.

With the settings we used, the average contributions to the final sensitivity values for the shoreline areas are: biological occurrences 49%, resource (human) use 20%, oil residence index 14%, archaeological sites 12%, communities 4% and special status areas (Ramsar sites) 1%. However, this is a simplification since the oil residence index value is a factor in the calculation of the value for biological occurrences, and thus also has an indirect influence on the sensitivity values.

20.2.3 The Selected Areas

To supplement the rather general mapping of shoreline sensitivity using the 50-km long shoreline areas, a number of small sensitive localities have been pinpointed as priority areas. A total of about 80 areas along the coast and within fjords have been selected as priority areas in the case of an oil spill situation. The basis for their selection is, compared to the coastline in general, that they are:

i) of high value either environmentally or for resource use;
ii) sensitive to oil spills; and
iii) of a size and form that may allow effective protection in an oil spill situation with a manageable amount of manpower and equipment.

The last criterion is important because it elucidates the rather limited possibilities to protect the coastline during a large oil spill and shows that tough prioritising is necessary.

20.2.4 Example of Indexing and Data Integration: Seabird Colonies

The major part of the biological information incorporated in the Atlas derives from databases maintained by NERI and the Greenland Institute of Natural Resources. As far as possible automatic selection and integration routines were developed to link these institutional databases and the Atlas database. One of the databases is a file on all seabird breeding colonies known in Greenland, with information on the birds, the sites and all survey results from the sites (Boertmann et al., 1996). A selection from the database was used in the Atlas. This selection was based upon the geographical range between 62° and 68° N and on the most comprehensive surveys, as many colonies have been surveyed several times. However, inferior colonies have been omitted and criteria for inclusion are listed in Table 20.1. Most colonies have a mixed species assemblage and the total number of colonies (with different geographical location) selected is 158.

The seabird colony data is used to produce a relative seabird abundance input value to the Seabird Oil Vulnerability Index scoring system which is integrated into the overall Shoreline Sensitivity Index. The seabird oil vulnerability index component takes into account the sensitivity to oil spills of the bird species both on an individual level and on a population level, as far as possible based on scientifically derived information on the characteristics of each species (Anker-Nilssen, 1987, Mosbech, 2000). These sensitivities are dependent on the behaviour and ecology of the birds, but also the distance to neighbouring colonies, which is a measure of the ability to re-colonise a colony. Moreover they take into account the status of the breeding population within the region, whether they are decreasing, increasing or stable, and finally their international conservation status (Mosbech et al., 1996).

20.2.4.1 Comments on some of the criteria

The breeding population of common eider in West Greenland has decreased seriously for a century, and within the Atlas region large and dense colonies have disappeared. Large breeding populations are mainly found dispersed in extensive archipelagos. To exclude sites with a few scattered nesting eiders the criterion for inclusion is ≥ 5 birds.

As gulls are only moderately sensitive to oil spills, only the largest colonies are included.

Black-legged kittiwake colonies with less than 50 pairs are excluded as they tend to be less stable over time.

Arctic terns usually breed in dense colonies on low islands. The population in West Greenland is generally decreasing. Small colonies of less than 30 pairs are excluded. Terns are moderately sensitive to oil spills, but colonies situated on low islands are very sensitive to disturbance e.g. from oil spill response activities.

All species of the family auks (alcids) are very sensitive to oil spills. This is caused by their behaviour as well as by their very low population turnover. Therefore, protection of their breeding sites is a high priority. Moreover, the breeding population of Brünnich's guillemot in West Greenland is seriously decreasing due to a very high hunting pressure, and the few breeding sites within the Atlas region are therefore all included.

20.2.4.2 Relative abundance of the species/species groups

Some of the species are pooled into groups, as they have a similar sensitivity to oil spills. The relative abundances applied to species and species groups appear in Table 20.2.

The relative abundance is the input from these site-specific shoreline species elements to the calculation of the sensitivity index for each shoreline area. Similar relative abundance values are calculated for all the biological and human use resources.

20.2.5 Example of Indexing and Data Integration: Archaeological and Historical Sites

Information on the archaeological and historical sites in Greenland has been collected for over 100 years and is contained in a database under development at the Greenland National Museum and Archives. There is information on about 1500 archaeological sites in the mapped area. All known prehistoric and historic sites are included in the present Atlas, but to protect the sites only the most basic information on each site is given e.g. the site type and dating of the site.

There are great differences in what is known about different archaeological sites in the Atlas area due to the source of the information. The major part of the information is derived from secondary sources (i.e. local informants) and may therefore be less accurate. Basic information on each site is often not reported and therefore the degree of sensitivity of each site is based on an estimate.

The sensitivity of each archaeological site is expressed on a scale 1-3;

1. Sites that are unlikely to be endangered.
2. Sites threatened either directly or indirectly. This group includes all sites close to the coastline that are presumed to represent value either as a historical source, as a recreational site or as a special historical highlight.

3. Threatened sites of significant importance that demand special measures be taken in case of a marine oil spill or other activities in connection to mineral or oil investigations and extraction.

Based on this assigned sensitivity each archaeological site is ranked on a scale from 0-5. For each segment (50-km shoreline area) these rankings are added and the results are grouped again on a scale from 0-5 similar to the relative abundance of the biological resources.

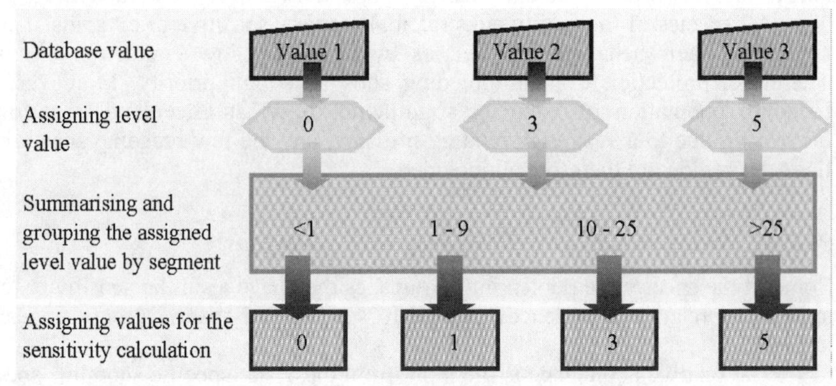

Figure 20.1 The Figure shows how the archaeological sites were given assigned values for the sensitivity calculation.

20.2.6 Integrating Local Knowledge

Local knowledge is included at several levels in the Atlas. Besides specific interview studies much information in the institutional databases is derived from locals helping researchers finding sites such as bird colonies and areas of archaeological significance. A specific interview study has been conducted to map the capelin (*Malotus villosus*) and lumpsucker (*Cyclopterus lumpus*) spawning and fishing areas and arctic char (*Salvelinus alpinus*) rivers (Nielsen et al., 2000). Capelin and lumpsucker spawn in the tidal zone and are therefore very vulnerable to oil spills. The study relied heavily on preliminary maps distributed for comments and the addition of supplementary information. Questionnaires and maps with preliminary information were sent out and returned with additional information, and these results integrated into new versions of the maps. Semi-directed interviews were conducted following the methodology of Huntington (1998). In the semi-directed interviews the maps were used to facilitate and guide the exchange of information among a group of local fishermen.

20.2.7 Community Consultation

A community consultation phase was carried out during the project. A draft version of the Atlas was mailed to local communities and user organizations. Later meetings were held with all municipal councils and most settlement councils as well as the local hunter and fishermen associations. The purpose of the community consultations was not only to verify the information in the Atlas, but also to allow for comments and discussion of the method we had used and the result of the sensitivity ranking. A computer with the GIS and a printer, which could print overheads, was brought along and used where specific thematic maps were needed.

Maps on overheads explicitly showing the human use ranking were presented and facilitated fruitful discussions from which much supplementary information came forward. Although the sensitivity mapping was positively received the level of abstraction and complexity in the index calculation sometimes lowered the enthusiasm in discussions of the final rankings. As an example an extremely important capelin spawning area could be in an area with low sensitivity ranking, because it takes many sensitive elements to raise the final ranking of the coast to high sensitivity. Here the use of small Selected Areas provides a better tool to accommodate local expectations and wishes.

In the settlements, where Inuit hunters are the majority, the layman – expert divide was very pronounced and personal contacts were of crucial importance for establishing a fruitful dialogue. In some settlements there was concern that the publication of maps with "their" Arctic char rivers would tempt townspeople to come fishing. It was therefore decided to restrict the publication of this information on the Internet until proper regulation to protect fishing rights is in place.

The advent of oil exploration in this area introduces a "low probability – high potential consequence" risk. The oil spill sensitivity Atlas is a valuable tool to minimise the risk. However, it is also important to communicate the realistic limitations of this tool.

20.2.8 Versatile Distribution of the Atlas

The users of the atlas are diverse with different backgrounds and qualifications. The primary target groups are government administrators (both national and municipal) who are involved in planning oil spill response; their counterparts in the oil industry; and oil spill responders in industry, navy and the local fire brigades. Since the target groups of this atlas are not necessarily familiar with GIS it was decided to distribute the final product in a form that is simple and quick to use and with an easy-to-use interface. The atlas with all text, maps, tables and images is distributed on CD and on the Internet as a number of linked PDF-documents, which can be read with the free Acrobat Reader application. Furthermore, all geographic data is available for users to pan, zoom and print seamless maps from the CD with the free GIS-viewer ArcExplorer. There has also been a demand for large plastic-coated map sheets and a limited number of printed copies of the atlas.

The PDF-document consists of 434 pages including 34 map areas covering the entire area at a scale of 1:250 000, suitable for printing on paper. Each map area is covered by two types of maps. One contains the shoreline sensitivity results (plate 20.1) and another the physical environment and logistics (plate 20.2). The PDF-document includes extra information for decision-makers e.g. aerial photos,

photos of specific bird colonies and sensitivity index calculations for each segment, accessible by hyperlinks.

Adobe Acrobat is not a GIS program, but is a good map viewing system. Distributing the Atlas with Acrobat gave us the ability to decide how the final maps were presented for the end user. Another advantage to using Acrobat was the map-producer's ability to decide at what scale the maps should be used. In this way there is no risk of misinterpretation of the spatial data caused by using them at a scale for which they were not designed. The end user must print the maps at the exact scale at which they were meant to be printed – this option is not available with the GIS-viewer ArcExplorer. ArcExplorer gives the end user the ability to view and use the maps and their information without having to learn the basics of a GIS.

20.3 CONCLUSION

The Atlas was produced over a 7-month period and the relatively short campaign worked well. The involved persons could see the end already when they started which helped the facilitation of enthusiasm among relevant participants.

The integration of institutional databases worked well, and the approach makes updates of the sensitivity Atlas and developments/adaptations to other purposes easier. The task initiated valuable contacts and discussions among database managers in the relevant institutions, as well as discussions with a focus on the research opportunities of integrated GIS analysis of georeferenced databases from various fields.

The PDF-format is a good choice for distributing an atlas with a mix of formatted text, tables, maps and images which should be connected with hyperlinks. The quality of the maps is better than what is possible with use of today's Internet mapping solutions. The same documents can be used for printing, CD-distribution and Internet-access. Since Adobe Acrobat is not a GIS, a GIS-viewer was necessary to enable users with no GIS experience to use the feature tables on the CD. ArcExplorer was found to be useful since the software is free of charge and comes with an extensive user guide.

The Selected Area concept helped pinpoint where to focus oil spill response and facilitated communication during community consultations because a single hot-spot could be highlighted based on one single very important item. The need for Selected Areas to supplement the general oil spill sensitivity mapping is related to the size of the coastal units. We chose 50-km units, like in the Lancaster Sound Atlas (Dickens et al., 1990), because of the vast area and the relatively limited amount of information. If the size of the coastal units is reduced to as little as 3 km on average, as is the case for Newfoundland in The Canadian Atlantic Region Sensitivity Mapping Program (Anonymous, 1999), the need for Selected Areas disappears.

The experience from community consultations and collection of local knowledge is that local communities need personal contact on a continuous basis to overcome the layman-expert divide. Continuity and build-up of confidence are important prerequisites for exchange of information and ideas. Some restriction of the publication of information obtained from local sources should be accepted.

20.4 THE STUDY TEAM

The project was carried out by the National Environmental Research Institute, the Geological Survey of Denmark and Greenland, the Greenland Institute of Natural Resources, the University of Copenhagen (Institute of Geography), the Greenland National Museum, The Greenland Secretariat of the Danish National Museum, Danish Meteorological Institute, AXYS Environmental Consulting Ltd. and SL Ross Environmental Research Ltd. The Danish Energy Agency funded the Atlas.

20.5 REFERENCES

Anker-Nilssen, T., 1987, *Metoder til konsekvensanalyser olje/sjøfugl*, (Trondheim: Viltrapport 44, Norsk Institutt For Naturforskning), pp. 114. In Norwegian.

Anker-Nilssen, T., 1994, *Identifikasjon og prioritering av miljøressurser ved akutte oljeutslipp langs norskekysten og på Svalbard*, (Norge: Norsk Institutt for Naturforskning, oppdragsmelding 310). In Norwegian.

Boertmann, D., Mosbech, A., Falk, K., and Kampp K., 1996, *Seabird Colonies in Western Greenland (60° - 79° 30′ N. lat.);* NERI Technical Report no. 170, (National Environmental Research Institute), pp. 148.

Dickins, D., Bjerkelund, I., Vonk, P., Potter, S., Finley, K., Stephen, R., Holdsworth, C., Reimer, D., Godon, A., Duval, W., Buist, I., and Sekerak, A., 1990, *Lancaster Sound region. A coastal atlas for environmental protection*, (Vancouver: D.F. Dickins Associates Ltd.)

Halls, J., Michel, J., Zengel, S., Dahlin, J., and Petersen, J., 1997, *Environmental Index Sensitivity Guidelines, version 2;* NOAA Technical Memorandum NOS ORCA 115, (Seattle:NOAA), pp. 79.

Huntington, H., 1998, Observations on the utility of the semidirective interview for documenting traditional ecological knowledge; *Arctic*, **51**, pp. 237-242.

Moe, K.A., Skeie, G.M., Brude, O.W., Lovas, C.M., Nedrebo, M., and Weslawski, J.M., 2000, The Svalbard intertidal zone: A concept for the use of GIS in applied oil sensitivity, vulnerability and impact analyses. *Spill Science & Technology Bulletin* **6** (2), pp. 187-206.

Mosbech, A., 2000, *Predicting Impacts of Oil Spills - Can Ecological Science Cope? A Case Study Concerning Birds in Environmental Impact Assessments*, University of Roskilde, National Environmental Research Institute, Department of Arctic Environment, pp. 129. Available online at URL: http://www.dmu.dk/1_viden/2_Publikationer/3_Ovrige/rapporter/PHD_AndersMosbech.pdf

Mosbech, A., Anthonsen, K.L., Blyth, A., Boertmann, D., Buch, E., Cake, D., Grøndahl, L., Hansen, K.Q., Kapel, H., Nielsen, S., Nielsen, N., Von Platen, F., Potter, S., and Rasch, M., 2000, *Environmental Oil Spill Sensitivity Atlas for the West Greenland Coastal Zone*, (Denmark: The Danish Energy Agency, Ministry of Environment and Energy) pp. 281 + appendix pp. 153. Available on CD-ROM, Print (limited distribution) and on the Internet http://Environmental-Atlas.dmu.dk

Mosbech, A., Dietz, R., Boertmann D., and Johansen, P., 1996, *Oil Exploration in the Fylla Area, An Initial Assessment of Potential Environmental Impacts*,

(Denmark: National Environmental Research Institute), NERI Technical Report no. 156, pp. 92 http://technical-reports.dmu.dk

Nansingh, P. and Jurawan, S., 1999, Environmental sensitivity of a tropical coastline (Trinidad, West Indies) to oil spills. *Spill Science & Technology Bulletin* **5**(2), pp. 161-172.

Nielsen, S.S., Mosbech, A., and Hinkler, J., 2000, *Fiskeriressourcer på det lave vand i Vestgrønland.- En interviewundersøgelse om forekomsten af lodde, stenbider og ørred; Danmarks Miljøundersøgelser.* Arbejdsrapport fra DMU nr. 118.

Table 20.1. Criteria for inclusion of seabird breeding colonies.

Species	**Criterion**	**No. of colonies meeting the criterion**	**No. of colonies included because other species meet their criterion (mixed colonies)**
Single species			
Northern fulmar	all colonies	3	-
Great cormorant	all colonies	34	-
Common eider	colonies with ≥5 individuals	31	42
Iceland gull	colonies with ≥500 individuals	10	32
Glaucous gull	colonies with ≥500 individuals	1	44
Unsp. glaucous/ Iceland gull	colonies with ≥500 individuals	1	3
Black-legged kittiwake	colonies with ≥50 individuals	38	7
Arctic tern	colonies with ≥30 individuals	23	7
Common guillemot	all colonies	5	-
Thick-billed murre	all colonies	5	-
Razorbill	colonies with ≥5 individuals	61	12
Black guillemot	colonies with ≥250 individuals	5	88
Dovekie	all colonies	1	-
Atlantic puffin	all colonies	17	-
Species combinations in mixed colonies (not meeting single species criterion)			
Razorbill and common eider	all colonies	1	-
Common eider and Arctic tern	all colonies	1	-

Table 20.2. Relative abundance of species/species groups in seabird breeding colonies. Add one point to relative abundance if colony/shoreline segment hold three or more alcid species and otherwise only reach a relative abundance of 3 or less. Add one point if colony/shoreline segment hold four or more gull and kittiwake species and otherwise only reach a relative abundance of 2 or less.

Species or species group:	**Alcids** (4 species with different occurrence in the Atlas region)				**Seaducks**	**Gulls**	**Kittiwake**	**Arctic tern**	**Tubenoses**	**Cormorants**	**Relative abundance**
Species in group:	Black guillemot	Razor-bill	Atlantic puffin	Thick-billed murre	Common eider	3 species			Northern Fulmar	Great Cormorant	
Number of individuals in colony:	1-100	1-20	1-5	1-10	1-50	1-200	1-100	1-50	1-200	1-20	1
	101-200	21-50	6-10	11-50	51-100	201-400	101-1000	51-200	201-1000	21-50	2
	201-5000	51-100	11-20	51-100	101-200	401-1000	1001-2000	201-1000	1001-2000	51-100	3
	501-1000	101-200	21-50	101-200	201-500	1001-2000	2001-10,000	1001-2000	2001-10,000	101-200	4
	>1000	>200	>50	>200	>500	>2000	>10,000	>2000	>10,000	>200	5

CHAPTER TWENTY-ONE

Environment Canada's Atlantic Sensitivity Mapping Program

André Laflamme, Stéphane R. Leblanc, and Roger J. Percy

21.1 INTRODUCTION

Canada's Atlantic Region along with other regions across the country have focused on providing consistent and standardized applications related to coastal mapping and data integration/generation during a drill or spill incident. This consistency is crucial if personnel are to be brought in from different regions, as they are immediately familiar with the process and terminology. In an effort to protect the environment and mitigate potential impacts, Environment Canada's Atlantic Sensitivity Mapping Program (ASMP) was designed to provide this level of support to environmental responders. The ASMP has become a very powerful tool providing a consistent terminology through the entire range of pre-spill planning, preparedness, and real-time response activities.

This paper will describe the scope, objectives, and current status of this mapping initiative and highlight recent developments in combining the full range of activities from data generation and decision development to the generation of sensitivity mapping. The desktop mapping application provides an easy-to-use approach to the manipulation, display and output of a wide range of technical and supporting data and information stored in various databases. In the development of its mapping program, Environment Canada relied on crucial partnerships with organizations willing to share data and expertise. Response managers and environmental responders now have access to sensitive resource information that normally would be difficult to collate and present in a map form under the pressures of a spill response.

The objective of developing and maintaining the best possible sensitivity mapping system is to provide planners and managers with the full range of information they require as part of pre-spill activities as well as resource protection recommendations at the time of a spill. The data and information are based on consistent sets of terms and definitions that describe the shore-zone character, the objectives and strategies for a specific response, and the methods by which those

0-41531-972-2/04/$0.00+$1.50

objectives may be achieved. These data are linked with other resource information in a GIS based system.

Standard or accepted terms, definitions, and shoreline segmentation procedures are already in place for describing the shore-zone character and shore-zone oiling conditions. In this program, a set of standardized objectives and strategy statements have been developed that can be entered easily into a database; these provide a better level of consistency than do phrases or sentences constructed by different recorders or evaluators. The suggested protection and treatment objectives and strategies are intended for consideration by the spill response management team. The actual type and volume of spilled oil, plus local environmental conditions and local priorities would be brought to bear on the decision process at the time of a spill. The suggested objectives and strategies provide a starting point and a framework for decision makers and planning and operations managers to discuss objectives and priorities. The concept of management by objectives provides a framework for decision-makers to set the goals of an operation at both the regional and a segment-by-segment level (Percy, LeBlanc, Owens, 1997).

The pre-spill database is integrated with the actual Sensitivity Mapping Program which is capable of displaying natural, cultural and man-made features vulnerable to oil spills. The computerized mapping system facilitates quick access and management of multiple data sets. A user-friendly interface allows queries and statistical analysis of data and display of graphical outputs. The system provides a tool for both planning and response; information can be accessed or modified using a laptop computer and real-time spill information or trajectory model outputs can be incorporated.

21.2 PARTNERSHIP

Following the Exxon Valdez spill in Alaska, the government of Canada realized the need for a system to provide access to sensitivity data for planning and response purposes. The Green Plan provided initial funding to develop, create and maintain a sensitivity mapping system to support environmental responders during marine spill incidents. Environment Canada was tasked to lead this project and was assigned the responsibility to gather and manage appropriate data sets from various agencies. Because of its mandate, Environment Canada has environmental emergency officers on duty on a 24/7 basis. Therefore, information must be quickly accessible in order to mitigate a potential impact on marine and coastal resources.

Without ready access to environmental data, the integrity of coastal and marine resources can be compromised during a spill incident if immediate action is not taken to protect them. Partnerships are crucial in order for environmental emergency responders to locate and identify sensitive resources at a spill site, especially within the first few hours/days of an incident. One of Environment Canada's foci is to approach and involve other federal, provincial, municipal agencies, private industry, local communities, etc. to make environmental data accessible. Most of the organizations involved during a spill incident are part of the Regional Environmental Emergencies Team know as REET.

REET has two main operating roles: *planning* and *response*. As part of the planning function, "team" members meet once a year to exchange scientific and technical information on such matters as contingency planning and spill response techniques. During this time, REET members also update and review their respective roles in any emergency response. In its response role, REET operates as a team of experts, advising the On-Scene-Commander, or OSC, in emergency situations. Chaired by Environment Canada, it is composed of scientific advisors, private contractors, community groups, etc.

21.3 GEOGRAPHIC APPLICATION

The coastal area covered by the Atlantic Region Sensitivity Mapping Program encompasses four provinces: these are New Brunswick, Nova Scotia, Prince Edward Island, and Newfoundland/Labrador. Approximately 12,500 unique shoreline segments covering more than 40,000 kilometres of coastline have been identified in Atlantic Canada. Labrador is the only area not presently covered by the shoreline classification; however growing interests and activities in this area will likely require the completion of the pre-spill database in the near future. Portions of the province of Quebec have been included in the mapping system: these are Chaleur Bay (on the north shore of the St. Lawrence River) and the Magdalene Islands, since these areas would likely be impacted by spills in the Atlantic Region. Despite the coastal applications of the mapping system, it also has the flexibility to cover the inland part of the Atlantic Provinces. Environmental data has been collected for the Maine and New Brunswick border. The Atlantic Region Sensitivity Mapping Program in conjunction with the Maine Department of Environmental Protection have agreed to exchange cross border information on coastal areas which can be used for planning and response during marine incidents that could impact both countries. As the information becomes available for inland areas, the mapping system will integrate the information in a format that is compatible with the existing data sets. Most of this information comes from federal, provincial, and municipal government, as well as local knowledge.

21.4 SHORELINE CLASSIFICATION AND PRE-SPILL DATABASE

The objective of the pre-spill database is to collate data and information that would be required and used by the spill response management team in the development of planning, priority and operations decisions. This database plays a fundamental role in the definition of resource protection priorities, and constitutes an introduction to the Shoreline Clean-up and Assessment Techniques (SCAT) process.

The database development procedure involves an initial segmentation of the shoreline followed by the creation of data templates for each segment. This process involves the use of various tools such as low-altitude videotape survey data, aerial photography, and pre-existing mapping materials to define sections of shoreline that have a uniform along-shore character. In Atlantic Canada, each segment has a unique two-letter prefix code followed by a sequential number

(Figure 21.1). The two-letter prefix is unique to one coastal area in Atlantic Canada which makes each code different (e.g.: Halifax Harbour has the following segment codes: HX-01 to HX-75).

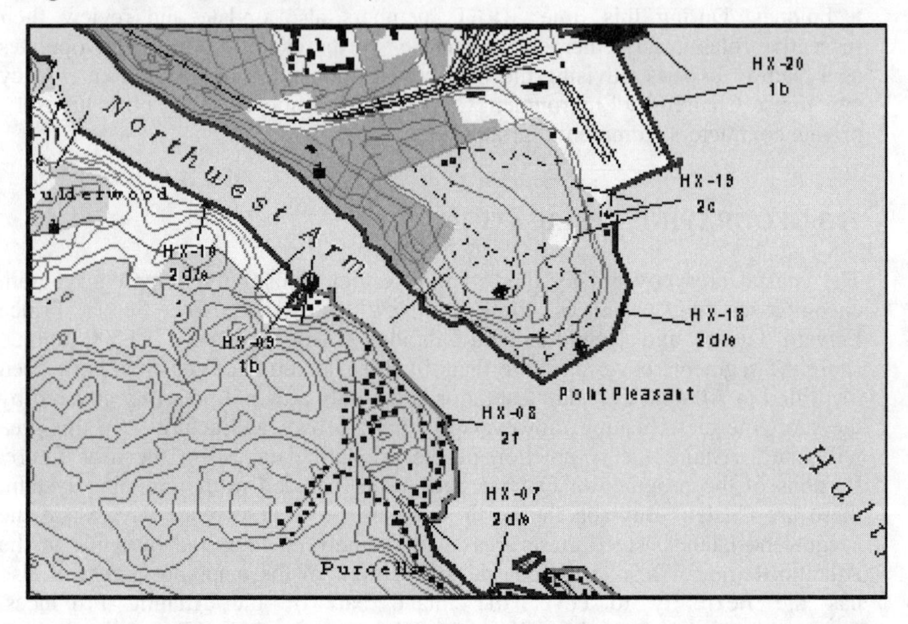

Figure 21.1 An example of the shoreline segmentation of the Northwest Arm area near Halifax, NS.

The description of the shore zone and the development of appropriate response strategies are presented in a systematic format based on four distinct templates: shore zone character, shoreline protection, shoreline treatment, and summary of response and requirements.

These templates contain a total of 143 different attributes which are unique for each shoreline segment. The Shore Zone Character template describes information such as shoreline material/type, nearshore environment, longshore current, oil traps and potential behaviour, resources at risk, etc. (Owens & Dewis, 1995). The shoreline material/type is further subdivided into five distinct categories: lower inter-tidal material, lower inter-tidal form, shoreline type (area located between the high and low tide mark), backshore material, and backshore form (Table 21.1)

The Shoreline Protection and Treatment/Cleanup Templates offer a variety of shoreline data including treatment and protection methods, objectives, strategies, and operational considerations. The last template is known as the Summary of Response Requirements. It is a summary of the protection and treatment templates and includes a response priority code (L = low, M = medium, H = high, VH = very high). The response priority code is defined based on the information available at the time of collection of the pre-spill database. Although it is a starting point in defining priorities, Environment Canada's Sensitivity Mapping Program is now in the process of incorporating other data sets in order to define a response priority code which will better reflect the actual resource inventory for a specific shoreline segment.

Table 21.1 Shoreline material/type found in Atlantic Canada

Lower ITZ Material	Lower ITZ Form	Backshore Form
anthropogenic concrete	anthropomorphic breakwater	anthropomorphic breakwater
anthropogenic wood	anthropomorphic pier/jetty	anthropomorphic bridge
anthropogenic riprap	anthropomorphic pilings	anthropomorphic causeway
bedrock resistant	anthropomorphic seawall	anthropomorphic road
bedrock unresistant	anthropomorphic bridge	anthropomorphic dyke
boulder	anthropomorphic wharf	anthropomorphic pier/jetty
cobble	beach	anthropomorphic wharf
mixed coarse with sand	cliff	anthropomorphic railway
marsh grass	dune	anthropomorphic seawall
Mud	platform	barrier beach
pebble	salt marsh	beach
Sand	tidal flat	cliff
	delta	dune
Backshore Material	low-islets	flat
anthropogenic asphalt		peat bog
anthropogenic concrete	**Shoreline Type**	platform
anthropogenic riprap	bedrock	salt marsh
anthropogenic wood	boulder beach	spit
bedrock resistant	man-made solid	wetland bog
bedrock unresistant	mixed sand-gravel beach	delta
mixed coarse with sand	mud tidal flat	
mixed coarse-no sand	pebble-cobble beach	
marsh grass	salt marsh	
peat	sand beach	
sand	sand tidal flat	

These templates use a knowledge-based concept, as data and recommendations are entered, in part, from knowledge and experience rather than from an objective analysis. The templates are described in detail by Owens and Dewis (1995). The shoreline protection and treatment or cleanup techniques that are recommended for each segment are derived from the Environment Canada Field Guide for the Protection and Cleanup of Oiled Shorelines (Owens, 1996).

The description of the physical character of the shore zone for each segment is broken down into the lower intertidal zone, the upper intertidal zone (which corresponds to the nine accepted standard shoreline types (Figure 21.2)) and backshore coastal character.

The shoreline type is a description of that area of the shore zone where oil is most likely to be stranded and the coastal character is described since this is the area in which backshore operations will stage and deploy resources. The description also includes identification of features that are likely to affect the behaviour of persistent oil, such as alongshore traps, potential boulder or riprap reservoirs, etc. Figure 21.3 gives an example of the Shore Zone Character template for a specific shoreline segment in the Atlantic Region.

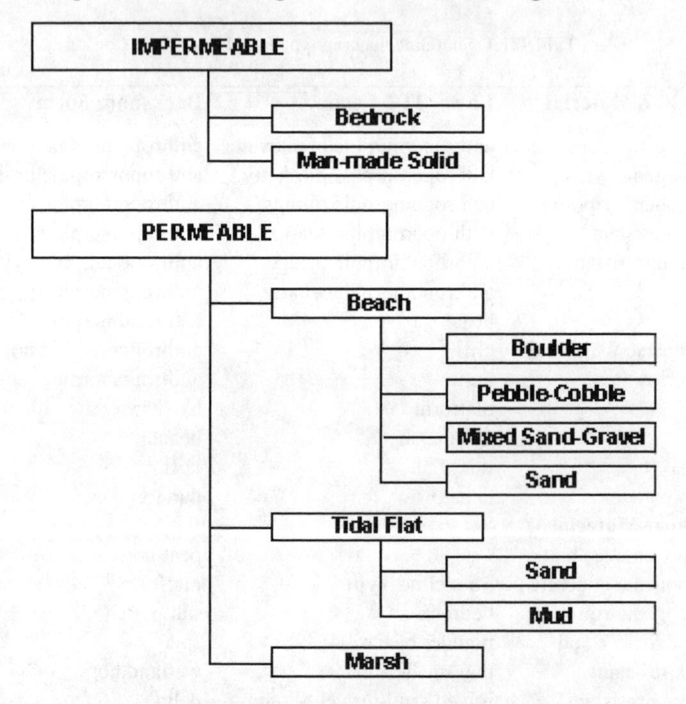

Figure 21.2 Nine standard shoreline types used to describe shore character.

Other areas outside Canada where the same shoreline classification approach has been applied include Hawaii (Honolulu-Waikiki), Russia (Sakhalin Island) and Alaska (Port Valdez). A number of countries around the world have shown interests in the Atlantic Region Sensitivity Mapping Program, including Bangladesh, Brazil, Spain, Israel, Chile, and France.

SHORE ZONE CHARACTER

REGION :ATL-4 **SEGMENT : HX-75**

SEGMENT LENGTH : **1.4 km**

SHORELINE MATERIAL/TYPE :

Lower ITZ material : **bedrock resistant**
Lower ITZ type : **platform**
Shoreline type : **bedrock**
Backshore material : **bedrock resistant**
Backshore type : **platform**

Permanent inlet : Cyclical inlet :
Inlet location changes : Inlet shape changes :
Inlet width : Number of channels :

NEARSHORE ENVIRONMENT :

Tidal range : Mean tide **1.5 (m)** Large tide **2.1 (m)**
Open, exposed coast ? **Yes**

PREDOMINANT LONGSHORE CURRENT/DRIFT DIRECTION: ?

OIL TRAPS AND POTENTIAL BEHAVIOUR :

Natural alongshore barrier (e.g. headland) **No**
Man-made alongshore barrier (e.g. wharf) **No**
Pebble/cobble - penetration potential **No**
Bay or re-entrant **No**
Riprap or boulder - reservoir potential **No**
Marsh meadow oiling potential during high water levels **No**
Sand/gravel - burial potential **No**
Overwash potential into lagoon/marsh **No**
Tidal lagoon or estuary **No**
Tidal inlet or channel **No**

RESOURCES AT RISK :

Birds (shore birds, ducks) **No**
Crustaceans or Mollusks **No**
Fish (nearshore only) **No**
Flora/Plant communities **No**
Mammals (marine) **No**
Agricultural **No**
Commercial or Industrial **No**
Harbour or Marina **No**
Recreation **No**
Residential **No**

Primary resources: **AESTH: scenic lookout**

Secondary resources:

Is the segment within a PAR ? **Yes**

INFORMATION SOURCES :

Topographic Map(s): 11 D/11
Hydrographic chart(s) : 4237
Videotape(s): GSC Open File # , Seg #58 (1989)

Figure 21.3 The shore zone character template is one of the four available in the pre-spill database. More than 143 attributes are available for each shoreline segment.

21.5 SENSITIVITY MAPPING SYSTEM

All computerized mapping systems require base map layers. Over the past years, Environment Canada purchased National Topographic Data Base (NTDB) digital maps from Natural Resources Canada - Centre for Topographic Information in Sherbrooke (Quebec). Three different scale were purchased: 1:50,000, 1:250,000 and 1:1,000,000. All three scales are used to represent specific information according to whether large- or small-scale maps are required. Once the base maps were visualised to reflect as much as possible the symbols and colours of the hard-copy versions, the digital maps were grouped to form one big region (Figure 21.4).

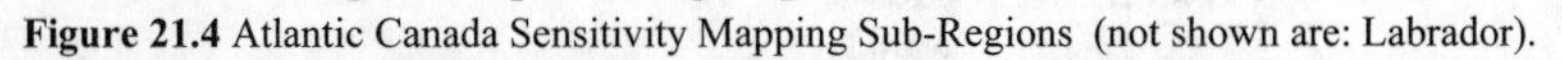

Figure 21.4 Atlantic Canada Sensitivity Mapping Sub-Regions (not shown are: Labrador).

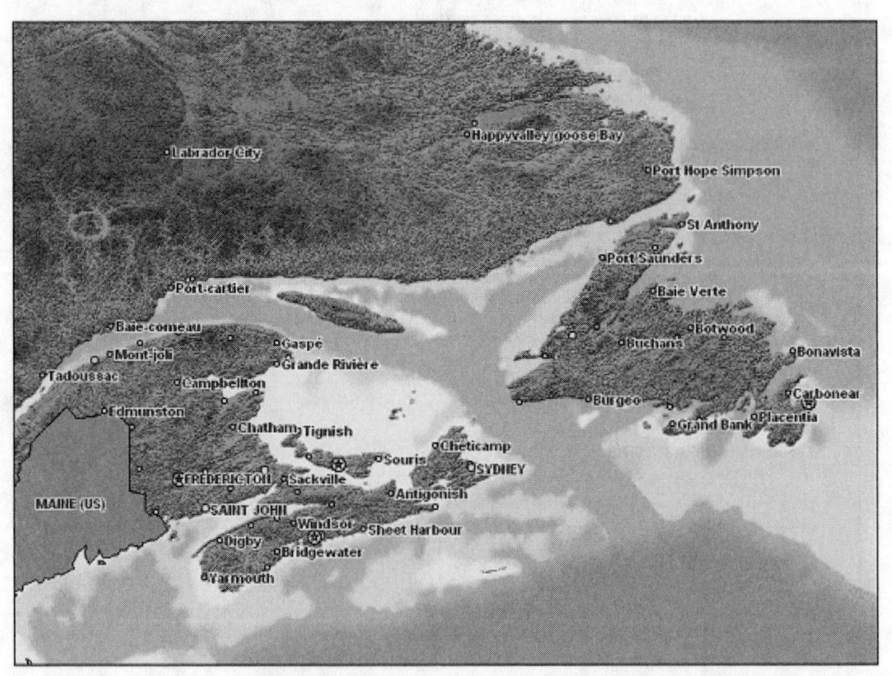

The shoreline classification and the pre-spill database constitute the most important components in the mapping system. The classification describes the physical aspects of the shoreline and provides useful information on protection and clean-up methods. In addition, the mapping system allows for the display of various databases such as birds, fish, shellfish, aquaculture sites, parks, archaeological sites, etc. (Table 21.2). With all this information available, the mapping system can provide a detailed report for any given area.

The user has the option of defining a buffer zone which can be used to determine the sensitive resources within an area; to calculate features such as length of shoreline or area affected; or to display data in a graphical form (Figure 21.5). This last example contains shoreline lengths within a polygon for shoreline type, backshore type, biological resources, and human use resources. The information can be displayed in the form of bar graphs or pie charts, and a detailed report on the affected resources may also be generated. The report contains information on various species or human-use resources, their sensitivity to oil spills and their seasonal vulnerability. A complete database of photographs related to sensitive birds, fish, shellfish, vegetation, and human use structures such as aerial photographs of small craft harbour can also be displayed. In addition to the Natural Resources digital base maps, the system can also display information from digital hydrographic charts or digital elevation models (Figure 21.6).

Table 21.2 Type of information and source (* - Available only during real incidents)

Database	Source
Pre-spill database (MS Access)	Environment Canada
Shoreline Classification	Environment Canada
1:50,000 NTDB maps	Natural Resources Canada
1:250,000 NTDB maps	Natural Resources Canada
Logistical/Operational Data	Response Organization
Fish	Fisheries & Oceans Canada
Shellfish	Fisheries & Oceans Canada
Mammal	Fisheries & Oceans Canada
Amphibian/Reptile	Fisheries & Oceans Canada
Vegetation	Fisheries & Oceans Canada
Small Craft Harbour	Fisheries & Oceans Canada
Whale Sanctuary	Fisheries & Oceans Canada
Fish Weir & Trap	Fisheries & Oceans Canada
Bird	Canadian Wildlife Services
National Wildlife Area	Natural Resources Canada
Wildlife Conservation Area	Natural Resources Canada
Ecological Reserve	Canadian Wildlife Services
Bird Sanctuary	Canadian Wildlife Services
Seabird Colony	Canadian Wildlife Services
Aquaculture Site	Provincial Government
Archaeological Site	Provincial / Federal Government *
Recreational Beach	Provincial Government
Provincial Park	Provincial Government
Municipal Park	Municipal Government
Sewage Outfall	Municipal Government
Water Intake	Municipal Government
Historic Site	Heritage Canada
Sewage Treatment Plant	Municipal Government
Recreational Fishing	Municipal Government
Salmon River	Provincial Government
Fish Processing Plant	Municipal Government
Hydrographic Chart	Canadian Hydrographic Services
Federal Properties	Treasury Boards Canada
National Historic Site	Heritage Canada
Federal Park	Heritage Canada
Aerial Shoreline Video	Geological Survey of Canada
Digital Elevation Model	Natural Resources Canada
Offshore Rigs Location	Environment Canada
Pulp & Paper Mills	Environment Canada

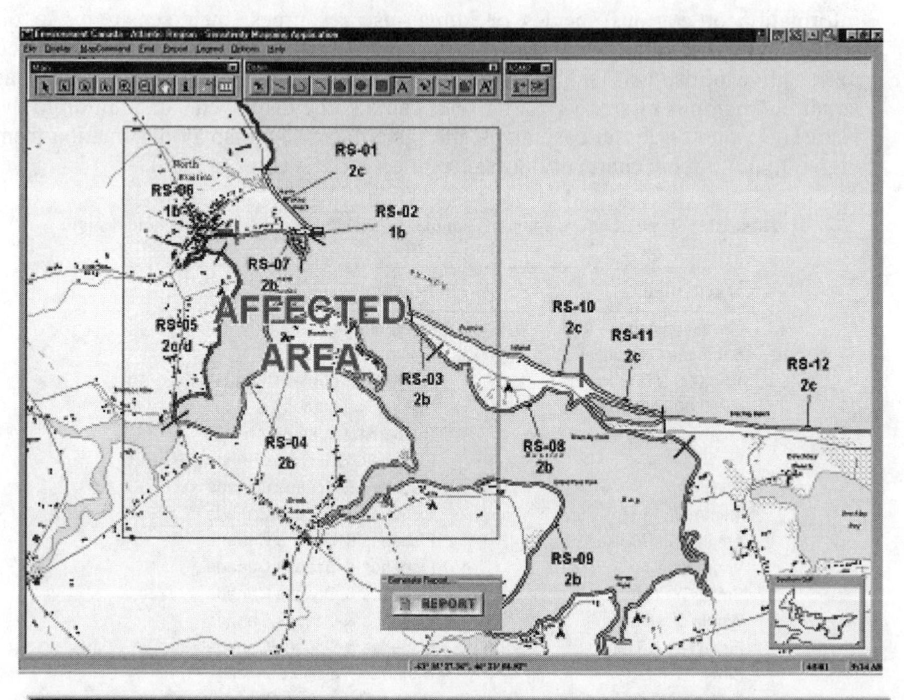
RS-01
2c
RS-06
1b
RS-02
1b
RS-07
2b
AFFECTED
AREA
RS-05
2c/d
RS-10
2c
RS-11
2c
RS-12
2c
RS-03
2b
RS-04
2b
RS-08
2b
RS-09
2b
REPORT

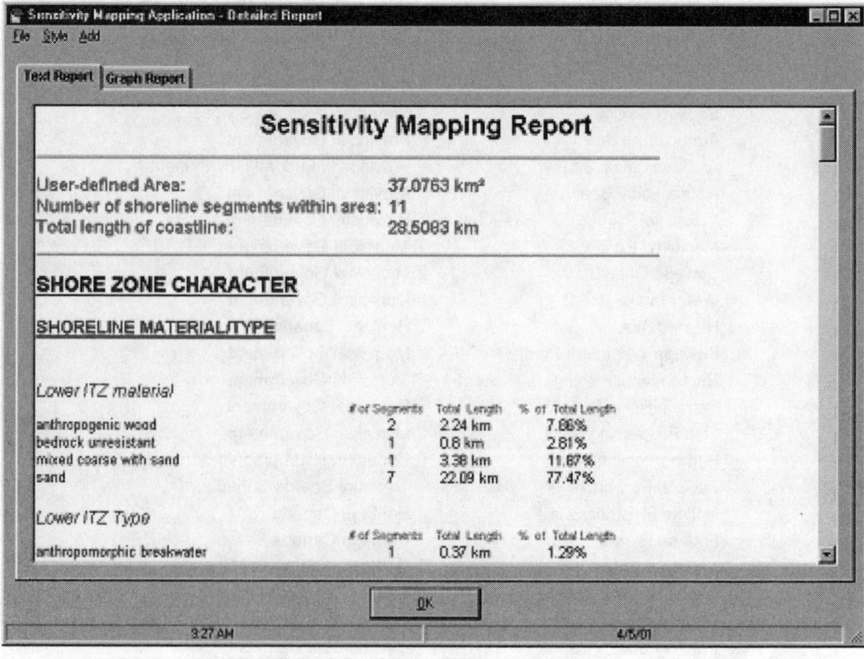
Sensitivity Mapping Application - Detailed Report
File Style Add
Text Report
Graph Report
Sensitivity Mapping Report
User-defined Area: 37.0763 km²
Number of shoreline segments within area: 11
Total length of coastline: 28.5083 km
SHORE ZONE CHARACTER
SHORELINE MATERIAL/TYPE
Lower ITZ material
# of Segments Total Length % of Total Length
anthropogenic wood 2 2.24 km 7.86%
bedrock unresistant 1 0.8 km 2.81%
mixed coarse with sand 1 3.38 km 11.87%
sand 7 22.09 km 77.47%
Lower ITZ Type
# of Segments Total Length % of Total Length
anthropomorphic breakwater 1 0.37 km 1.29%
OK
3:27 AM
4/5/01

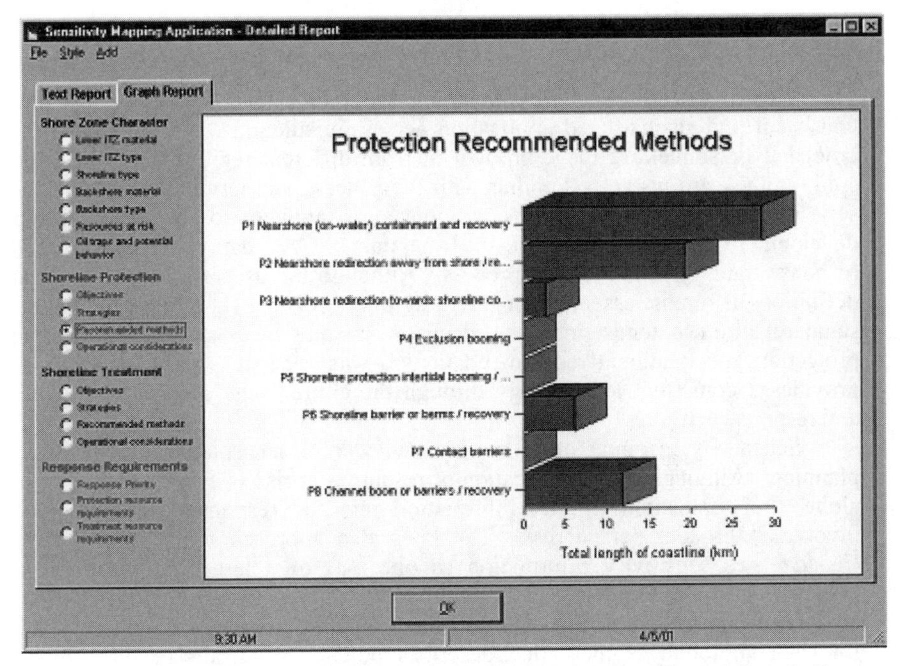

Figure 21.5 a,b,c One of the strengths of the mapping system is the ability to define a buffer zone and extract resource information within that area. The program can generate reports in text and graphical form.

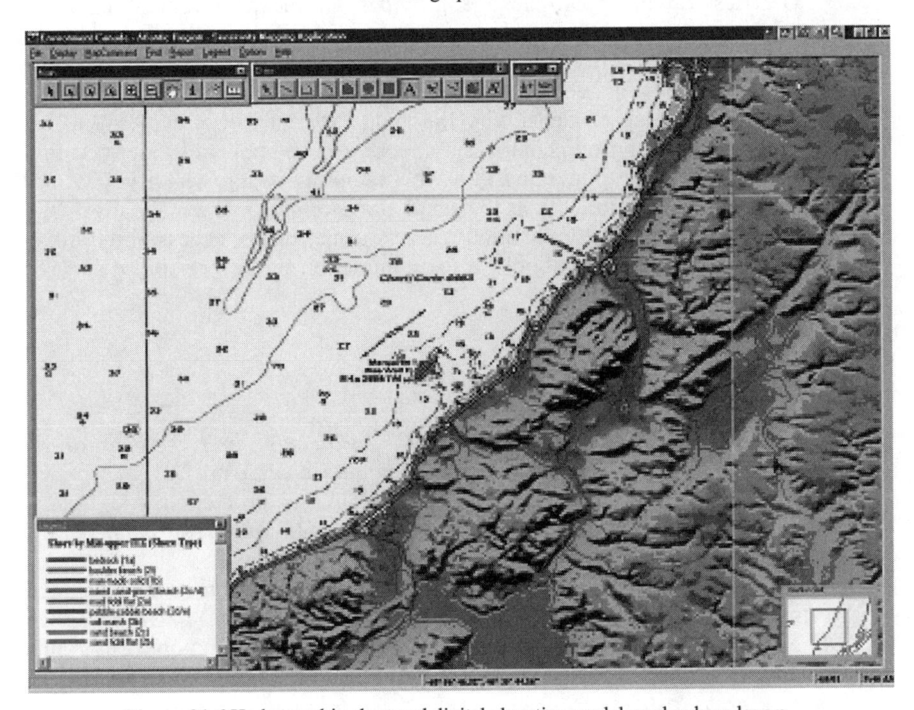

Figure 21.6 Hydrographic chart and digital elevation model used as base layers.

21.6 SUMMARY

The Atlantic Region Sensitivity Mapping Program's main focus is to provide a consistent and standardized application across jurisdictions. This consistency is crucial if personnel are to be brought in from different regions during a drill or spill, as they are instantly familiar with the process and terminology. The same sets of standard terms and conditions are applied to pre-spill database development, sensitivity and pre-spill mapping, SCAT data generation, and the response management decision process. Although the use of standard terms and definitions is almost essential for the description of oiled shorelines, the value of standardization extends into the decision process with regard to shoreline protection and treatment recommendations. This concept is very powerful as it provides a consistent terminology through the entire range of pre-spill planning and response activities.

Sensitivity mapping often has been viewed as an isolated activity in spill planning. Although the identification of resources at risk is a study that can stand alone, the application of that information into the response decision process involves a broader perspective. The integrated approach for spill response in Canada uses sensitivity information as one part of a larger body of relevant information that can and should be used by the decision team in the formulation of an appropriate response. For example, the pre-spill database includes information for each shoreline segment that describes potential oil traps or oil behaviour considerations, possible operational constraints, as well as the resources that are at risk.

The desktop mapping package provides an inexpensive and easy-to-use approach to data management and manipulation. The data can be updated quickly and easily with new information as this becomes available, or during a spill incident, and this new information can be presented immediately in map form.

The Atlantic Region Sensitivity Mapping Program keeps evolving to better serve the needs of environmental responders. All components described in this document will soon be integrated in a web-mapping format which will offer the same features and functionality as the stand alone version. The main advantage of designing a web mapping application resides in its broader accessibility and ease of use. Future developments include the completion of the pre-spill database for Labrador.

21.7 REFERENCES

Owens, E.H., 1996a, *Field Guide for the protection and cleanup of oiled shorelines*, 2nd Edition (Dartmouth: Environment Canada), pp. 201.

Owens, E.H., 1996b, The management of shoreline protection and treatment operation. In *Proceedings 19th Arctic and Marine Oil Spill Programme (AMOP) Technical Seminar* (Edmonton: Environment Canada).

Owens, E.H. and Dewis, W.S., 1995, A pre-spill shoreline protection and shoreline treatment data base for Atlantic Canada. In *Proceedings 18th Arctic and Marine Oil Spill Programme (AMOP) Technical Seminar* (Edmonton: Environment Canada).

Percy, R.J., Leblanc, R.S. and Owens, E.H., 1997, An integrated approach to shoreline mapping for spill response planning in Canada. In *Proceedings 20th Arctic and Marine Oil Spill Programme (AMOP) Technical Seminar* (Edmonton: Environment Canada).

Epilogue: Meeting the Needs of Integrated Coastal Zone Management

Jennifer L. Smith and Darius J. Bartlett

The emergence of Integrated Coastal Zone Management (ICZM) represents a paradigm shift for a range of practitioners who work in the complex, dynamic area where land meets sea. The structure and implementation of geomatics technologies has been strongly affected by this shift, and GIS/RS practitioners have stepped up to meet the information needs of ICZM by creating coastal information systems featuring increased rigour, openness, and useability.

As a result of this forcing, coastal GIS/RS is increasingly differentiating itself from the marine sciences and emerging as a unique discipline. In 1999, Wright wrote: "it may be fair to say that marine applications of GIS have been more in the realm of basic science whereas coastal applications, due in part to the intensity of human activities, have encompassed both basic and applied science, as well as policy and management." This move towards an integrated approach is being realized in parallel with similar shifts in the approach of other actors in the coastal zone, including scientists, managers and planners. The principles of Integrated Coastal Zone Management make explicit what coastal GIS/RS practitioners have known for years: that the coastal zone is a uniquely complex system that requires new and innovative management approaches.

Here we outline the nature of this paradigm shift and the ways in which the authors in this volume have tackled the challenges of applying GIS to the integrated management of the coastal zone.

The principles of Integrated Coastal Zone Management

A thorough discussion of the inception and principles of Integrated Coastal Zone Management can be found in Cicin-Sain (1993) and Cicin-Sain and Knecht (1998). The following draws heavily upon these sources.

ICZM has been described as a process for decision-making: it should be continuous, iterative, and should recognize the contributions of stakeholders and the natural dynamism, both physical and ecological, of the coastal environment.

A primary goal of ICZM is to overcome the compartmentalized approach to managing coastal resources by harmonizing the decisions of diverse jurisdictions and levels of government. In particular ICZM aims to bridge the traditional divide between management of the land and of the water. ICZM, therefore, is also about building institutions that facilitate this integration.

ICZM is founded on principles of sustainable development, recognizing that the coastline is the fount of resources of great value to human communities and

0-41531-972-2/04/$0.00+$1.50

that these resources should be managed in ways that conserve their value for future generations. In addition to the extractive resources (such as fishing, hunting, aggregate extraction, and harvesting of algae), ICZM aims to recognize and conserve those resources that provide ecosystem services, such as seagrass beds that provide nursery habitat and mangroves that act as a defense against storm surge and erosion.

The outcome of these principles is a planning methodology that employs an ecosystem approach to management (considering interconnected elements of the ecosystem) and incorporates adaptive management to deal with uncertainty, variability, and change.

The role of GIS/RS in ICZM

The principle function of geomatics technologies in the coastal milieu has traditionally been that of a tool for converting data into information, which is in turn either published as basic science or fed into the balance sheet of the rational decision-making process. With the emergence of ICZM, however, the role of GIS/RS has evolved. In response to the ICZM imperative, coastal geographic information systems are increasingly integrative of scientific and socioeconomic elements, increasingly dynamic, temporal and predictive, and are often the tool that effectively brings stakeholders together.

GIS is often defined as a synergy of hardware, software, data, and people (ESRI 1990). Clearly, the changing role of the technology is accompanied by a changing role for the people who build and operate coastal information systems.

This volume contains a diverse collection of chapters, all of which address the application of GIS/RS and the role of its practitioners in the various facets of an integrated approach to coastal zone management. The authors in this book have begun the process of finding solutions to the challenges presented by ICZM, and are working towards these goals on several fronts.

Contributing to the capacity requirements of ICZM

Cicin-Sain and Knecht (1998) list four areas of critical capacity required for ICZM. In defining "technical capacity" the authors make direct reference to the role of coastal databases and information systems in gathering information on coastal ecosystems and processes, on the patterns of human use that occur in the coastal zone, and on the ongoing effectiveness of management activities. But GIS/RS also has the potential to contribute to the other areas of required capacity outlined by Cicin-Sain and Knecht: authors in this volume show how coastal spatial databases can assist in the formulation, implementation and enforcement elements of the *legal and administrative capacity* required for ICZM. For example, Both Pan and Lindsey *et al.* outline the use of GIS to fulfil legistlative imperatives; the TIS developed by Bourcier has the potential to provoke harmonisation of diverse jurisdictions and regulatory regimes. Some authors have outlined their contribution to the *human resources capacity* of coastal management initiatives, including bringing together multidisciplinary teams and fostering public awareness of and involvement in coastal issues. To this end, Bartlett and Sudarshana examine issues surrounding technology transfer; the standardised approach to risk mapping developed by Laflamme *et al.* facilitates transfer of workers from one Canadian province to another in response to environmental emergencies; Nwilo describes programs that have facilitated exchanges and training for coastal zone managers in Africa; and many authors in this volume have found avenues for disseminating and interpreting their work to interested stakeholders. And, although GIS are typically expensive to implement, they can sometimes bring economies and recognition to coastal management initiatives, thus enhancing the *financial capacity* of the project in the long-term. Initiatives for advancing geomatics in the coastal zone, like those of Webster *et al.* and Nwilo, have attracted funding with spinoffs for coastal communities.

Integrating data from diverse sources

If the worth of a GIS is determined by how faithfully it represents the real world, creating a servicable GIS is challenging indeed when data of sufficient resolution, coverage and quality are difficult to acquire and unite for reasons of protectionism, conflict and variable format or standards. The ICZM imperative of integration across jurisdictions and institutions makes this issue doubly significant in the coastal zone. Authors in this volume, along with virtually every other coastal GIS practitioner, face this challenge daily and are working to find strategies that yield information systems that meet the needs of ICZM. Bourcier, Laflamme *et al.* and Mosbech *et al.*, among many others in this book, have addressed the challenge of collecting and harmonizing data within a distributed, fractured, multi-agency and multi-jurisdictional region (the Seine estuary, the Eastern Canadian coastal zone and the Greenlandic coastal zone, respectively). Gomm considers issues specific to data integration across the land-sea divide, and Gourmelon writes about the process of coordinating a large number of institutions in the creation of a large-scale GIS.

Sometimes the greatest challenge to harmonising datasets is finding them. Longhorn describes the development of Coastal Spatial Data Infrastructures which, at their best, facilitate the formulation of standards, the sharing of data, and, ultimately, a more collaborative, open and connected network of coastal GIS applications.

Even in areas where standards and data-sharing protocols are in force, the increasingly rich and diverse database of many coastal regions leaves to GIS practitioners the task of determining the best available representation of coastal phenomena, as in Li *et al.,* who compared different data sources to arrive at the most accurate depiction of the coastline or Hwang and Ku, who outlined methods for choosing remotely sensed data at a resolution that best meets the needs of a coastal modelling excercise.

The work of all of these authors shows that geography can provide the scale and scope that is needed to unite diverse interest groups. Issues of data integration appear to be casting GIS developers as a uniting force and GIS as a nexus in the project of ICZM.

Adapting tools and technologies to serve the stakeholders of ICZM

We are now faced with the task of expressing the methods and results of what was, until recently, a specialized and technical discipline to a diverse group of scientists, managers, regulators, legislators, resource users and members of the public. These people are increasingly not just data *consumers* but also data *users*, who both demand and deserve to peer inside the black box of GIS technology and GIS-based decision support. GIS practitioners in this volume have responded by creating innovative visualisations, interfaces and forums for participation in the course of applying GIS to the coastal zone.

Involving stakeholders in both the formulation and the output of coastal GIS goes beyond the technical and presents intercultural and interpretive challenges. Bartlett and Sudarshana touch on the pervasiveness of Western metaphors in coastal GIS and the need to consider the cultural background of the stakeholders or collaborators. On Alaska's Pribilof Islands, Lindsay *et al.* collaborated across jurisdictions and cultures to construct a GIS that represents archeological features, important traditional resources, local placenames and the legacy of destructive industries. In another Arctic environment, Mosbech *et al.* gathered traditional knowledge to create an oil spill sensitivity database that considers resources of importance to local people. Both of these projects required that the researchers constructing the GIS consider the needs, concerns, and sensitivities of those people who will ultimately be affected by decisions based upon the outcomes of the exercise. Jude describes an exercise in visualizing the outcome of a coastal development: he applies GIS and RS in a way that makes the effects of change real and explicit to a non-technical audience.

Broad new user groups are now working directly with the data and software that has traditionally been the domain of experts. The Internet has excercised its all-pervasive influence here: Laflamme *et al.*, Green and King and Macharia describe web applications that bring GIS to a broader audience when and where information is needed. Longhorn discusses Coastal Spatial Data Infrastructures, that aid in making data accessible to a greater cross-section of users and in supporting the wider popularization of desktop GIS. Planning for consultation, infrastructures and dissemination is now seen as a critical phase in any coastal GIS initiative, and coastal GIS practitioners must keep the needs of stakeholders and future users in mind.

Improving tools and technologies

Because the land-sea interface is a complex ecotone, the scale of biophysically distinct areas in the coastal zone often necessitates a spatial resolution that has traditionally been difficult or expensive to acquire. GIS/RS practitioners working in the coastal zone were until recently either limited to a few data sources (such as analogue photogrammetry or bathymetric and topographic mapping by traditional survey and sounding means), or required to adapt the scale and resolution of their work to the quality of readily available remotely sensed data, i.e. satellite data at a 30m or greater resolution, often with poor depth penetration, particularly for turbid temperate waters. The challenge of acquiring high-quality data has been somewhat ameliorated by the increasing availability of quick-deploying hyperspectral sensors like CASI (Compact Airborne Spectrographic Imager) and active laser sensors like LiDAR (Light Detection and Ranging) or active satellite sensors with specialised uses in the marine environment (such as RadarSat). Both Pan and Webster document their experiences and the technical challenges they faced in acquiring and applying data from these sensors, and both authors make explicit the great advances possible with such high (spatial and/or spectral) resolution data. Of particular importance to ICZM is the value of these fast-deploying sensors for monitoring change and perhaps even emergencies in the dynamic and heterogeneous coastal zone. High resolution data is one more piece in the puzzle of building rigorous, accurate and precise models of coastal zone processes.

Adapting the technology to represent complex dynamic systems

Tracking and analyzing spatio-temporal variation is a problem that has received much attention from geographers and GIS practitioners. In the coastal zone, it is compounded by the other challenges described here. Practitioners are faced with the need to deal with (and differentiate between) error and dynamism in an inherently dynamic environment. Bruce provides a detailed outline of the types of error and uncertainty encountered in coastal zone GIS and the various methods that have been developed to help practitioners quantify and work with this uncertainty. Li *et al.* describe methods for defining 'the coastline' in such a way that the variability of this elusive entity is accounted for. Schupp *et al.* have measured historical shoreline change in order to learn about future change. Webster *et al.* describe how dynamic coastal processes at different time scales (extreme tidal range, seasonal variability, and sea level rise) entered into the decisions that were made in the process of acquiring high resolution imagery and elevation data in a coastal area, then uses these data to plan for future changes. If change is the currency of ICZM, then developing data structures and modes of analysis that integrate dynamism into coastal GIS is critical.

Modelling for decision support

Applying GIS/RS in ICZM presents the challenge of using new technology in a real-life context, where models and findings are fed into a legislative and regulatory application and are subjected to not only academic but also governmental, public and vested private sector scrutiny. Coastal GIS practitioners have acknowledged the need to continue to make their models more robust at the same time as they become more complex and comprehensive. Populus *et al.* and Mayerle and Toro, for example, both undertook to model complex hydrodynamic processes with the explicit goal of using these models for decision support. Euan-Avila *et al.* considered land-based activities and their socio-economic drivers in modeling effects on water quality.

Conclusions

Researchers and practitioners are increasingly taking an interdisciplinary, integrative approach to applying GIS/RS in the coastal zone. As the body of work in this area grows, so too does our understanding of coastal processes and the role of humans in perturbing and managing these systems. GIS has much to contribute to Integrated Coastal Management, and GIS practitioners have many more challenges to face in constructing truly integrative coastal spatial information systems.

REFERENCES

Cicin-Sain, B., 1993, Sustainable development and integrated coastal management. *Ocean & Coastal Management*, **21** (1-3), pp.11-43.

Cicin-Sain, B. and Knecht, R. 1998, *Integrated Coastal and Ocean Management: Concepts and Practices,* (Washington and California: Island Press).

ESRI, 1990, *Understanding GIS: The ARC/INFO Method*, (Redlands, CA: Environmental Systems Research Institute).

Wright, D. J., 2000, Down to the Sea in Ships: The Emergence of Marine GIS. In *Marine and Coastal Geographic Information Systems*, edited by D.J. Bartlett and D. J.Wright, (London: Taylor and Francis).

Index

0-41531-972-2/04/$0.00+$1.50

D

E